Computational Methods for Approximation of Large-Scale Dynamical Systems

Computational Methods for Approximation of Large-Scale Dynamical Systems

Mohammad Monir Uddin

CRC Press
Taylor & Francis Group
Boca Raton London New York

CRC Press is an imprint of the
Taylor & Francis Group, an **informa** business
A CHAPMAN & HALL BOOK

CRC Press
Taylor & Francis Group
6000 Broken Sound Parkway NW, Suite 300
Boca Raton, FL 33487-2742

Printed on acid-free paper

International Standard Book Number-13: 978-0-8153-4803-0 (Hardback)

Library of Congress Cataloging-in-Publication Data

Names: Uddin, Mohammad Monir, 1978- author.
Title: Computational methods for approximation of large-scale dynamical systems / Mohammad Monir Uddin.
Description: Boca Raton : CRC Press, Taylor & Francis Group, 2019. | Includes bibliographical references and index.
Identifiers: LCCN 2019011935| ISBN 9780815348030 (hardback : alk. paper) | ISBN 9781351028622 (ebook)
Subjects: LCSH: Differentiable dynamical systems.
Classification: LCC QA614.8 .U33 2019 | DDC 515/.39--dc23
LC record available at https://lccn.loc.gov/2019011935

Visit the Taylor & Francis Web site at
http://www.taylorandfrancis.com

and the CRC Press Web site at
http://www.crcpress.com

About the Book

This book discusses computational techniques for the model reduction of large-scale sparse linear time invariant (LTI) continuous-time systems. The book emphasizes the model reduction of descriptor systems, i.e., the systems whose dynamics obey differential-algebraic equations. For the first-order and second-order LTI systems, we show first-order-to-first-order, second-order-to-first-order and second-order-to-second-order reduction techniques. For the first-order-to-first-order or second-order-to-first-order models, order reduction by the *balanced truncation* (BT) and the *iterative rational Krylov algorithm* (IRKA) methods are discussed. For the second-order-to-second-order reduction (i.e., structure preserving model reduction), alongside BT, this book discusses another technique called *projection onto the dominant eigenspace of the Gramian* (PDEG). We also discuss the *low-rank alternating direction implicit* (LR-ADI) iteration and the issues related to solving the Lyapunov equation of large-scale sparse LTI systems to compute the low-rank Gramian factors. Note that the low-rank Gramian factors are important ingredients for BT and PDEG-based model reductions. The efficiency and capability of the methods are confirmed by numerical experiments with the data of real-world models.

Contents

Acknowledgments

Personal thanks

It is with great pleasure that I would like to express first my heartfelt appreciation for my teacher **Prof. Peter Benner** (Director of the Max Planck Institute (MPI) for Dynamics of Complex Technical Systems, Magdeburg) who introduced me to the highly fascinating academic and research area of model order reduction. I was supervised in my master's and Ph.D. theses by Prof. Peter Benner with his invaluable guidance as well as financial support. I am also thankful for the experts on the topics of this book and my teachers, colleagues and research students for their valuable suggestions, motivation and cordial help in different stages to complete this book. Among them are:

Dr. Mian Ilyas Ahmad
NUST
Islamabad, Pakistan

Dr. Kapil Ahuja
Indian Institute of Technology (IIT)
Indore, India

Prof. Athanasios C. Antoulas
Rice University
Houston, Texas, USA

Prof. Christopher Beattie
Virginia Tech
Blacksburg, USA

Dr. Tobias Breiten
Karl-Franzens-Universität Graz
Graz, Austria

Dr. Xin Du
Shanghai University
Shanghai, China

Prof. Serkan Gugercin
Virginia Tech
Blacksburg, USA

Shazzad Hasan
North South University
Dhaka, Bangladesh

Prof. Matthias Heinkenschloss
Rice University
Houston, Texas, USA

Dr. Mohammad Sahadet Hossain
North South University
Dhaka, Bangladesh

Mohammad Sumon Hossain
University of Groningen
Groningen, The Netherlands

Dr. Burkhard Kranz
Fraunhofer IWU
Chemnitz, Germany

Dr. Patrick Kürschner
MPI Magdeburg
Magdeburg, Germany

Prof. Volker Mehrmann
TU Berlin
Berlin, Germany

Prof. Timo Reis
Universität Hamburg
Hamburg, Germany

Dr. Jens Saak
MPI-Magdeburg
Magdeburg, Germany

Dr. René Schneider
TU Chemnitz
Chemnitz, Germany

Nuren Shams
North South University
Dhaka, Bangladesh

Prof. Boris Shapiro
Stockholm University
Stockholm, Sweden

Prof. Tatjana Stykel
Universität Augsburg
Augsburg, Germany

Dr. Matthias Voigt
TU Berlin
Berlin, Germany

Dr. Heiko K. Weichelt
MPI Magdeburg
Magdeburg, Germany

Institutional support

Fraunhofer Institute for Machine Tools and Forming Technology IWU
Dresden, Germany

International Max Planck Research School
Magdeburg, Germany

Max Planck Institute for Dynamics of Complex Technical Systems
Magdeburg, Germany

North South University
Dhaka, Bangladesh

Stockholm University
Stockholm, Sweden

TU Cheminitz
Cheminitz, Germany

Preface

Before implementing new ideas or decisions in different disciplines of science, engineering and technology, an experiment is required. The conventional approach of this experiment would require a laboratory with a range of specialized equipment; this can be an expensive approach to demonstrate a concept. The alternate approach to exploring scientific ideas to test their validity, one that is less expensive and often easier to apply than performing real-world experiments, is through computer simulation. For simulation, one needs to convert a physical model into a mathematical model. In real-world applications, the mathematical models are often represented by linear time invariant (LTI) continuous-time systems. In many cases, these systems are subject to additional algebraic constraints which lead to differential-algebraic equations (DAEs) or descriptor systems. These descriptor systems are in either first- or second-order form. The mathematical models are generated in various ways. In many applications, the systems are obtained by finite element (FEM) or finite difference (FDM) discretization. In order to model a system accurately, a sufficient number of grid points must be generated because many geometrical details must be resolved. Sometimes, physical systems consist of several bodies and each body is composed of a large number of disparate devices. Therefore, the mathematical models become more detailed and different coupling effects must be included. In either case, the resulting systems are typically very *large* and *sparse*; they might often be well-structured as well.

A *large-scale*[1] system leads to additional memory requirements and enormous computations. Large-scale systems also prevent frequent simulations, which is often required in certain applications. Sometimes, the generated systems are too large to store due to limited computer memory. To circumvent these complexities, reducing the size of the systems is unavoidable. The method to reduce a higher dimensional model to a lower one is called *model order reduction* (MOR). The fundamental aim of MOR is to replace the high-dimensional dynamical systems by substantially lower-dimensional systems, while the response of the original system is approximated to the highest possible extent. In certain cases, some important features such as stability, passivity, definiteness, symmetry and so forth of the original system must be preserved in the reduced systems.

[1]The notion of large is always changing with the increasing capabilities of computational hardware.

Currently, MOR is considered as an indispensable subject in the different branches of science, engineering and technology. Moreover, MOR has crucial applications in industries. The demand for model reduction is increasing every day because modern science and technology heavily rely on simulation-based works. It has also become an interesting and rewarding area of research in both academics and industries. Over the last few decades, many research articles, technical reports, Ph.D. theses as well as some texts were written on MOR.

For the MOR of LTI systems, various techniques have been developed and are available in the literature. In a broad sense, all the methods can be classified into one of two categories, namely,

→ *Gramian*-based methods and

→ *moment matching*-based methods.

The Gramian-based methods include *optimal Hankel norm approximation*, *singular perturbation approximation*, *dominant subspaces projection*, *frequency weighted balanced truncation*, *dominant pole algorithm* and *balanced truncation* (BT). On the other hand, moment matching can be implemented efficiently via *rational interpolatory projection* methods. In this case, the prominent algorithm is IRKA, i.e., iterative rational Krylov algorithm.

Among all the aforementioned methods, balanced truncation (BT) and the interpolatory method via IRKA are currently the most commonly used techniques for large-scale dynamical systems. The system theoretic method, balanced truncation, has an a priori *error bound* which means, for a given system, the method can generate a best approximate system with respect to a given tolerance. Besides this, balanced truncation preserves the stability of the original systems, i.e., if the given system is stable, the method ensures a stable reduced system. Although these important properties make balanced truncation superior to the other methods, the main disadvantage of this method is its requirement to solve two continuous-time algebraic Lyapunov equations for the original model which requires substantial computational resources. On the other hand, the recently developed, interpolatory projection method via IRKA is interesting to the model reduction community since it is computationally efficient. It requires only matrix-vector products or linear-system solves and does not require solving the Lyapunov equations. Unfortunately, this method has neither an a priori error bound nor guaranteed stability preservation.

Both BT and IRKA are well developed for the standard or generalized LTI systems. Recently, their idea has also been extended to large-scale descriptor systems. Over the last few decades, many research articles, technical reports and Ph.D. theses have been published on MOR for both LTI standard/generalized and descriptor systems. Although there are some texts, in general, all of them are mainly focused on either standard or generalized LTI systems. This book emphasizes model reduction for LTI continuous-time descriptor systems of first-order and second-order form. First-order-to-first-order and second-order-to-first-order reduction techniques are discussed using the BT and IRKA methods. In the literature, second-order-to-second-order model reduction is known as structure preserving MOR. Along with BT, this book also discusses another technique named PDEG (projection onto the dominant eigenspace of the Gramian). This new method is considered superior to the other two since it preserves some required properties of the original system.

For implementing BT- and PDEG-based model reduction, the most expensive task is the computation of two Gramian factors by solving two continuous-time algebraic Lyapunov equations. In recent time, several efficient approaches have been proposed in literature; these exploit the fact that often all the coefficient matrices are sparse and the number of inputs and outputs is very small compared to the number of degrees of freedoms (DoFs). The alternating direction implicit (ADI)-based method low-rank (LR-)ADI iteration is the most useful for a large-scale sparse dynamical system. The book considers this technique and the related issues are discussed elaborately for solving the Lyapunov equations arising from large-scale sparse LTI continuous-time systems. The theoretical results of each topic are tested by numerical experiments using data of some benchmark model examples. These models are also introduced in the book.

There are *three* parts in this book:

→ Part I is preliminaries. Its purpose is to establish notations and introduce important concepts that will be used in the next parts of the book. This part reviews important definitions, theorems and properties of linear algebra and control systems. Low-rank solutions of Lyapunov equations for LTI continuous-time generalized systems are discussed here. Additionally, model reduction of generalized state space and second-order standard systems are introduced here. Each chapter introduces key ideas and proceeds to solve them in a step-by-step method, finally summarizing it all in the form of a precise algorithm. Available softwares are introduced (if necessary) and some exercise problems are also included for practice. If necessary, numerical experiments are illustrated to confirm the efficiency of the proposed algorithms.

→ The main focus of this book is Part II, which is about MOR of LTI continuous-time descriptor systems. The techniques discussed in Part I are generalized here for descriptor systems. There are three types of descriptor systems: index 1, index 2 and index 3. For each index, the systems can either be in their first-order or second-order form. This part provides algorithms of BT and IRKA for the various indexed descriptor systems of first-order form. For the second-order descriptor systems, BT and IRKA methods are discussed to find second-order-to-first-order reduction. On the other hand, for second-order-to-second-order reduction, besides BT, PDEG methods are also discussed. As was the case for Part I, the methods are discussed on the basis of a key idea with the algorithms being mathematically provided in a step-by-step system and summarized into a short and precise form in the end. In each case, the algorithm is tested by applying it to real-world problems. Numerical results are discussed for showing the efficiency and capabilities of the developed algorithms and for its comparison with other prominent algorithms.

→ Part III contains the two appendices of the book. Appendix A presents several model examples which are used for the numerical experiments throughout the different chapters of this book. Appendix B contains MATLAB®codes for selected algorithms from Part I. The ideas of these codes can be generalized to implement the algorithms presented in Part II.

The book contains 12 chapters. A brief introduction of the chapters is as follows.

- ⊙ Chapters 1 and 2 briefly review some fundamental theorems and definitions and presents some algorithms from linear algebra and control systems, respectively. The goal is to establish some notations and fundamental results that will be used in the later chapters of this book.

- ⊙ The low-rank alternating direction implicit (LR-ADI) iteration is discussed in Chapter 3 for solving the Lyapunov equations of generalized state space systems to compute its low-rank Gramian factors. This iterative method is also used in later chapters for computing the low-rank factor of the Gramians of the different types of models which are considered in this book.

- ⊙ Chapter 4 studies model reduction methods of generalized LTI continuous-time systems. This chapter discusses the fundamental properties of balanced truncation and iterative rational Krylov algorithm (IRKA) and establishes algorithms of both the methods for the underlying system.

- ⊙ Model reduction of second-order standard system is considered in Chapter 5. Second-to-first-order reduction is implemented by BT and IRKA, while the second-to-second-order reduction was realized by the BT and PDEG methods.

- ⊙ Chapter 6 introduces some special types of descriptor systems which are studied in the next chapters. Structured descriptor systems including their indices, solvability condition, transfer function and spectrum are briefly discussed in this chapter.

- ⊙ Model reduction of first-order index 1, index 2 and index 3 systems are discussed in Chapters 7, 8 and 10, respectively. Both the balanced truncation and IRKA methods are discussed in each chapter elaborately. The idea of Chapter 8 is generalized in Chapter 9 for the model reduction of unstable first-order index 2 system. The LR-ADI algorithms are also developed in each chapter for computing the low-rank Gramian factors of the respective DAE system.

⊙ Chapters 11 and 12 are concerned with the model reduction of second-order descriptor systems. Model reduction of second-order index 1 systems is discussed in Chapter 11, while Chapter 12 is concerned with model reduction of second-order higher index (i.e., index 3) DAE systems. Both chapters consider second-order-to-first-order and second-order-to-second-order reduction methods of the underlying DAE systems. Both chapters also discuss LR-ADI iteration for the low-rank solutions of second-order index 1 and index 3 DAE systems.

Some noteworthy key features are as follows:

- This book provides a step-by-step guideline for the readers. We begin with the preliminaries that provide all the background knowledge necessary to understand the later chapters. In each chapter, the topics are discussed both theoretically and mathematically. Each chapter provides necessary mathematical algorithms, sufficient examples, numerical experiments and related exercises. With the combination of this book and its supplementary materials, the reader will gain a sound understanding of the topic.

- The software codes based on our algorithms, using MATLAB are provided in the book. The solutions of the exercises, data sets for experiments and electronic version of the software are provided in the book's website.

- For the numerical tests, in most cases, real-world data sets are used. In some chapters, the models that are used are obtained from industries and research institutes and they are not found in other literature.

- The methodologies discussed in this book are updated and efficient.

This book is intended for assisting students of advanced bachelor's, master's or Ph.D. levels and the researchers in academics and industries. Students from different disciplines like engineering (including but not limited to electrical, mechanical, civil and computer science), mathematics, system and control theory, biological science, economics and any other area where mathematical modeling is used for simulation, analysis, optimization and control may find this book helpful. While the basic contents of this book can be useful to the advanced bachelor's degree-level students in any discipline, it would also be very helpful to the researchers in academics and industries alike.

MATLAB and Simulink are registered trademarks of The MathWorks, Inc. For MATLAB and Simulink product information, please contact:

The MathWorks, Inc.
3 Apple Hill Drive
Natick, MA, 01760-2098 USA
Tel: 508-647-7000
Fax: 508-647-7001
E-mail: info@mathworks.com
Web: https://www.mathworks.com

How to buy: https://www.mathworks.com/store Find your local office: https://www.mathworks.com/company/worldwide

Author

Dr. Mohammad Monir Uddin received his PhD in Mathematics from the Max Planck Institute (MPI) for Dynamics of Complex Technical Systems, Germany in 2015. During his PhD Research (2011–2015), he was a member of the International Max Planck Research School (IMPRS) for Advanced Methods in Process and Systems Engineering Magdeburg, Germany (2012–2015). Dr. Uddin completed his second M.Sc. in Applied Mathematics from Stockholm University, Sweden in 2011. He completed his B.Sc. (2003) and M.Sc. (2005) in Mathematics from the University of Chittagong, Bangladesh. Additionally, he worked as a research fellow in the Department of Mathematics at Stockholm University and also in the Department of Mathematics at Chemnitz University of Technology, Germany. Currently, Dr. Uddin is an Assistant professor in the Department of Mathematics and Physics at North South University, Bangladesh. He also has teaching and research experience at various Universities in home and abroad.

List of Acronyms and Symbols

Acronyms:

ADI	alternating direction implicit
ARE	algebraic Riccati equation
ASS	adaptive spindle support
BT	balanced truncation
BIPS	Brazilian interconnected power system
CBM	clamped beam model
CDMS	constrained damped mass-spring
CDP	CD player
CLE	controllability Lyapunov equation
CTCO	constrained triple chain oscillator
DAE	differential-algebraic equation
DoFs	degrees of freedom
FDM	finite difference method
FEM	finite element method
GPARE	generalized projected algebraic Riccati equation
HSV	Hankel singular value
IRKA	iterative rational Krylov algorithm
ISS	international space station
JCF	Jordan canonical form
LR-ADI	low-rank alternating direction implicit
LR-SRM	low-rank square root method
LTI	linear time invariant
MNA	modified nodal analysis
MOR	model order reduction
ODE	ordinary differential equation
OLE	observability Lyapunov equation
PDE	partial differential equation
PDEG	projection onto dominant eigenspace of the Gramian
ROM	reduced order model
SEI	supersonic engine inlet
SFM	stabilizing feedback matrix
SLICOT	subroutine library in systems and control theory
SLNSM	semi-discretized linearized Naviar-Stokes model
SLSM	semi-discretized linearized Stokes model
SMWF	Sherman-Morrison-Woodbury formula
SOFOR	second-order-to-first-order reduction
SPD	symmetric positive definite
SPMOR	structure preserving model order reduction
SRM	square root method
SVs	singular values
SVS	singular value decomposition
TCO	triple chain oscillator
TF	transfer function

Symbols:

$\mathbb{R}$	field of real numbers
$\mathbb{C}$	field of complex numbers
$\mathbb{R}^{m\times n}$	set of all real matrices
$\mathbb{C}^{m\times n}$	set of all complex matrices
$\mathbb{R}^n$	vector space of real n components
$\mathbb{C}^n$	vector space of complex n components
$\mathbb{C}^-$	left complex half plane
$\mathbb{C}^+$	right complex half plane
$:$	such that
$m \times n$	m by n
A^T	transpose of A
A^*	complex conjugate transpose of A
A^{-1}	inverse of A
M_c	controllability matrix
M_o	observability matrix
$\in$	belongs to
$a_{i,j}$	an element of a matrix at i-th row and j-th column
$\Lambda(A)$	Eigen spectrum of A
$\lambda_i(A)$	i- eigenvalue of matrix A
$A - \lambda E$	generalized matrix pencil
$\sigma_{max}(A)$	largest singular value of A
$\sigma_{min}(A)$	smallest singular value of A
Σ	diagonal matrix with singular values
$\mathrm{Null}\,(A)$	null space of A
$\mathrm{img}(A)$	image of A
$\ker\,(A)$	kernal of A
$\mathrm{span}\,(A)$	all linear combination of columns of A
$\|A\|_2$	2-norm of A
$\|A\|_F$	Frobenius norm of A
$\hat{A}$	reduced-order matrix of A
$\mathcal{L}$	Laplace's transformation
$\delta(t)$	dirac delta function
$u_{step}(t)$	unit step function
$\{\mu_i\}_{i=1}^J$	ADI shift parameters
$\{\alpha_i\}_{i=1}^J$	interpolation points
$A \otimes B$	Kronecker product of the matrices A and B
σ	singular value
$\mathcal{K}_m$	m dimensional Krylov subspace
$\mathrm{range}\,(A)$	range of A
$\mathrm{colsp}\,(A)$	column space of A
$\mathrm{tr}\,(A)$	trace of A
$a \ll b$	a is very very smaller than b
$a \neq b$	a is not equal to b
$a \geq b$	$a - b \geq 0$
$a \leq b$	$a - b \leq 0$
$X \approx Y$	X is approximate to Y
ϵ_{mach}	machine precision
$\mathcal{R}(X_i)$	Lyapunov residual at i-th iteration
τ	tolerance for stopping LR-ADI iteration
$\overline{\alpha}$	complex conjugate of α
$\mathrm{Re}\,(\alpha)$	real part of α
$\mathrm{Im}\,(\alpha)$	imaginary part of α
$\det\,(A)$	determinant of A
$\mathrm{rank}\,(A)$	rank of A
N^ν	nilpotent matrix N with nilpotency ν
$\|.\|_{\mathcal{H}_2}$	$\mathcal{H}_2$ norm
$\|.\|_{\mathcal{H}_\infty}$	$\mathcal{H}_\infty$ norm
$G(s)$	transfer function (matrix)
$G(\mathrm{j}\omega)$	frequency response

Part I

PRELIMINARIES

Chapter 1

Review of Linear Algebra

1.1 Introduction

Linear algebra, or more specifically numerical linear algebra, plays an important rule in linear control theory. The purpose of this book is to help its readers gain a better understanding of the different techniques of model reduction for linear dynamical systems. As a prerequisite, the readers are expected to have adequate knowledge regarding linear algebra. For the reader's convenience, this chapter quickly goes through some basic concepts from linear algebra that will be crucial to the understanding of the material that will be covered later in the book. We will briefly review some fundamental theorems and definitions and present some algorithms that are essential to linear algebra from previous literature. The primary goal is to establish some notations and fundamental results that will be used in the later chapters.

Note that detailed discussion of a topic and proof of theorems are omitted since the details can be found in the reference literature. Most of the materials of this chapter are available in numerical linear algebra textbooks; e.g., [10, 57, 59, 79, 102, 139, 142, 144, 179].

1.2 Matrices

A matrix is a rectangular array of numbers (either real or complex) arranged in rows and columns. The numbers in the array are called the entries in the matrix. The size or order of a matrix is described in terms of the number of rows (horizontal lines) and columns (vertical lines) it contains. A matrix with only one column is called a *column matrix* (or a column vector) and a matrix with only one row is called *row matrix* (or a row vector).

In general, an $m \times n$ matrix can be written as

$$A = \begin{bmatrix} a_{11} & a_{12} & \cdots & a_{1n} \\ a_{21} & a_{22} & \cdots & a_{2n} \\ \vdots & \vdots & \ddots & \vdots \\ a_{m1} & a_{m2} & \cdots & a_{mn} \end{bmatrix} = \left[a_{ij}\right]; \begin{cases} i = 1, \cdots, m \\ j = 1, \cdots, n \end{cases} \tag{1.1}$$

where $\begin{pmatrix} i \\ j \end{pmatrix} = \begin{pmatrix} 1, 2, \cdots, m \\ 1, 2, \cdots, n \end{pmatrix}$. Here, a_{ij} is an element of A located in the i-th row and j-th column. If all the elements of A are real, then it is said to be a real matrix and may be written as $A \in \mathbb{R}^{m\times n}$. Similarly, $A \in \mathbb{C}^{m\times n}$ represents a complex matrix with at least one of its elements being a complex number.

Some useful matrices

***Square and rectangular matrices*:** An $m \times n$ matrix $A = \left[a_{ij}\right]$ is called *square* if $m = n$, i.e., number of rows is equal to the number of columns. If $m \neq n$, then the matrix is called rectangular. If $m > n$, then the matrix is *thin rectangular matrix* and if $m < n$, then the matrix is *fat rectangular matrix.*

***Diagonal matrix*:** A square matrix $A = \left[a_{ij}\right]$ is called diagonal if all the elements except the diagonal are zeros.

***Identity matrix*:** A diagonal matrix $A = \left[a_{ij}\right]$ is called identity if $a_{ij} = 1$ when $i = j$. We often denote an identity matrix by I.

***Triangular matrix*:** A square matrix $A = \left[a_{ij}\right]$ is called upper triangular if $a_{ij} = 0$ when $i > j$. Similarly, if $a_{ij} = 0$ when $i < j$, then the matrix is called lower triangular.

Transpose and conjugate transpose of a matrix: The transpose of $A \in \mathbb{R}^{m \times n}$ is another $n \times m$ matrix A^T obtained by interchanging the rows and columns in A. On the other hand, the *conjugate transpose* or *Hermitian transpose* of $A \in \mathbb{C}^{m \times n}$ is $A^* \in \mathbb{C}^{n \times m}$ obtained from A by taking the transpose and then taking the complex conjugate of each of the entries.

Symmetric matrix and Hermitian matrix: A real matrix $A \in \mathbb{R}^{n \times n}$ is called symmetric if $A = A^T$. If A is symmetric, then $A - A^T = 0$ (zero matrix). On the other hand, a complex matrix $A \in \mathbb{C}^{n \times n}$ is called Hermitian matrix (or self-adjoint matrix) if it is equal to its own conjugate transpose, that is the element in the i-th row and j-th column is equal to the complex conjugate of the element in the j-th row and i-th column, for all indices i and j.

Positive (semi-) definite matrices: A symmetric (or Hermitian) matrix A is called positive definite if the scalar $z^T Az$ (or $z^* Az$) is strictly positive for every non-zero column vector $z \in \mathbb{R}^n$.

Inverse matrix: An $n \times n$ matrix A is said to be invertible if there exists another $n \times n$ matrix B such that $AB = BA = I$. Inverse of A is denoted by A^{-1}. An invertible matrix is often called nonsingular matrix. The determinant of such matrix is non-zero. If a matrix is not invertible, then it is called singular matrix.

Orthogonal and unitary matrices: A matrix $A \in \mathbb{R}^{n \times n}$ is called orthogonal if $A^T A = AA^T = I$, i.e., the transpose is equal to its inverse. Similarly, complex square matrix A is unitary if its conjugate transpose A^* is also its inverse, that is, if $AA^* = A^* A = I$.

Sparse matrix and dense matrix: A matrix with special structure that has relatively few non-zero entries is called *sparse matrix*. A sparse matrix is a matrix that allows special techniques to take advantage of the large number of zero elements and their locations. Usually, standard discretization of partial differential equations leads to large and sparse matrices. More details about sparse matrices, their properties, their representations, and the data structures that are used to store them can be found in [143].

Nilpotent matrix: A square matrix N is called nilpotent if for some positive integer ν, $N^\nu = 0$ (zero matrix). The smallest ν is called the nilpotency of the matrix.

Block matrix: Suppose an $(m + n) \times (m + n)$ is partitioned as:

$$A = \begin{bmatrix} A_{11} & A_{12} \\ A_{21} & A_{22} \end{bmatrix},$$

where A_{11}, A_{12}, A_{21} and A_{22} are the $m \times m$, $m \times n$, $n \times m$ and $n \times n$ submatrices, respectively. Then A is called block matrix. A diagonal matrix where all diagonal entries are submatrices is called *block diagonal matrix.*

1.3 Vector space and subspace

Vector space and subspace are very useful and fundamental to linear algebra. This section briefly discusses both of these topics and others related to them.

Definition 1 (Vector space). *A nonempty set of vectors V in which two operations; addition (i.e., associating two elements $u, v \in V$ by $u+v$) and scalar multiplication (i.e., associating an element $u \in V$ and any scalar $k \in \mathbb{R}$ by ku) are defined is called a vector space but only if V satisfies the following axioms.*

(a) *There is a vector 0 in V called* zero vector, *such that $u + 0 = u$ for any vector $u \in V$.*
(b) *If $u, v \in V$, then $u + v \in V$.*
(c) *If $u \in V$, then there exists $-u \in V$ such that $u + (-u) = (-u) + u = 0$.*
(d) *If $u \in V$, then for any scalar k, $ku \in V$.*
(e) *For $u, v \in V$, $u + v = v + u$.*
(f) *For $u, v \in V$ and any scalar k, $k(u + v) = ku + kv$.*
(g) *For scalars k, l and any vector $v \in V$, $(k + l)u = ku + lu$.*
(h) *For scalars k, l and any vector $v \in V$, $(kl)u = k(lu)$.*
(i) *For $u, v, w \in V$, $(u + v) + w = u + (v + w)$.*
(j) *For the unit scalar 1 and $u \in V$, $1u = u$.*

According to the definition, any *Euclidean n-space* ($\mathbb{R}^n$), where $n = 1, 2, \cdots$ is an example of a vector space.

Definition 2 (Subspace). *A subset W of a vector space V is called subspace if it is itself a vector space under the addition and scalar multiplication defined in V.*

Any *Euclidean n-space* ($\mathbb{R}^n$), where $n = 1, 2, \cdots$ is a subspace since any $\mathbb{R}^n$ is subset of itself and it is a vector space. The following observation is useful to identify a subspace.

Theorem 1. *Suppose V is a vector space and W is a nonempty sub-set of V. Then W is a subspace of V if and only if the following conditions are satisfied.*

- *If $u, v \in W$, then $u + v \in W$.*
- *For any vector $u \in W$ and any scalar k, $ku \in W$.*

Definition 3 (Linear combination). *A vector w is called a linear combination of the vectors $v_1, v_2, \cdots, v_r$ if it can be expressed as*

$$w = x_1v_1 + x_2v_2 + \cdots + x_rv_r,$$

where $x_1, x_2, \cdots, x_r$ are scalars.

If $S = \{v_1, v_2, \cdots, v_r\}$ is a set of vectors in a vector space V, then the subspace W of V consisting of all linear combinations of the vectors in S is called the space spanned by W. To indicate that W is the space spanned by the vectors in the set $S = \{v_1, v_2, \cdots, v_r\}$, we write

$$W = \text{span}\,(S) \quad \text{or} \quad W = \text{span}\,\{v_1, v_2, \cdots, v_r\}$$

Theorem 2. *If $S = \{v_1, v_2, \cdots, v_r\}$ are vectors in a vector space V, then* span (S) *is a subspace of V.*

Definition 4 (Linearly independent). *Let $S = \{v_1, v_2, \cdots, v_n\}$ be a nonempty set of vectors. If the homogeneous linear system*

$$Ax = 0 \tag{1.2}$$

where $A = \begin{bmatrix} v_1 & v_2 & \cdots & v_n \end{bmatrix}$ and $x = \begin{bmatrix} x_1 \\ x_2 \\ \vdots \\ x_n \end{bmatrix}$ has only trivial solution, then S is called a linearly independent set, i.e., the vectors in S are linearly independent. If (1.2) has nontrivial solution also, then S is called a linearly dependent set, i.e., the vectors in S are linearly dependent.

Equation (1.2) reveals that if the vectors in S are linearly independent, then the linear combination of the vectors $v_1, v_2, \cdots, v_n$, i.e.,

$$x_1v_1 + x_2v_2 + \cdots + x_rv_n \tag{1.3}$$

is zero (vector), where $x_1 = 0$, $x_2 = 0, \cdots$, $x_n = 0$. A nonzero vector space V is called finite-dimensional if it contains a finite set of vectors $\{v_1, v_2, \cdots, v_n\}$ that forms a basis.

Definition 5 (Basis). *Let V be a finite-dimensional vector space, and a linearly independent set of vectors $S = \{v_1, v_2, \ldots, v_n\}$ is called basis for V if*

- *S is linearly independent and*
- span $(S) = V$.

If no such set exists, V is called infinite dimensional. In addition, we shall regard the zero vector space to be finite dimensional.

Theorem 3. *If* $S = \{v_1, v_2, \cdots, v_n\}$ *is a basis for a vector space* V*, then every vector* $v \in V$ *can be expressed in the form*

$$v = c_1v_1 + c_2v_2 + \cdots + c_nv_n$$

in exactly one way.

Range and null space of a matrix

Two important subspaces of a matrix are its range and null space. For $A \in \mathbb{R}^{m\times n}$, these are defined as

Definition 6 (Range). *The range of* $A \in \mathbb{R}^{m\times n}$ *is denoted by* Range (A) *and defined as follows:*

$$\text{Range}\,(A) = \{Ax; \textit{for all } x \in \mathbb{R}^n\}.$$

Range of A *is the same as the column space of* A *which is spanned by the columns of* A *and it is denoted by* colsp (A)*. It is a subspace of* $\mathbb{R}^m$*. The dimension of column space (or range) of* A *is called rank of the matrix* A*, i.e.,* $\text{rank}\,(A) = \dim\,(\text{colsp}\,(A))$.

> For an $n \times n$ matrix A if $\text{rank}\,(A) = n$, then it is called *full-rank matrix.* If $\text{rank}\,(A) < n$, then A is called *rank deficient matrix.*

Definition 7 (Null space). *The solution space of a homogeneous linear system,* $Ax = 0$ *is called null space of* A*. It is denoted by* Null (A) *and defined as*

$$\text{Null}\,(A) = \{x : Ax = 0\}.$$

Null space of A *is also known as Kernal of* A *and denoted by* $\ker(A)$*. The dimension of a null space is known as its* nullity.

1.4 Orthogonalization and Gram-Schmidt process

Orthogonalization is the process of finding a set of orthogonal vectors that span a particular subspace. Formally, starting with a linearly independent set of vectors $\{v_1, \cdots, v_r\}$ in an inner product space (most commonly the Euclidean space $\mathbb{R}^n$), orthogonalization results in a set of orthogonal vectors $\{u_1, \cdots, u_r\}$ that generate the same subspace as the vectors $\{v_1, \cdots, v_r\}$. Every vector in the new set is orthogonal to each other with the new set and

Algorithm 1: Modified Gram-Schmidt process.

Input : $v_1, v_2, \cdots, v_r$ from $\mathbb{R}^n$.
Output: Orthogonal set of vectors $\tilde{v}_1, \tilde{v}_2, \cdots, \tilde{v}_r$.

```
for i = 1 : r do
    for j = 1 : i − 1 do
        v_i = v_i − (ṽ_j^T v_i)/(ṽ_j^T ṽ_j) ṽ_j.
    ṽ_i = v_i/‖v_i‖_2.
```

Algorithm 2: Update modified Gram-Schmidt process.

Input : $v_1, v_2, \cdots, v_r$ from $\mathbb{R}^n$ and ϵ (substantially small).
Output: Orthogonal set of vectors $\tilde{v}_1, \tilde{v}_2, \cdots, \tilde{v}_r$.

```
for i = 1 : r do
    for j = 1 : i − 1 do
        v_i = v_i − (ṽ_j^T v_i)/(ṽ_j^T ṽ_j) ṽ_j.
    if ‖v_i‖_2 ≤ ε then
        delete ṽ_i.
    else
        ṽ_i = v_i/‖v_i‖_2.
```

the old set having the same linear span. In addition, if we want all the resulting vectors to be unit vectors, then the procedure is called orthonormalization. In the literature this procedure of orthogonalization is known as *Modified Gram-Schmidt process*, which is summarized in Algorithm 1. Note that the vectors $\tilde{v}_i$, where $i = 1, 2, \cdots, r$ are normalized, i.e., the 2-norm of each $\tilde{v}_i$ is 1. Hence $Q = [\tilde{v}_1, \tilde{v}_2, \cdots, \tilde{v}_r]$, is an orthonormal matrix. Sometimes, after orthogonalization of all the previous vectors $\tilde{v}_j$, $\|v_i\|_2$ becomes zero or close to zero. This indicates v_i and v_{i-1} are linearly dependent. If it happens, then $\tilde{v}_i$ does not provide any new information to the subspace or vector space and thus can be deleted. The updated version of the *Modified Gram-Schmidt process* is summarized in Algorithm 2.

1.5 Krylov subspace and Arnoldi process

Krylov subspace plays an important role in this book. To perform interpolatory projection based model reduction and in the eigenvalue problems, we need to form basis of different dimensional Krylov subspaces. The Arnoldi

---

Algorithm 3: Arnoldi process for orthonormal basis of Krylov subspace.

Input : $A \in \mathbb{R}^{n\times n}, b \in \mathbb{R}^n$.

Output: Orthogonal set of vectors $v_1,\ v_2,\ \cdots,\ v_m$.

$v_1 = \frac{b}{\|b\|_2}$.

for $i = 1 : m$ **do**

  $w = Av_i$.

  for $j = 1 : i$ **do**

    $h = v_j^T w$.

    $w = w - hv_j$.

  if $\|w\|_2 \leq \epsilon$ *(a tolarence)* **then**

    stop.

  else

    $v_i = \frac{w}{\|w\|_2}$.

---

process is considered as one of the efficient methods to form the basis of Krylov subspace. This section presents an algorithm for creating the basis of Krylov subspace using Arnoldi iterations.

Definition 8 (Krylov subspace). *Consider $A \in \mathbb{R}^{n\times n}$, $b \in \mathbb{R}^n$ and a set of linearly independent vectors,*

$$V_m = \{b, Ab, \cdots, A^{m-1}b\}. \tag{1.4}$$

A subspace which is spanned by V_m is called m dimensional Krylov subspace *associated with A and b and it can be defined as*

$$\mathcal{K}_m(A, b) = \text{span}\left(V_m\right). \tag{1.5}$$

The explicit formulation of Krylov basis V_m in (1.4) is not suitable for numerical computations. As m increases, the vector $A^m b$ always converges to an eigenvector belonging to a dominant eigenvalue. This implies that the vectors in V_m become more and more linearly dependent. To avoid these effects, one should choose a basis of a better nature, for example an orthonormal basis. However, in this case we can follow the modified Gram-Schmidt procedure introduced above. Exploiting the idea of modified Gram-Schmidt procedure to form the orthonormal basis of Krylov subspace is known as Arnoldi procedure. In fact, the Arnoldi procedure applies the Gram-Schmidt procedure to transform the vectors in V_m into orthonormal set of vectors $\{v_1, v_2, \cdots, v_m\}$ which form a basis of m-dimensional Krylov subspace $\mathcal{K}_m(A, b)$. The Arnoldi procedure is summarized in Algorithm 3.

1.6 Eigenvalue problem

Matrix eigenvalues and eigenvectors are the roots of many mathematical and engineering problems. In this book, classification of dynamical systems (e.g., standard or descriptor) and their analysis such as stability, unstability etc. are based on the eigenvalues. In the interpolation-based model reduction to compute the interpolation points, tangential direction and computing ADI shift parameters for solving Lyapunov equations (see the later chapters) we need to compute eigenvalues of a matrix. This section briefly discusses an efficient computation of eigenvalues of large matrices.

Definition 9. *For a matrix $A \in \mathbb{C}^{n\times n}$, a nonzero vector $x \in \mathbb{C}^n$ is called eigenvector and the scalar $\lambda \in \mathbb{C}$ is called eigenvalue if they satisfy*

$$Ax = \lambda x. \tag{1.6}$$

Equation (1.6) is known as an *eigenvalue problem.* The equation

$$\det(A - \lambda I) = 0, \tag{1.7}$$

where I is an $n \times n$ identity matrix is called the *characteristic polynomial.* The set of all solutions $\{\lambda_i\}_{i=1}^n$ is called *eigen-spectrum* or *spectrum* and it is denoted by $\Lambda(A)$.

Computing the full spectrum of a large matrix A by solving Equation (1.7) is infeasible. However, we can apply the Krylov-based Arnoldi process to compute all (or some) eigenvalues of a large matrix A efficiently. The following theorem provides its basic insights.

Theorem 4. *Let the columns of $V_{m+1} = [v_1, v_2, \cdots, v_{m+1}] \in \mathbb{R}^{n\times(m+1)}$ form an orthogonal basis for $\mathcal{K}_{m+1}(A, v_1)$, then there exists an $(m+1)\times m$ unreduced upper Hessenberg matrix:*

$$\widehat{H}_m = \begin{bmatrix} h_{11} & h_{12} & \cdots & h_{1m} \\ h_{21} & h_{22} & \cdots & h_{2m} \\ & \ddots & \ddots & \vdots \\ & & & h_{m,m} \\ & & & h_{m+1,m} \end{bmatrix} \tag{1.8}$$

such that

$$AV_m = V_{m+1}\widehat{H}_m. \tag{1.9}$$

Conversely, a matrix V_{m+1}, with orthonormal columns satisfies a relation of the form in (1.9) only if the columns of V_{m+1} form an basis for $\mathcal{K}_{m+1}(A, v_1)$.

This theorem justifies the following definition:

Definition 10 (Arnoldi decomposition). *Let the column of $V_{m+1} \in \mathbb{R}^{n \times m+1}$ form an orthogonal basis. If there exists a Hessenberg matrix $\widehat{H}_m \in \mathbb{R}^{m+1 \times m}$ of the form in (1.8) so that*

$$AV_m = V_{m+1}\widehat{H}_m. \tag{1.10}$$

Then (1.10) is called an (unreduced) Arnoldi decomposition of order m.

By a suitable partition of $\widehat{H}_m$, we can rewrite (1.10) as

$$AV_m = \begin{bmatrix} V_m & v_{m+1} \end{bmatrix} \begin{bmatrix} H_m \\ h_{m+1,m}e_m^T \end{bmatrix} = V_m H_m + h_{m+1,m} v_{m+1} e_m^T. \tag{1.11}$$

Excepting the rank one perturbation from (1.11) we get

$$AV_m = V_m H_m. \tag{1.12}$$

One can write (1.12) as follows:

$$H_m = V_m^T A V_m. \tag{1.13}$$

Note that H_m is a projection of A onto the Krylov subspaces $\mathcal{K}_m(A, v_1)$. As a result, its eigenvalues are related to those of A and are called *ritzvalues* of A.

Definition 11 (Rayleigh-quotient and ritzvalue). *Let $A \in \mathbb{R}^{n \times n}$ and let the columns of $V_m \in \mathbb{R}^{n \times m}$ be orthonormal. The $m \times m$ matrix $H_m = V_m^T A V_m$ is called* Rayleigh-quotient, *an eigenvalue λ of H_m is called a* ritzvalue. *If v is an eigenvector of H_m associate with λ, then $V_m v$ is called a* ritzvector *belonging to λ.*

Algorithm 4 shows how to compute $(m+1) \times m$ upper Hessenberg $\widehat{H}$ (as in (1.8)) and $n \times (m+1)$ orthonormal matrix V_{m+1} (as defined in (1.9)) by applying the Arnoldi procedure onto $A \in \mathbb{C}$.

Generalized eigenvalue problem

Following the above discussion now, let us turn to the generalized eigenvalues and eigenvectors. Consider, an ordered pair (A, E), where $A, E \in \mathbb{C}^{n \times n}$, a non-zero vector $x \in \mathbb{C}^n$ and $\alpha, \beta \in \mathbb{C}$ and both not zero, such that

$$\alpha Ax = \beta Ex. \tag{1.14}$$

If $\alpha \neq 0$, then

$$Ax = \lambda Ex, \tag{1.15}$$

where $\lambda = \frac{\beta}{\alpha}$. The scalar λ is called *eigenvalue or characteristic value* of the pair (A, E) and x is an *eigenvector* corresponding to the eigenvalue λ.

Algorithm 4: Arnoldi process for computing upper Hessenberg matrix.

Input : $A \in \mathbb{R}^{n \times n}, b \in \mathbb{R}^n$.
Output: $\widehat{H}$ (as in (1.8)) and $V_{m+1} = [v_1, \cdots, v_{m+1}]$.

$v_1 = \frac{b}{\|b\|_2}$.
for $i = 1 : m$ **do**
  $w = Av_i$
  for $j = 1 : i$ **do**
    $h_{j,i} = v_j^T w$.
    $w = w - h_{ji} v_j$.
  if $\|w\|_2 \leq \epsilon$ **then**
    stop.
  else
    $h_{i+1,i} = \|w\|_2$.
    $v_i = \frac{w}{h_{i+1,i}}$.

Algorithm 5: Arnoldi process for ritzvalues of A.

Input : $A \in \mathbb{R}^{n \times n}, b \in \mathbb{R}^n$.
Output: Some large magnitude ritzvalues of A.

Compute Hessenberg matrix $\widehat{H}$ using Algorithm 4.
Compute some m eigenvalues of $\widehat{H}(1 : m, 1 : m)$.

These eigenvalues are *finite*. If (1.14) holds with $\alpha = 0$ and $\beta \neq 0$, then (A, E) has *infinite* eigenvalues, i.e., the eigenvalues of (A, E) associated with β are infinite. This is the *generalized eigenvalue* problem (see [10, 102, 142, 179]). Many eigenvalue problems that arise in applications are most naturally formulated as generalized eigenvalue problems. In the case $E = I$ (identity matrix), we obtain the standard eigenvalue problems.

Theorem 5. *Let $A, E \in \mathbb{C}^{n \times n}$ and $\lambda \in \mathbb{C}$ be non-zero.*

- *λ is an eigenvalue of (A, E) iff $\frac{1}{\lambda}$ is an eigenvalue of (E, A).*
- *∞ is an eigenvalue of (A, E) iff E is a singular matrix.*
- *∞ is an eigenvalue of (A, E) iff 0 is an eigenvalue of (E, A).*
- *If E is nonsingular, the eigenvalues of (A, E) are exactly the eigenvalues of $E^{-1}A$.*

Definition 12 (Matrix pencil)**.** *Let, $A, E \in \mathbb{C}^{n \times n}$ the expression $A - \lambda E$ with indeterminant λ, where $\lambda \in \mathbb{C}$ is called* matrix pencil *(or* eigen pencil*). We denote this by $\mathcal{P}(\lambda)$.*

The terms matrix pair or matrix pencil are used more or less interchangeably, i.e., if any non-zero vector x is an eigenvector of the pencil $(A - \lambda E)$, it is also called an eigenvector of the pair (A, E). Note that, $\lambda \in \mathbb{C}$ is an eigenvalue of (A, E) iff $A - \lambda E$ is singular [179] that is to say

$$\det(\lambda E - A) = 0. \tag{1.16}$$

This is known as the *characteristic equation* of the pair (A, E), where the function $\Delta(\lambda) = (\lambda E - A)$ is the *characteristic polynomial* with a degree equal to or less than n.

Definition 13 (Singular and regular pairs)**.** *A matrix pair (A, E), where $A, E \in \mathbb{C}^{n \times n}$, is called* singular *if its characteristic polynomial, i.e., $\det(\lambda E - A)$ is identically zero. Otherwise, it is called* regular *pair*

Remark 1. *If E is nonsingular, the eigenvalues of (A, E) are equal to the eigenvalues of $E^{-1}A$. Therefore, for computing the generalized eigenvalues or ritzvalues, we can use Algorithm 5 to form the upper Hessenberg by just replacing the input A by $E^{-1}A$.*

1.7 Matrix factorizations

Matrix decomposition is essential for accelerating a computational task associated with the matrix. Depending on the nature, we can decompose a matrix different ways. In the following subsections, we briefly introduce some important types of matrix factorizations which will be frequently used in the next chapters of this book.

1.7.1 Eigen decomposition

Let A be an $(n \times n)$ diagonalizable matrix with n linearly independent eigenvectors $\{s_1, \cdots, s_n\}$. Then a factorization of A is,

$$A = S\Lambda S^{-1},$$

where S is an $n \times n$ matrix whose i-th column is the eigenvector s_i of A and $\Lambda = \operatorname{diag}(\lambda_1, \cdots \lambda_n)$ is the diagonal matrix whose diagonal elements are the corresponding eigenvalues. This is called eigen decomposition or sometimes spectral decomposition. If none of the eigenvalues of A are zero, then A is *nonsingular* and its inverse is given by

$$A^{-1} = S\Lambda^{-1}S^{-1}, \tag{1.17}$$

where Λ^{-1} is obtained by inverting all of its diagonal entries. If A is a Hermitian matrix or is symmetric, then its eigenvalues are real and the eigenvectors can be chosen such that they are orthogonal to each other. Thus a symmetric matrix A can be decomposed as

$$A = S\Lambda S^T \tag{1.18}$$

where S is an orthogonal matrix, and Λ is a diagonal matrix whose entries are the eigenvalues of A.

1.7.2 Singular value decomposition

Let m and n be arbitrary; we do not require $m \geq n$. Given $A \in \mathbb{C}^{m\times n}$, not necessarily of full rank, the *singular value decomposition (SVD)* of A is a factorization

$$A = U\Sigma V^*, \tag{1.19}$$

where $U \in \mathbb{C}^{m\times m}$, $V \in \mathbb{C}^{n\times n}$ are unitary, and $\Sigma \in \mathbb{R}^{m\times n}$ is diagonal. In addition, it is assumed that the diagonal elements σ_j $(j = 1, 2, \cdots, k)$ of Σ are nonnegative and in nonincreasing order; i.e.,

$$\sigma_1 \geq \sigma_2 \geq \cdots \sigma_k \geq 0,$$

where $k = \min(m, n)$. This SVD is referred to as *full SVD* by some authors. On the other hand, if $U_1 = (u_1 \cdots u_n) \in \mathbb{C}^{m\times n}$, $\Sigma_1 = \operatorname{diag}(\sigma_1, \sigma_2, \cdots, \sigma_n)$ then $A = U_1\Sigma_1 V^*$. This factorization is known as *thin SVD* of A. The thin SVD is also referred to as the *economic* SVD.

1.7.3 LU decomposition

Let A be an $n \times n$ matrix. The LU factorization decomposes the matrix A into two factors, an upper triangular matrix U and a lower triangular matrix L with proper row and/or column orderings or permutations, i.e., an LU factorization of A is

$$A = LU.$$

The LU factorization with partial pivoting returns a permutation matrix P (which reorders either the rows or the columns of A) such that

$$PA = LU,$$

where U is an upper triangular matrix and L is a lower triangular matrix. On the other hand, the LU factorization with full pivoting involves both row and column reordering, i.e., it returns a row permutation matrix P and column permutation matrix Q such that

$$PAQ = LU,$$

where L, U are defined as above.

1.7.4 Cholesky decomposition

Let A be a symmetric positive definite matrix. Then A can be decomposed as

$$A = LL^T$$

where L is a lower triangular matrix and L^T is the transpose of L. This is known as Cholesky decomposition. If A is a symmetric positive semi-definite matrix, this decomposition is also possible. If A is a Hermitian positive definite (or semidefinite) matrix, then we can write the Cholesky decomposition of A as

$$A = LL^*$$

where L^* denotes the conjugate transpose of L.

1.7.5 QR decomposition

The QR decomposition of an $n \times n$ matrix A yields an orthogonal (unitary if A is complex) matrix $Q \in \mathbb{R}^{n\times n}$ and an upper triangular matrix $R \in \mathbb{R}^{n\times n}$, such that

$$A = QR.$$

If A is an $m \times n$ matrix with $m \geq n$, then the QR factorization of A is

$$A = QR = Q\begin{bmatrix} R_1 \\ 0 \end{bmatrix} = \begin{bmatrix} Q_1 & Q_2 \end{bmatrix}\begin{bmatrix} R_1 \\ 0 \end{bmatrix} = Q_1 R_1.$$

where Q_1 is an $m \times n$ orthogonal matrix and R_1 is an $n \times n$ upper triangular matrix.

1.7.6 Schur decomposition

The Schur decomposition of an $n \times n$ complex matrix is as follows

$$A = QUQ^*$$

where Q is a unitary matrix and U is a quasi-triangular Schur matrix. If A is real, then Q is an orthogonal matrix.

1.8 Vector norms and matrix norms

Definition 14 (Vector norm). *Let* $u = \left[u_1^T, \cdots, u_n^T\right]^T$ *be a vector in* $\mathbb{C}^n$. *Then we define the vector* $p-$ *norm of* u *as*

$$\| u \|_p = \left(\sum_{i=1}^{n} \| u_i \|^p \right)^{\frac{1}{p}}, \qquad \textit{for } 1 \leq p < \infty.$$

In particular, 1-norm and 2-norm can be defined by, respectively,

$$\| u \|_1 = \sum_{i=1}^{n} |u_i| \quad \text{and} \quad \| u \|_2 = \sqrt{\sum_{i=1}^{n} |u_i|^2}.$$

Likewise, the ∞-norm of the vector u is defined by

$$\| u \|_\infty = \max_{1 \leq i \leq n} | u_i | .$$

Definition 15 (Matrix norm). *Let $A = [a_{ij}] \in \mathbb{C}^{m \times n}$; then the matrix norm induced by a vector p-norm is defined as*

$$\| A \|_p = \sup_{u \neq 0} \frac{\| Au \|_p}{\| u \|_p},$$

where u is a vector in $\mathbb{C}^n$. The matrix norms induced by vector p-norms are sometimes called induced p-norms.

In particular, the induced matrix 1-norm can be computed by

$$\|A\|_1 = \max_{1 \leq j \leq n} \| a_j \|_1; \quad a_j \text{ is the } jth \text{ columns of } A.$$

One of the most important types of norm of A is 2-norm which can be computed by

$$\|A\|_2 = \sqrt{\text{maximum eigenvalues of } (A^* A)} = \sqrt{\lambda_{max}(A^* A)}. \tag{1.20}$$

Another important and frequently used matrix norm is *Frobenius norm* defined by

$$\|A\|_F = \left(\sum_{i=1}^{m} \sum_{j=1}^{n} | a_{ij} |^2 \right)^{1/2}, \tag{1.21}$$

where a_{ij} is an element of A located in the i-th row and j-th column.

1.9 Some important definitions and theorems

Invariant Subspaces: In general, a subspace $S \subset \mathbb{C}^n$ is called *invariant* for the transformation A, or *A-invariant*, if $Ax \in S$ for every $x \in S$. In other words, S is invariant for A means that the image of S under A is contained in S; $AS \subset S$. For example, 0, $\mathbb{C}^n$, $\ker(A)$, $\text{img}(A)$ are all A-invariant subspaces.

Projection matrix: A projector or projection matrix P is a square matrix that satisfies

$$P = P^2. \tag{1.22}$$

Such a matrix is also said to be *idempotent.* If P is a projector, $(I - P)$ is also a projector since

$$(I - P)^2 = I - 2P + P^2 = I - 2P + P = I - P.$$

$I - P$ is called *complementary* projector to P. Let $V = \text{Range}\,(P)$, then P is projector onto V. On the other hand, if $\{v_1, v_2, \cdots, v_r\}$ is a basis of V and

$$V_r = \begin{bmatrix} v_1 & v_2 & \cdots & v_r \end{bmatrix},$$

then $P = V(V^TV)^{-1}V^T$ is a projector onto V.

Definition 16 (Orthogonal and oblique projectors). *Let S_1 and S_2 be two orthogonal subspaces of $\mathbb{C}^m$ such that $S_1 \cap S_2 = \{0\}$ and $S_1 + S_2 = \mathbb{C}^m$, where $S_1 + S_2$ denote the span of S_1 and S_2, i.e., the set of vectors $s_1 + s_2$ with $s_1 \in S_1$ and $s_2 \in S_2$. If a projector projects onto a subspace S_1 along a subspace S_2, then it is called* orthogonal projector. *Otherwise, the projector is called* oblique projector.

Orthogonal projection is known as Galerkin projection, and oblique projection is known as Petrov-Galerkin-type projection.

Definition 17 (Spectral projector). *If V is an A-invariant subspace corresponding to a subset of A's spectrum, then we call P a spectral projector.*

Lemma 1. *If $P = P^T$, then P is an orthogonal projector, otherwise P is an oblique projector.*

Lemma 2. *Let V be another r-dimensional subspace of $\mathbb{R}^n$ with the basis matrix $V_r = \begin{bmatrix} v_1, v_2, \cdots, v_r \end{bmatrix}$. If W is another r-dimensional subspace of $\mathbb{R}^n$ and $W_r = \begin{bmatrix} w_1, w_2 \cdots w_r \end{bmatrix}$ is a basis matrix for W, then $P = V_r(W_r^TV_r)^{-1}W_r^T$ is an oblique projector onto V along W.*

Low-rank approximation of a matrix: Any matrix can be approximated by its low-rank matrix. This result is referred to as the matrix approximation theorem of *Eckart-Young-Mirsky theorem.*

Theorem 6 (Eckart-Young-Mirsky theorem). *Let the SVD of $A \in \mathbb{R}^{m\times n}$ $(m \leq n)$ be $A = U\Sigma V$. The best rank-r approximation to A is given by*

$$\hat{A} = \sum_{i=1}^{r} \sigma_i u_i v_i^T, \tag{1.23}$$

where u_i *and* v_i *denote the ith column of* U *and* V*, respectively. Then the approximation errors in spectral norm (2-norm) and Frobenius norm are respectively given by,*

$$\|A - \hat{A}\|_2 = \sigma_{r+1} \quad \textit{and} \quad \|A - \hat{A}\|_F = \sqrt{\sum_{i=r+1}^{m} \sigma_i^2}.$$

We can compress the images based on this theorem (see the Exercises).

Image compression by SVD

A digital image with $m \times n$ pixels can be represented by a matrix $A \in \mathbb{R}^{m \times n}$, where $A(i,j)$ contains color information of pixel (i,j). Then the memory requirements of this image (in single precision) is $4.m.n$ bytes. By taking the SVD of A as in (1.23) if we save $u_1, \cdots, u_r, \sigma_1 v_1, \cdots, \sigma_r v_r$, then the memory requirement is $4r.(m+n)$.

If a matrix has rank deficiency, then by applying this theorem we can approximate the matrix by its low-rank factors. For example, suppose the rank (numerical rank) of the matrix $A \in \mathbb{R}^{n \times n}$ is r. Then the SVD of the matrix is

$$A = U\Sigma V^T = \begin{bmatrix} U_r & U_{n-r} \end{bmatrix} \begin{bmatrix} \Sigma_r & \\ & \Sigma_{n-r} \end{bmatrix} \begin{bmatrix} V_r^T \\ V_{n-r}^T \end{bmatrix}$$

where $\Sigma = \operatorname{diag}(\Sigma_1, \Sigma_2) = \operatorname{diag}(\sigma_1, \cdots, \sigma_r, \sigma_{r+1}, \cdots, \sigma_n)$. Now truncating the negligible singular values $\sigma_{r+1}, \cdots, \sigma_n$, we can write

$$A \approx U_r \Sigma_r V_r^T = (U_r \Sigma^{\frac{1}{2}})(V_r \Sigma^{\frac{1}{2}})^T = RL^T,$$

where R and L are called low-rank factors of A. If the matrix A is symmetric and positive definite, then $L = R$.

Definition 18 (Numerical rank). *The matrix* $A \in \mathbb{R}^{n \times n}$ *has* numerical-rank r *if the singular values of* A *which are* $\sigma_1, \cdots, \sigma_n$ *satisfy*

$$\sigma_1 \geq \sigma_2 \geq \sigma_r > \epsilon \geq \sigma_{r+1} \geq \cdots \geq \sigma_n,$$

where ϵ *is an error tolerance.*

Numerical rank computation

To compute the numerical rank of a matrix A, count the singular values which are larger than ϵ (given tolerance). If this number is r, then we call numerical rank of A is r under the tolerance ϵ.

The Jordan canonical form of a matrix: For an $n \times n$ matrix A, if there exists a nonsingular matrix U such that

$$U^{-1}AU = J = \begin{bmatrix} J_1 & & \\ & \ddots & \\ & & J_k \end{bmatrix},$$

where

$$J_j = \begin{bmatrix} \lambda_j & 1 & & \\ & \lambda_j & \ddots & \\ & & \ddots & 1 \\ & & & \lambda_j \end{bmatrix}; \quad j = 1, \cdots, k$$

are $p_j \times p_j$ matrices and $p_1 + \cdots + p_k = n$, then J is called Jordan canonical form (JCF) of A. The matrices J_j are called Jordan blocks and for each $j = 1, \cdots, k$, λ_j is the eigenvalues of A with multiplicity p_j.

Kronecker product and vec-operation:

Definition 19 (Kronecker product). *Consider an $m \times n$ matrix A as defined in (1.1) and B be any $p \times q$ matrix. The Kronecker product of A and B is defined as:*

$$A \otimes B = \begin{bmatrix} a_{11}B & a_{12}B & \cdots & a_{1n}B \\ a_{21}B & a_{22}B & \cdots & a_{2n}B \\ \vdots & \vdots & \ddots & \vdots \\ a_{m1}B & a_{m2}B & \cdots & a_{mn}B \end{bmatrix}.$$

The matrix $A \otimes B$ is an $mp \times nq$ matrix. If the matrices A and B are invertible, then $A \otimes B$ is also invertible and $(A \otimes B)^{-1} = A^{-1} \otimes B^{-1}$.

Definition 20 (Vec-operation). *Consider an $m \times n$ matrix A as defined in (1.1). The vec-operation on A converts the matrix into a vector as follows:*

$$\operatorname{vec}(A) = \left[a_{11}, \cdots, a_{m1}, a_{12}, \cdots, a_{m2}, \cdots, a_{n1}, \cdots, a_{mn}\right]^T.$$

Schur complement: Suppose A is a block matrix with the submatrices A_{11}, A_{12}, A_{21} and A_{22}, such that

$$A = \begin{bmatrix} A_{11} & A_{12} \\ A_{21} & A_{22} \end{bmatrix}.$$

If A_{22} is invertible, then the *Schur complement* of A_{22} of the matrix A is defined by

$$A/A_{22} = A_{11} - A_{12}A_{22}^{-1}A_{21}.$$

Similarly, if A_{11} is invertible, then the Schur complement of A_{11} of the matrix A is defined by $A/A_{11} = A_{22} - A_{21}A_{11}^{-1}A_{12}$. If the matrix A is sparse due to invertibility of A_{22} (or A_{11}), the Schur complement A/A_{22} (or A/A_{11}) is dense. It can be proved (see, e.g., the Exercises in this chapter), all the eigenvalues of the Schur complement A/A_{22} (or A/A_{11}) coincide to the eigenvalues of A. These properties will be used in the model reduction of first-order and second-order index 1 descriptor systems. It can also be proved that if the matrices A and A_{11} are nonsingular, then A/A_{22} is also nonsingular and

$$A^{-1} = \begin{bmatrix} A_{11}^{-1} + A_{11}^{-1}A_{12}(A/A_{22})^{-1}A_{21}A_{11}^{-1} & -A_{11}^{-1}A_{21}(A/A_{22})^{-1} \\ -(A/A_{22})^{-1}A_{21}A_{11}^{-1} & (A/A_{11})^{-1} \end{bmatrix}.$$

See, e.g., [191] for further details and more interesting properties of Schur complements.

Sherman-Morrison-Woodbury (SMW) formula: Suppose A is an $n \times n$ invertible matrix and B, C and D are the matrices with sizes $n \times k$, $k \times k$ and $k \times n$, respectively. The inversion of $A + BCD$ can be computed as:

$$(A + BDC)^{-1} = A^{-1} - A^{-1}B(D^{-1} + CA^{-1}B)^{-1}CA^{-1}.$$

This formula is known as *Sherman-Morrison-Woodbury* [185, 186] formula or just *Woodbury* formula. It is a generalized version of the *Sherman-Morrison formula* [150]. In the case of $D = I$ (identity matrix) the formula becomes

$$(A + BC)^{-1} = A^{-1} - A^{-1}B(I + CA^{-1}B)^{-1}CA^{-1}.$$

This formula is very efficient in computing $(A + BC)^{-1}$ (see, e.g., the Exercises).

1.10 Some useful MATLAB functions

Function name	Description
`rand, eye, zeros, diag`	create different types of matrices
`sprand, speye, sparse, spdiags`	create different types of sparse matrices
`whos`	provide the information (size, bytes, class, sparsity, etc.) of the variable
`eig`	compute all eigenvalues and eigenvectors of a matrix or matrix-pencil
`eigs`	compute a few eigenvalues and eigenvectors of a matrix or matrix-pencil
`svd`	singular value decomposition
`lu`	lu factorization of a matrix
`chol`	cholesky decomposition of a matrix
`qr`	qr factorization of a matrix
`schur`	schur decomposition of a matrix
`norm`	matrix or vector norm
`normest`	compute the matrix 2-norm
`rank`	matrix rank computation
`det`	determinant of a matrix
`cond`	compute the condition number
`condest`	1-norm condition number estimate
`\, gmres, bicg`	solve linear system
`orth`	form an orthonormal basis for the range of a matrix
`null`	form an orthonormal basis for the nullspace of a matrix
`plot`	linear plot

`loglog`
 	
`semilogx (semilogy)`
	log-log scale plot using logarithmic scaled for both the x-axis and y-axis
	log-log scale plot using logarithmic scaled for x- (y-) axis

`spy`	visualize the sparsity pattern
`nnz`	number of nonzero matrix elements
`reshape`	convert a Matrix to a vector
`vec2mat`	convert a vector to a matrix
`kron`	kronecker product
`imshow, image`	display image, display image from array
`imread`	read image from graphics file

■ Exercises:

1.1 Load two matrices as $A =$ `gallery`('normaldata', [60, 60], .0); and $B =$ `bucky`;. Show the sparsity pattern of the matrices. Also, find the memory requirements for both matrices.

1.2 Apply the appropriate MATLAB commands to the matrices A and B from Exercise 1.1 to find the following identities:

   (i) Discuss the invertibility using the determinant.

   (ii) Compute the eigenvalues and plot them.

   (iii) Find the rank of the matrices and discuss whether they have full-rank or not.

   (iv) Find the singular values and plot them using `semilogy` command.

1.3 Use appropriate MATLAB commands to find (where possible) the eigenvalue, singular value, LU and Cholesky factorizations of the matrices A and B from Exercise 1.1. If any of the factorizations is not possible, then discuss the reasons.

1.4 Consider a matrix $A =$ `bucky` and suppose rank $(A) = n$. Applying the SVD onto A, construct a matrix A_r whose rank is r. Consider $r = 5, 20$ and 30; for each case show that $\|A - A_r\|_2 = \sigma_{r+1}$, where σ_{r+1} denotes the $r + 1$st singular value of A.

1.5 Are the columns of the matrix A from Exercise 1.1 orthogonal? If not using Algorithm 1 or 2, orthogonalize the columns of A and show that $A^T A = AA^T = I$.

1.6 Consider a linear system $Ax = b$. Load A and b using the MATLAB commands: `load west0479`, $A =$ west0479; and $b =$ `sum`$(A, 2)$;. Then, solve the linear system as follows:

   (i) $x = A\backslash b$;

   (ii) $[L, U, P] =$ `lu`(F); $x = U\backslash(L\backslash(P * b))$;

   (iii) $[L, U, p] =$ `lu`$(F,' vector')$; $x = U\backslash(L\backslash(b(p, :))$;. Now compare the procedures in terms of accuracy and efficiency.

1.7 We know that Algorithm 3 generates the orthogonal basis of a m Krylov-subspace $\mathcal{K}_m(A, b)$. Considering A and b from Exercise 1.6, form the basis of $\mathcal{K}_{50}(A, b)$, $\mathcal{K}_{100}(A, b)$ and $\mathcal{K}_{500}(A, b)$.

1.8 Compute all the eigenvalues of $A = bucky$ using $d_o =$ `eig(full`(A)`)`. By applying Algorithm 5, compute 20 ritzvalues (d_l) with large magnitude and 30 ritzvalues with small magnitude (d_s). Now plot d_o, d_l and d_s on a single figure and show the comparisons.

1.9 Suppose a matrix $A \in \mathbb{R}^{n\times n}$ is partitioned as

$$A = \begin{bmatrix} J_1 & J_2 \\ J_3 & J_4 \end{bmatrix},$$

with $J_1 \in \mathbb{R}^{k\times k}$, $J_2 \in \mathbb{R}^{k\times n-k}$, $J_3 \in \mathbb{R}^{n-k\times k}$ and $J_4 \in \mathbb{R}^{n-k\times n-k}$ ($k < n$). A matrix $A_s = J_1 - J_2 J_4^{-1} J_3$ is called Schur complement of A w.r. to block J_4. Show that all the eigenvalues of A_s coincide with the eigenvalues of A. For the matrix $A =$ `bucky` show this identity.

1.10 Write a MATLAB routine which reads in a matrix containing image data, computes a best rank-k approximation of this matrix and then displays the approximation error (w.r. to the 2-norm), original and compressed image as well as required memory for storing the images. Run the program by loading the data (MATLAB commands): `load clown`. Test different values of k and empirically determine the smallest value of k which can be used without a visible loss of accuracy.

1.11 Show that the solution of the linear system $(A + BC^T)x = b$, where $A \in \mathbb{R}^{n\times n}$ $B, C \in \mathbb{R}^{n\times k}$ and $b \in \mathbb{R}^n$ can be obtained by solving the linear system

$$\begin{bmatrix} A & B \\ C^T & -I \end{bmatrix} \begin{bmatrix} x \\ \gamma \end{bmatrix} = \begin{bmatrix} b \\ 0 \end{bmatrix},$$

for x. Then show that x can also be computed as:

$$x = A^{-1}b - A^{-1}B(I + CA^{-1}B)^{-1}CA^{-1}b.$$

The last expression can be obtained by applying the SMW (Sherman-Morrison-Woodbury) formula. Construct the matrices A, B and C and a vector b using MATLAB `sprand` command (for example, you can consider $n = 1000$, $k = 100$). Now using MATLAB\(backslash), find x in all cases mentioned above and discuss their efficiency.

Chapter 2

Dynamic Systems and Control Theory

2.1 Introduction

The central topic of this book is the model reduction of LTI (linear time invariant) continuous-time systems which is considered as a branch of control theory. In order to develop the algorithms for the model reduction of dynamical control systems, one needs to exploit basic concepts of control theory. This chapter reviews some important topics of control theory from the previous literature. We briefly introduce some basic theorems and definitions, important tools and essential notations and present the results that will be used throughout the remainder of this book. Note that we have intentionally excluded detailed discussion of any particular topic and proof of the theorems, since details are available in the references. The materials which are discussed here are available in any book on linear control systems. See, e.g., [51, 80, 104, 108, 118, 153, 155, 193, 195]

2.2 A brief introduction of dynamical control systems

A dynamical control system is a system that evolves with time according to some fixed rule. In general, such a rule can be stated as

$$E\dot{x}(t) = f(x(t), u(t), t); \quad x(t_0) = x_0, \tag{2.1}$$

where, $E \in \mathbb{R}^{n\times n}$ is a constant matrix, $x(t) \in \mathbb{R}^n$ is the vector of states, $u(t) \in \mathbb{R}^p$ is the vector of control inputs and f is a continuously differentiable function. The dimension or the order of the system is determined by n, i.e., the number of states. The dot notation denotes ordinary differentiation with respect to t. For a given initial condition $x(t_0) = x_0$ and the input $u(t)$ the state at any future time, $x(t_1)$, can be determined by integrating (2.1) from t_0 to t_1. The relationship given in (2.1) is very general and can be used to describe a wide variety of different systems. If the function f does not depend explicitly on time, then system (2.1) is called time-invariant and it is a simplification of the time-variant system. This is often a very reasonable assumption, since the underlying physical laws themselves do not typically depend on time. For time-invariant systems, the parameters or coefficients of the function, f, are constant. The states and the control may, however, still be time dependent. A time-invariant system can be written as

$$E\dot{x}(t) = f(x(t), u(t)); \quad x(t_0) = x_0 \tag{2.2}$$

The time-invariant dynamical system (2.2) can be either linear or nonlinear. A linear time-invariant (LTI) continuous-time system can often be represented in matrix vector form as

$$E\dot{x}(t) = Ax(t) + Bu(t); \quad x(t_0) = x_0 \tag{2.3}$$

where $A \in \mathbb{R}^{n\times n}$ and $B \in \mathbb{R}^{n\times p}$ are the constant matrices known as system and input matrices, respectively. In reality, physical systems, for the most part, are nonlinear. In that case, however, for purposes like analysis, simulation and optimization, the nonlinear system can be converted into linear system. Such conversion techniques can be found in various books of control theory (see, e.g., [51, 80, 108, 118, 153, 155], also the following examples). This book only discusses the LTI continuous-time system. In the following section, we introduce the properties of such system which will be used throughout the coming chapters.

2.3 Representations of LTI dynamical systems

In the above section, we have introduced a LTI dynamical system. In various disciplines and for various purposes they are represented in different ways. Here we introduce two commonly used representations.

2.3.1 Generalized state-space representation

A generalized state-space linear time invariant (LTI) continuous-time system can be represented as

$$E\dot{x}(t) = Ax(t) + Bu(t); \quad x(t_0) = x_0,\, t \geq t_0, \tag{2.4a}$$
$$y(t) = Cx(t) + D_a u(t), \tag{2.4b}$$

where (2.4a) is known as system equation and (2.4b) is known as output equation in which the time dependent vector $y(t) \in \mathbb{R}^m$ denotes the measurement output. Output equation is necessary because often there are state variables which are not directly observed or are otherwise not of interest. The output matrix, $C \in \mathbb{R}^{m\times n}$, is used to specify which state variables are available for use by the controller. Also, often, there is no direct feed-forward in which case $D_a \in \mathbb{R}^{m\times n}$ is the zero matrix. The system in (2.4) is called single-input/single-output (SISO) system if $m = p = 1$; otherwise, it is known as multi-input/multi-output (MIMO) system.

- In the SISO system, $B \in \mathbb{R}^{n\times 1}$ and $C \in \mathbb{R}^{1\times n}$.
- In the MIMO system, $B \in \mathbb{R}^{n\times p}$ and $C \in \mathbb{R}^{m\times n}$, where either p or m, or both are greater than 1.

If E is an identity matrix (i.e., I), then the system (2.4) is known as a standard LTI continuous-time system. However, if E is an invertible matrix, then the system (2.4) can be converted into a standard system

$$\dot{x}(t) = A_s x(t) + B_s u(t); \quad x(t_0) = x_0,\, t \geq t_0, \tag{2.5a}$$
$$y(t) = Cx(t) + D_a u(t), \tag{2.5b}$$

where $A_s = E^{-1}A$ and $B_s = E^{-1}B$.

For convenience, we denote system (2.4) by $\Sigma_{gss} := (A, B, C, D_a, E)$.

2.3.2 Transfer function representation

The state-space representation discussed in the above section is essentially a time domain representation of the system. It is possible to convert a system's time-domain representation into a complex-domain representation by applying the *Laplace transformation.* This representation is also known as Laplace-domain representation. In this representation the governing differential equations are converted into algebraic equations and hence simulation and analysis of the system becomes easier.

Definition 21 (Laplace Transformation). *The Laplace transform of a time domain function, $f(t)$, is defined below:*

$$F(s) = \mathcal{L}\{f(t)\} = \int_0^\infty e^{-st} f(t)dt$$

where the parameter $s = \sigma + j\omega$ is a complex frequency variable.

We apply the Laplace transformation on the system (2.4). In this case, we consider that $X(s)$, $Y(s)$ and $U(s)$ are the Laplace transformations[1] of $x(t)$, $y(t)$ and $u(t)$, respectively. Then the system in (2.4) becomes:

$$sEX(s) - Ex(0) = AX(s) + BU(s), \tag{2.6a}$$

$$Y(s) = CX(s) + D_aU(s). \tag{2.6b}$$

Equation in (2.6a) implies

$$X(s) = (sE - A)^{-1}Ex(0) + (sE - A)^{-1}BU(s). \tag{2.7}$$

Inserting this identity into (2.6b), we obtain

$$Y(s) = C(sE - A)^{-1}Ex(0) + C(sE - A)^{-1}BU(s) + D_aU(s). \tag{2.8}$$

If the initial condition is $x(0) = 0$, then (2.8) can be represented as

$$Y(s) = \mathrm{G}(s)U(s), \tag{2.9}$$

where

$$\mathrm{G}(s) = \mathrm{C}(sE - A)^{-1}B + D_a. \tag{2.10}$$

is called the *transfer function* of the system.

Definition 22. *Transfer function (TF) is the input-output relation or the ratio of the input to the output of a dynamical system in the complex-domain. For the dynamical system (2.5), the TF is denoted by (2.10). To find a TF, we always consider the system's initial condition to be zero.*

[1] $\mathcal{L}\{\dot{x}(t)\} = sX(s) - X(0)$, $\mathcal{L}\{x(t)\} = X(s)$ and $\mathcal{L}\{u(t)\} = U(s)$

In the SISO case, TF of the system is just a rational function. On the other hand, for the MIMO systems, G(s) of (2.4) is the $p \times m$ matrix and can be written as

$$\mathrm{G}(s) = \begin{bmatrix} \mathrm{G}_{11}(s) & \mathrm{G}_{12}(s) & \cdots & \mathrm{G}_{1m}(s) \\ \mathrm{G}_{21}(s) & \mathrm{G}_{22}(s) & \cdots & \mathrm{G}_{2m}(s) \\ \vdots & \vdots & \cdots & \vdots \\ \mathrm{G}_{p1}(s) & \mathrm{G}_{p2}(s) & \cdots & \mathrm{G}_{pm}(s) \end{bmatrix}, \tag{2.11}$$

where $\mathrm{G}_{il} = C(i,:)(sE - A)^{-1}B(:,l) + D_a(i,l)$ with $i = 1, 2, \cdots, p$ and $l = 1, 2, \cdots, m$. Transfer function is an important tool to analyze dynamical control systems. In fact, the behavior of the dynamical systems can be fully characterized by its transfer function.

Definition 23. *The transfer function $G(s)$ as defined in (2.10) is called* proper *if* $\lim_{s\to\infty} G(s) < \infty$ *and* strictly proper *if* $\lim_{s\to\infty} G(s) = 0$. *Otherwise, $G(s)$ is called* improper.

2.4 System responses

In order to analyze a dynamical control system, we need to solve the system to find the output response. The output response of a system is often known as the system response. Two types of response, namely time domain and frequency domain, are most commonly used by control system engineers for analyzing the systems. In the following subsections, we give their brief introductions.

2.4.1 Time response

The time-domain response represents how the state of a dynamic system changes in time when subjected to a particular input. Based on different input, we have different types of responses from the system. To find time-domain responses, the system (2.4) first needs to be rewritten into system (2.5). Multiplying both sides of (2.5a) by $e^{-A_s t}$ we obtain

$$e^{-A_s t}\dot{x}(t) = e^{-A_s t}A_s x(t) + e^{-A_s t}B_s u(t),$$

which can be represented as

$$\frac{d}{dt}\left[e^{-A_s t}x(t)\right] = e^{-A_s t}B_s u(t).$$

Integrating both sides of this equation from t_0 to t,

$$\begin{aligned}
\left[e^{-A_s t}x(t)\right]_{t_0}^{t} &= \int_{t_0}^{t} e^{-A_s \tau} B_s u(\tau) d\tau \\
e^{-A_s t}x(t) - e^{-A_s t_0}x(t_0) &= \int_{t_0}^{t} e^{-A_s \tau} B_s u(\tau) d\tau \\
x(t) = e^{A_s(t_0 - t)}x(t_0) &+ \int_{t_0}^{t} e^{A_s(t-\tau)} B_s u(\tau) d\tau. \qquad (2.12)
\end{aligned}$$

Now plugging $x(t)$ into (2.5b) yields

$$y(t) = Ce^{A_s(t-t_0)}x(t_0) + \int_{t_0}^{t} Ce^{A_s(t-\tau)} B_s u(\tau) d\tau + D_a u(t). \qquad (2.13)$$

Assuming $t_0 = 0$, we obtain the time domain response of the system (2.4) as

$$y(t) = Ce^{A_s t}x(0) + \int_{t_0}^{t} Ce^{A_s(t-\tau)} B_s u(\tau)\tau + D_a u(t), \qquad (2.14)$$

where the matrix exponential $e^{A(t-t_1)}$ is called *state transition matrix.*

Definition 24 (Matrix exponential). *The matrix exponential* e^{At} *as defined above has the form*

$$e^{At} = \sum_{k=0}^{\infty} \frac{(At)^k}{k!}.$$

In general, the system's time response is measured by the output response $y(t)$ for a particular initial condition $x(0)$ and the control input $u(t)$. Based on different types of inputs, different types of system responses are defined. Two important system responses are the impulse response and the step response; these are commonly used to analyze the characteristics of LTI systems. The impulse response of a system is the output of the system if $x(0) = 0$ and an impulse function is used as the input.

Definition 25 (Impulse function). *The impulse function, which is often called* dirac delta function, *is denoted as follows*

$$\delta(t) = \begin{cases} 0, & \text{if } t \neq 0 \\ \text{undefined}, & \text{if } t = 0 \end{cases}$$

Inserting $x(0) = 0$ and $u(t) = \delta(t)$, (2.14) implies

$$\begin{aligned}
y(t) &= \int_{t_0}^{t} \left(Ce^{A_s(t-\tau)}B + D\delta(t-\tau)\right) u(\tau)\, d\tau, \\
&= \int_{t_0}^{t} h(t-\tau)u(\tau)\, d\tau, \qquad (2.15)
\end{aligned}$$

where the matrix $h(t) = Ce^{A_s t}B + D_a\delta(t)$ is the *impulse response matrix*. In fact, the impulse response matrix relates the input to the output in the time domain. The (i, j) element of $h(t)$ is responsible for the *i-th* output at time t due to the *j-th* input (of unit impulse) of the system, while all other inputs are considered zero.

Another common and frequently used tool to analyze the system is the *step response*. Typically, this is the first characteristic to be analyzed for a newly designed system. Similar to the impulse response, the step response of a system is the output of the system when a unit step function is used as the input.

Definition 26 (Unit step function). *The unit step function is defined as*

$$u_{step}(t) = \begin{cases} 0, & \textit{if } t < 0 \\ 1, & \textit{if } t \geq 0 \end{cases}$$

When subjected to the step input, the system will initially have an undesirable output. This is called the *transient response*. The transient response occurs because a system is approaching its final output value. If the time goes to infinity, system response is called *steady-state response*. The steady-state response of the system occurs when the transient response has ended. The amount of time to take place the transient response is known as the rise time. The amount of time it takes for the transient response to end and the steady-state response to begin is known as the settling time. It is common practice for systems engineers to try and improve the step response of a system. In general, it is desirable to have the transient response reduced, the rise and settling times be shorter, and the steady-state to approach a desired "reference" output. Using the MATLAB command `step`, we can compute the step response of the system (2.4). However, for a large-scale system, we can apply the *implicit Euler method* that is summarized in Algorithm 6 for computing the step response of the system (2.4).

2.4.2 Frequency response

In the above subsection, we have shown the complex-domain representation of the LTI continuous-time system. In this representation the $\mathrm{G}(s)$ where, $s \in \mathbb{C}$, is the input-output relation of the system. For $s = \mathrm{j}\omega$, where $\omega \in \mathbb{R}$ is called the frequency, the output response

$$Y(\mathrm{j}\omega) = \mathrm{G}(\mathrm{j}\omega)U(\mathrm{j}\omega), \tag{2.16}$$

is called *frequency response* of the LTI continuous-time system. The matrix $\mathrm{G}(\mathrm{j}\omega)$ is known as *frequency response* matrix. For the LTI continuous-time system (2.4), the frequency response matrix is defined by

$$\mathrm{G}(\mathrm{j}\omega) = C(\mathrm{j}\omega E - A)^{-1}B + D_a. \tag{2.17}$$

Algorithm 6: Time-domain (step) response.

Input : E, A, B, C, D_a, t_{max}, N.
Output: stp

Compute $h = \frac{t_{max}-0}{N}$
Compute LU decomposition of $(E - h * A)$ as

$$[L, U, P, Q] = lu(E - h * A)$$

for $i = 1 : N$ **do**
  $x = Q * (U \backslash (L \backslash (P * (E * x + h * B))))$
Find $stp = C * x + D_a$

Algorithm 7: Singular value of the frequency response.

Input : E, A, B, C, D_a, $\{\omega_{min}, \omega_{max}\}$, N.
Output: SV

Compute $s = N$ logarithmically equally spaced points between $10^{\omega_{min}}$ and $10^{\omega_{max}}$.
for $i = 1 : N$ **do**
  $G = C(1\mathrm{j} * E - A)^{-1}B$
  if *SISO* **then**
    $SV = | G |$.
  else
    $SV =$ `max(svds(`G`))`.

Because there is no integral term in the frequency response, its computation is much easier than that of the time domain response. In fact, the frequency response matrix plays a key role for the frequency response of the dynamical system. Therefore, for the analysis of the LTI dynamical system we use magnitude and phase of the frequency response matrix. The frequency response is similar to the transfer function, except for the fact that it is the relationship between the system output and input in the complex Fourier domain, not the Laplace domain. The MATLAB bode diagram can compute magnitude and phase of the frequency response of a dynamical system (2.4). The graph of the magnitude of the frequency response is often called a *sigma plot.* Algorithm 7 can efficiently generate the singular values (SVs) of the frequency response for a large-scale dynamical system (2.4).

2.5 System Gramians

Two special matrices, namely controllability Gramian and observability Gramian, play important roles in the control theory. Among many applications, they are extremely useful in the model reduction of large-scale dynamical systems. The Gramian-based MOR methods are in general based on the principle of the system *controllability Gramian* and *observability Gramian.* In the following subsections, we briefly introduce the Gramians including their properties.

2.5.1 Controllability Gramian

For the LTI continuous-time system (2.4), the controllability Gramian can be defined as

$$P_t = \int_0^t \mathrm{e}^{E^{-1}At}(E^{-1}B)(E^{-1}B)^T \mathrm{e}^{A^T E^{-T} t}\,\mathrm{d}t.$$

If the system (2.4) is stable, i.e., all the eigenvalues of the matrix pencil (A, E) are in $\mathbb{C}^{-1}$, then the controllability Gramian of the system can be defined as

$$P = \int_0^\infty \mathrm{e}^{E^{-1}At}(E^{-1}B)(E^{-1}B)^T \mathrm{e}^{A^T E^{-T} t}\,\mathrm{d}t. \tag{2.18}$$

Since $P = P^T$, the controllability Gramian is symmetric. Multiplying (2.18) left side by v^T (v is a non-zero vector) and right side by v we have,

$$v^T P v = \int_0^\infty v^T \mathrm{e}^{E^{-1}At}(E^{-1}B)(E^{-1}B)^T \mathrm{e}^{A^T E^{-T} t} v\,\mathrm{d}t.$$

Since matrix exponential is always nonsingular and $E^{-1}BB^T E^{-T}$ is positive (semi-)definite, we can say that P is positive (semi-)definite. The following theorem shows that P can be computed by solving the continuous-time algebraic Lyapunov equation [73]

$$APE^T + EPA^T = -BB^T \tag{2.19}$$

Theorem 7. *For a stable system (2.4), the symmetric positive (semi-)definite matrix P, defined in (2.18) is the unique solution of the Lyapunov equation in (2.19).*

Proof. First we want to prove that P is the solution of the given Lyapunov equation. To prove it, we modify the Lyapunov equation (2.19) by premultiplying E^{-1} and postmultiplying by E^{-T}

$$E^{-1}AP + PA^T E^{-T} = -E^{-1}BB^T E^{-T}. \tag{2.20}$$

If $P = \int_0^{\infty} \mathrm{e}^{E^{-1}At}(E^{-1}B)(E^{-1}B)^T \mathrm{e}^{A^T E^{-T} t}\,dt$ is the solution of (2.19), it should satisfy (2.20). Now

$$\begin{aligned} E^{-1}AP + PA^T E^{-T} &= \int_0^{\infty} E^{-1}A\mathrm{e}^{E^{-1}At}(E^{-1}B)(E^{-1}B)^T \mathrm{e}^{A^T E^{-T} t}\,\mathrm{d}t + \\ &\quad \int_0^{\infty} \mathrm{e}^{E^{-1}At}(E^{-1}B)(E^{-1}B)^T \mathrm{e}^{A^T E^{-T} t} A^T E^{-T}\,\mathrm{d}t \\ &= \int_0^{\infty} \frac{d}{dt}\left(\mathrm{e}^{E^{-1}At} E^{-1}BB^T E^{-T} \mathrm{e}^{A^T E^{-T} t}\right). \end{aligned}$$

Since the system is stable, the eigenvalues of the matrix $\Lambda(A, E)$ or $\Lambda(E^{-1}A) \in \mathbb{C}^{-1}$. Therefore, $e^{E^{-1}At} \to 0$ as $t \to \infty$ and (2.20) holds.

□

Remark 2. *In [194], the authors use the frequency domain representations of these integrals*

$$P = \frac{1}{2}\int_{-\infty}^{\infty} (\imath\omega E - A)^{-1}BB^T(-i\omega E^T - A^T)^{-1}\,d\omega$$

to extend the definition of the Gramians to systems with no poles on the imaginary axis.

2.5.2 Observability Gramian

The observability Gramian of the system (2.4) can be defined as

$$Q_t = \int_0^{t} \mathrm{e}^{E^{-T}A^T t} E^{-T}C^T C E^{-1} \mathrm{e}^{AE^{-1}t}\,\mathrm{d}t.$$

In the case of a stable system, the observability Gramian of the system can be defined as

$$Q = \int_0^{\infty} \mathrm{e}^{E^{-T}A^T t} E^{-T}C^T C E^{-1} \mathrm{e}^{AE^{-1}t}\,\mathrm{d}t. \tag{2.21}$$

Like the controllability Gramian, Q is also symmetric and positive (semi-)definite. It can also be shown that Q is the solution of the Lyapunov equation [73]

$$A^T QE + E^T QA = -C^T C, \tag{2.22}$$

Remark 3. *In [194], the authors use the frequency domain representations of these integrals*

$$Q = \frac{1}{2\pi} E^T \left\{ \int_{-\infty}^{\infty} (\imath\omega E^T - A^T)^{-1} C^T C (-\imath\omega E - A)^{-1} \, d\omega \right\} E \tag{2.23}$$

to extend the definition of the Gramians to the systems with no poles on the imaginary axis.

2.5.3 Physical interpretation of the Gramians

The controllability and observability Gramians also have an interpretation from a physical point of view [77]. For the system (2.4) consider the following two relations

$$J_c = \min_u \int_{-\infty}^{0} u^*(t)u(t)dt, \quad x(0) = x_0, \, t \leq 0, \tag{2.24a}$$

$$J_o = \int_{0}^{-\infty} y^*(t)y(t)dt, \quad u(t) = 0, \, x(0) = x_0, \, t \geq 0, \tag{2.24b}$$

where J_c defines the minimum required energy to drive the system from its zero state to state x_0 and J_o is the obtained energy observed at the output under the zero input and the initial condition x_0. The functionals J_c and J_o can be determined from

$$J_c = x_0^* P^{-1} x_0 \tag{2.25}$$

and

$$J_o = x_0^* Q x_0. \tag{2.26}$$

where P and Q are the controllability and observability Gramians of the system. The relation in (2.25) says that any state $x_0 = x(t)$ that lies in an eigenspace of P^{-1} corresponding to large eigenvalues requires more input energy to control. Since the eigenvectors of P^{-1} corresponding to large eigenvalues are equal to the eigenvectors of P with small eigenvalues, it can be said that the state $x_0 = x(t)$ is difficult to control if it lies in an eigenspace of P corresponding to a small eigenvalue. Likewise, from (2.26) it can be said that the state that lies along one of the eigenvectors of Q with small eigenvalues is difficult to observe. We can assume that the states that are difficult to control and observe are less important. The balancing based MOR methods are based on identifying and truncating the less important states from the systems. We will revisit these issues in Chapter 4.

2.6 Controllability and observability

Controllability and *observability* are basic concepts in control theory; they are useful tools for solving many problems in system theory. The applications of these concepts can be found in [55, 80, 110, 115, 155, 195]. The ideas of controllability and observability of the system also play crucial roles in the MOR methods.

Definition 27. *The system in (2.4) is said to be controllable in $t_0 \leq t \leq t_f$, if there exists an admissible input $u(t)$ such that the system can be driven from its initial state $x(t_0)$ to any final state $x(t_f)$.*

To explain the idea of controllability, at the time $t_0 = 0$, let $x_0 = 0$. Then the relation in (2.12) yields [155]

$$\begin{aligned} x(t_f) &= \int_0^{t_f} e^{A_s(t_f-\tau)} B_s u(\tau) d\tau, \\ &= \int_0^{t_f} \left\{ I + A_s(t_f-\tau) + \frac{{A_s}^2}{2!}(t_f-\tau)^2 + \cdots \right\} B_s u(\tau) d\tau, \\ &= B_s \int_0^{t_f} u(\tau) d\tau + A_s B_s \int_0^{t_f} (t_f-\tau) u(\tau) d\tau + \\ & \quad A_s^2 B_s \int_0^{t_f} \frac{(t_f-\tau)^2}{2!} u(\tau) d\tau + \cdots . \end{aligned} \tag{2.27}$$

We see in (2.27) that $x(t_f)$ is the linear combination of $B_s, A_s B_s, \cdots, A_s^{n-1} B_s$. Therefore, it can be said that a final state $x(t_f)$ is controllable iff the controllability matrix

$$M_c = \begin{bmatrix} E^{-1}B & (E^{-1}A)E^{-1}B & \cdots & (E^{-1}A)^{n-1}E^{-1}B \end{bmatrix}$$

has full rank. The system (2.4) is said to be controllable if every state of the system is controllable, i.e., the controllability matrix has full rank [110], i.e., $\operatorname{rank}(M_c) = n$.

Definition 28. *The system in (2.4) is said to be observable in $t_0 \leq t \leq t_f$, if for a given input $u(t)$ the initial state $x(t_0)$ can be uniquely determined from the given output $y(t)$.*

Observability is the dual concept of the controllability. Analogous to controllability, one can notice that the system (2.4) is observable if the observability matrix

$$M_o = \begin{bmatrix} C \\ CE^{-1}A \\ \vdots \\ CE^{-1}A^{n-1} \end{bmatrix}$$

is nonsingular, i.e., the matrix M_o has full rank.

2.7 Stability

Stability is another important characteristic of the LTI system. It has many applications in control theory. In this book, we will exploit this characteristic in the solution of matrix equation and model reduction of the large-scale systems.

Definition 29 (Stable and unstable systems). *The dynamical system (2.4) is said to be asymptotically stable if all the eigenvalues of the matrix pencil* $\mathcal{P}(\lambda) = (\lambda E - A)$*, for any* $\lambda \in \mathbb{C}$*, have negative real part. If any of the eigenvalues has positive real part the system is unstable.*

Note that the eigenvalues of the matrix pencil are discussed in the previous chapter. We can view the eigenvalues of the system on the complex plane $\mathbb{C}$. In this case, if all the eigenvalues lie in the left complex half-plane (i.e., $\mathbb{C}^-$), the system will be *asymptotically stable*. However, if any of the eigenvalues lies in the right half-plane (i.e., $\mathbb{C}^+$), then the system is unstable. The stability of the system can also be defined from the TF, of the system. In the zero-pole representation of TF, if all the poles of the system have a negative real part, then the system is stable. If any of the poles has a positive real part, then the system is unstable. The stability of the system can also be found from the following observation:

Theorem 8. *For a given system*

$$E\dot{x}(t) = Ax(t), \quad x(0) = x_0,$$

the trajectory (i.e., the solution) of the system will be $x(t) \to 0$ *as* $t \to \infty$ *if all the eigenvalues of the matrix pencil* $\mathcal{P}(\lambda) = (\lambda E - A)$ *have a negative real part.*

Proof. If $u(t) = 0$ and $t_0 = 0$, then the equation in (2.12) leads to solution of the given system and can be written as

$$x(t) = e^{E^{-1}At}x_0.$$

Using similarity transformation, the matrix $E^{-1}A \in \mathbb{R}^{n\times n}$ can be put into Jordan canonical form as

$$E^{-1}A = \mathrm{TJT}^{-1} = \mathrm{T}\,\mathrm{diag}\,(\mathrm{J}_1, \mathrm{J}_2, \ldots, \mathrm{J}_k)\,\mathrm{T}^{-1},$$

where $\mathrm{J}_i \in \mathbb{C}^{n_i\times n_i}$ is the *Jordan block* with the eigenvalue λ_i ($\sum_{i=1}^{k} n_i = n$), and $\mathrm{T} \in \mathbb{R}^{n\times n}$ is the matrix with the basis of the eigenspace corresponding to the eigenvalues. Using the properties of the matrix exponential, we have

$$e^{E^{-1}At} = \mathrm{T}\,\mathrm{diag}\left(e^{\mathrm{J}_1 t}, e^{\mathrm{J}_2 t}, \ldots e^{\mathrm{J}_k t}\right)T^{-1}.$$

□

If all the eigenvalues have negative real parts, then $e^{J_i t} \to 0$ if $t \to \infty$. The stability of the system (2.4) can also be confirmed by the following observation.

Theorem 9. *The solution of the Lyapunov equation (2.20) is symmetric positive (semi-) definite if and only if all the eigenvalues of the system (2.4) have a negative real part.*

Proof. Consider that v is an eigenvector corresponding to the eigenvalue of the matrix pencil $\mathcal{P}(\lambda) = \lambda E - A$. Now premultiplying and postmultiplying (2.20) by v^* and v, respectively, we obtain

$$\begin{aligned} v^* E^{-1} A P v + v^* P A^T E^{-T} v &= -v^* E^{-1} B B^T E^{-T} v \\ \bar{\lambda} v P v + v^* P \lambda v &= -v^* E^{-1} B B^T E^{-T} v \\ (\bar{\lambda} + \lambda) v^* P v &= -v^* E^{-1} B B^T E^{-T} v. \end{aligned}$$

□

Since both $E^{-1}BB^TE^{-T}$ and P are symmetric positive (semi-)definite, the third line is true if $\bar{\lambda} + \lambda < 0$, i.e., $\mathrm{Re}(\lambda) < 0$. To prove the theorem, conversely follow the proof in Theorem 7. The stability of the system also relates with the controllability and observability of the system.

Theorem 10. *Let P and Q be the symmetric positive (semi-)definite solutions of the Lyapunov equations (2.19) and (2.22), respectively. If the system (2.4) is stable, then it is controllable and observable.*

If the system (2.4) is not stable, we can find a matrix $K \in \mathbb{R}^{p\times n}$ such that the closed-loop system

$$E\dot{x}(t) = (A - BK)x(t) + Bu(t); \quad x(t_0) = x_0,\ t \geq t_0, \tag{2.28a}$$
$$y(t) = Cx(t) + Du(t), \tag{2.28b}$$

is stable. This is called feedback stabilization. In this stabilization technique, finding the *stabilizing feedback matrix* K is a challenging task. One of the popular methods to stabilize the system is called *Riccati-based feedback stabilization.* In this method we need to solve the generalized continuous-time algebraic Riccati equation (GCARE)

$$A^T XE + EXA - EXBB^T XE = -C^T C, \tag{2.29}$$

for X. The feedback stabilization matrix, K, then can be formed by

$$K = B^T X. \tag{2.30}$$

Computation of K based on the Riccati equation can be found in any control system book under the topic *linear quadratic regulator* (LQR) problem.

Algorithm 8: System Hankel singular values (SHSVs).

Input : E, A, B, C.
Output: σ_h

1 Compute P by solving controllability Lyapunov equation (2.20).
2 Compute Q by solving controllability Lyapunov equation (2.22).
3 Compute the Cholesky decomposition of P and Q as

$$P = RR^T \quad \text{and} \quad P = LL^T.$$

4 Find $\sigma_h = \texttt{svd}(R^T L)$.

2.8 System Hankel singular values

The system Hankel singular values, or more simply Hankel singular values (HSVs), play a crucial role in the balancing-based model reduction that we will see later. In general, the HSVs of the system are the singular values of the Hankel operator (see, e.g., [7]). In [77], Glover shows that the system's HSVs are the positive square roots of the eigenvalues of the product of the controllability and observability Gramians, that is to say

$$\sigma_h = \sqrt{\lambda(PQ)} = \sqrt{\lambda(QP)}, \tag{2.31}$$

where λ denotes the eigenvalues. Since the controllability Gramian and the observability Gramian are symmetric positive definite, they always have the Cholesky decomposition:

$$P = RR^T \quad \text{and} \quad Q = LL^T. \tag{2.32}$$

It can be shown that (see, e.g., [109, 163])

$$\begin{aligned}\sigma_h =& \sqrt{\lambda(PQ)}\\ =& \sqrt{\lambda(RR^T LL^T)}\\ =& \sqrt{\lambda((R^T L)^T (R^T L))}\\ =& \sigma(R^T L),\end{aligned}$$

where σ_h denotes the singular values of $R^T L$. This means, the HSVs of the systems are the singular values of the product of the two Gramian factors. To compute the system's HSVs, we therefore use the Gramian factors of the systems. A procedure to compute the SHSVs is summarized in Algorithm 8.

2.9 Realizations

The transfer function ($G(s)$) of the system (2.4) is defined in (2.10). Typically, the set of matrices (E, A, B, C, D_a) is called the realization of the LTI system (2.4) under the transfer function $G(s)$. We know that the TF of the LTI system is invariant under *state-space transformations* or *coordinate transformations* [23]. For instance, if we replace $x(t)$ in (2.4) with

$$\tilde{x}(t) = Tx(t), \tag{2.33}$$

where the nonsingular matrix T is a coordinate transformation [23], we obtain a transformed system in which

$$(E, A, B, C, D_a) \Leftrightarrow (TET^{-1}, TAT^{-1}, TB, CT^{-1}, D_a). \tag{2.34}$$

The invariance of the transfer function under coordinate transformations of the system can be shown by

$$\begin{aligned}\tilde{\mathrm{G}}(s) =&(CT^{-1})(sTET^{-1} - TAT^{-1})^{-1}(TB) + D_a\\ =&C(sE - A)^{-1}B + D_a = G(s).\end{aligned}$$

Therefore, we can say that $(TET^{-1}, TAT^{-1}, TB, CT^{-1}, D_a)$ in another realization of the same transfer function $\mathrm{G}(s)$. Since the input/output relation of a system is not changing under coordinate transformations, a system may have infinitely many realizations. Among them there exist realizations where the dimension (r) of the system is minimum or the system consists of a minimum number of degree of freedoms (DoFs). This number r is called the *McMillan degree* of the system.

Definition 30. *A realization $(E_r, A_r, B_r, C_r, D_a)$ of the transfer function $G(s)$ in (2.10) of McMillan degree r is called a minimal realization.*

A state-space realization of a transfer function $\mathrm{G}(s)$ is minimal iff the system is controllable and observable. Note that although the McMillan degree is unique, the coordinate transformations lead to many minimum realizations of the same system. Fundamentally, the concept of MOR is to find a realization of a given system where the dimension of the system is as small as possible. We will study this in the next chapter.

2.10 The $\mathcal{H}_2$ norm and $\mathcal{H}_\infty$ norm

In the $\mathcal{H}_2$ and $\mathcal{H}_\infty$ control problems, the $\mathcal{H}_2$ norm and $\mathcal{H}_\infty$ norm are well known and play important roles to solve many problems. Both the norms are extremely useful in the model reduction of large-scale dynamical systems. The deviation of original and the reduced order models can be measured by measuring the $\mathcal{H}_2$ norm or the $\mathcal{H}_\infty$ norm. In the following subsections, we briefly introduce the two norms.

2.10.1 The $\mathcal{H}_2$ norm

The $\mathcal{H}_2$ norm is often known as the energy of the system obtained from the impulse input.

Definition 31 ($\mathcal{H}_2$ norm). *For a asymptotically stable system $(E, A, B, C, 0)$, the $\mathcal{H}_2$ norm is defined as*

$$\|G\|_{\mathcal{H}_2} = \sqrt{\frac{1}{2\pi}\int_{-\infty}^{\infty} \operatorname{tr}\left(G(j\omega)^* G(j\omega)\right) d\omega}. \tag{2.35}$$

Alternatively, the $\mathcal{H}_2$ norm can also be defined as

$$\|\mathrm{G}\|_{\mathcal{H}_2} = \sqrt{\int_0^{\infty} \operatorname{tr}\left(h^T(t)h(t)\right) dt}, \tag{2.36}$$

where $h(t) = Ce^{E^{-1}At}E^{-1}B$ is the impulse response matrix. This relation can be obtained from (2.35) by applying Parseval's theorem [2]. Rewriting relation (2.36) we obtain

$$\begin{aligned}\|\mathrm{G}\|_{\mathcal{H}_2} &= \sqrt{\operatorname{tr}\left(B^T E^{-T}\underbrace{\int_0^{\infty} e^{A^T E^{-T}t} C^T C e^{E^{-1}At} dt}_{=:Q} E^{-1}B\right)},\\ &= \sqrt{\operatorname{tr}\left(B_s^T Q B_s\right)},\end{aligned} \tag{2.37}$$

where $B_s = E^{-1}B$ and Q is the solution of the observability Lyapunov equation (2.22). Similarly, considering P as the solution of the controllability Lyapunov equation (2.19) the $\mathcal{H}_2$ norm can be defined by

$$\|\mathrm{G}\|_{\mathcal{H}_2} = \sqrt{\operatorname{tr}\left(CPC^T\right)}. \tag{2.38}$$

Note that the expression in (2.35) and (2.36) are the theoretical concepts of the $\mathcal{H}_2$ norm. Practically, the $\mathcal{H}_2$ norm of the system (2.4) can be computed from either expression (2.37) or (2.38). To compute the $\mathcal{H}_2$ norm of the system (2.4), we always consider $D_a = 0$. This is because the system should be strictly proper for computing the $\mathcal{H}_2$ norm. A procedure to compute the $\mathcal{H}_2$ norm of the system (2.4) is summarized in Algorithm 9.

Algorithm 9: $\mathcal{H}_2$ norm computation.

Input : E, A, B, C.
Output: h_2

1. Compute P by solving controllability Lyapunov equation (2.19) or compute Q by solving observability Lyapunov equation (2.22).
2. $h_2 = \sqrt{\mathrm{tr}\,(CPC^T)}$ or $\sqrt{\mathrm{tr}\,(B^TQB)}$

2.10.2 The $\mathcal{H}_\infty$ norm

For a given sinusoidal input with a particular frequency ω and unit magnitude, the largest possible output of a stable system can be measured by $|G(j\omega)|$. The $\mathcal{H}_\infty$ norm essentially provides the largest possible amplification of a stable system over the entire range of frequencies of a unit sinusoidal input. Hence, we can define the $\mathcal{H}_\infty$ norm as follows

Definition 32. *Let (2.4) be a stable SISO system. The transfer function of the system $G(s)$ is given by (2.10). Then the $\mathcal{H}_\infty$ norm of the system is defined by*

$$\|G\|_{\mathcal{H}_\infty} = \sup_{\omega \in \mathbb{R}} |G(j\omega))|\,, \tag{2.39}$$

where sup *denotes the maximum upper bound of the function $G(j\omega)$.*

In the case of a MIMO, the $\mathcal{H}_\infty$ norm is defined as follows

$$\|\mathrm{G}\|_{\mathcal{H}_\infty} = \sup_{\omega \in \mathbb{R}} \sigma_{max}(G(j\omega)), \tag{2.40}$$

with σ_{max} denoting the maximum singular value of $G(j\omega)$. To compute the $\mathcal{H}_\infty$ norm we can follow the procedure given below.

- Select a set of frequencies $\{\omega_1, \omega_2, \cdots \omega_N\}$.
- For $k = 1, 2, \cdots, N$ compute

  $$\mathcal{N}_k = \sigma_{\max}(G(j\omega_k)) \quad \text{or} \quad \lambda_{\max}(G^*(j\omega_k)G(j\omega_k))$$

  where σ_{max} denotes the maximum singular value and λ_{max} denotes the maximum eigenvalue.
- Evaluate the $\mathcal{H}_\infty$ norm as

  $$\|\mathrm{G}\|_{\mathcal{H}_\infty} = \max_{1 \le k \le N} \{\mathcal{N}_k\}$$

The complete procedure is summarized in Algorithm 10.

Algorithm 10: $\mathcal{H}_\infty$ norm computation.

Input : E, A, B, C, D_a, $\{\omega_{min}, \omega_{max}\}$, N.
Output: h_∞

Compute $\omega = N$ logarithmically equally spaced points between $10^{\omega_{min}}$ and $10^{\omega_{max}}$.
for $i = 1 : N$ **do**
    $G = C(\mathrm{j}\omega(i) * E - A)^{-1}B$
    if *SISO* **then**
        $\mathcal{N}_i = \mid G \mid$.
    else
        $\mathcal{N}_i = \mathtt{max}(\mathtt{svds}(G))$.
Evaluate:

$$h_\infty = \max_{1 \le i \le N} \{\mathcal{N}_i\}$$

2.11 Some useful MATLAB functions

Function name	Description
`expm`	compute matrix exponential
`ss`	construct state-space model or convert model to state space
`rss`	generate randomized continuous-time state-space models
`dss`	create generalized or descriptor state-space models
`step`	step response of dynamic systems
`stepplot`	plot step response of linear systems
`impulse`	impulse response of dynamic systems
`impulseplot`	plot impulse response of linear systems
`tf`	construct transfer function or convert to transfer function of LTI dynamical system
`freqresp`	frequency response of dynamic systems
`sigma`	singular value plot of the frequency response of the dynamic system
`frd`	constructs or converts to frequency response data model
`bode`	bode frequency response of dynamic system
`hsvd`	computes the Hankel singular values of linear systems
`hsvplot`	plots the Hankel singular values of an LTI model
`sigma`	singular value plot of dynamic systems
`norm`	compute $\mathcal{H}_2$ norm and $\mathcal{H}_\infty$ norm of LTI systems
`balreal`	Gramian-based balancing of state-space realizations
`gram`	controllability and observability Gramians
`lyap`	solve continuous-time Lyapunov equations
`ctrb`	compute the controllability matrix
`obsv`	compute the observability matrix
`care`	solve continuous-time algebraic Riccati equations

■ Exercises:

2.1 Consider an LTI continuous-time system:

$$(A, B, C, D) = \left(\begin{bmatrix} 0 & 1 \\ -5 & -1 \end{bmatrix}, \begin{bmatrix} 0 \\ 1 \end{bmatrix}, \begin{bmatrix} 1 & 0 \end{bmatrix}, 0 \right).$$

Find the transfer function $G(s)$ of the system manually. Then compute a frequency response $(G(\text{j}\omega))$ at the frequency $\omega = 2$. Hence compute the $\mathcal{H}_\infty$ norm of the system, i.e., $|G(\text{j}\omega)|$.

2.2 For the LTI continuous-time system as given in Exercise 2.1, write the corresponding (controllability and observability) Lyapunov equations and solve them analytically for computing the controllability Gramian and observability Gramian. Using these Gramians find the $\mathcal{H}_2$ norm of the system and then verify your results with the Gramians obtained by using MATLAB commands `gram`.

2.3 Use appropriate MATLAB commands to find controllability and observability Gramians, controllability and observability matrices, and eigenvalues of the system:

$$(A, B, C, D) = \left(\begin{bmatrix} -1 & 1 & 0 \\ 0 & -4 & -2 \\ 0 & 2 & 0 \end{bmatrix}, \begin{bmatrix} 1 \\ 0 \\ 0 \end{bmatrix}, \begin{bmatrix} 1 & 1 & 1 \end{bmatrix}, 0 \right).$$

Then discuss the controllability, observability and stability of the systems.

2.4 Writing sys = ss(A,B,C,D), create an object sys representing the LTI continuous time model. You can compute a minimal realization sysm=ss(sys,'min') if the realization of sys is not minimal. Using the model in Exercise 2.3, discuss whether the system is minimal or not. If it is not minimal, find sysm and plot the step response of both sys and sysm on a single figure to compare them.

2.5 Plot the bode diagram using MATLAB command `bode` for the model presented in Exercise 2.3. In the figure, the upper sub figure shows the magnitude of the frequency responses of the system over a certain frequency range. Show that this magnitude of the frequency response can be obtained by using the command `sigma` which is also known as the sigma plot of the system. Find the sigma plot of the system over different frequency ranges, e.g., $[0, 10^4]$, $[10^{-4}, 10^4]$ and $[10^{-4}, 0]$.

2.6 By applying Algorithm 7, find the singular values (s) of the frequency response of the model CD-player using a frequency interval $\omega \in [10^{-2}, 10^4]$ and 500 logarithomically distributed sample points. Now graph s using the MATLAB command `semilogx`(ω, s) and `loglog`(ω, s).

2.7 By applying Algorithm 6, find the step response (s) of the model CD-player using a time interval $t \in [0, 100]$ and 500 linearly distributed sample points. Now graph s using the MATLAB command `plot`(t, s).

2.8 Apply Algorithm 8 to compute the system Hankel singular values (SHSVs) (σ_s) of the model ISS. You can also compute the SHSVs (σ_m) by using the MATLAB command `hsvd`. Now graph both sets of the SHSVs on a single figure by using the MATLAB command `semilogy`$(\sigma_s, 'r*', \sigma_m, 'bo')$.

2.9 Based on the discussion in Subsection 2.10.1 (Equation (2.38)), write a MATLAB routine which will compute the $\mathcal{H}_2$ norm of an LTI continuous time system (2.4). Test your program to compute the $\mathcal{H}_2$ norm by using the ISS model.

2.10 Based on the discussion in Subsection 2.10.2, write a MATLAB routine which will compute the $\mathcal{H}_2$ norm of a LTI continuous time system (2.4). Test your program to compute the $\mathcal{H}_\infty$ norm by using the data of the ISS model. You can use a frequency interval $\omega \in [10^{-2}, 10^4]$ with 500 logarithomically distributed sample points.

2.11 Write a MATLAB routine which will check the stability of an LTI continuous time system (2.4). If the system is unstable, compute a feedback matrix K by solving CARE (using MATLAB command `care`) which will stabilize the system. Test the validity of your program by applying to the system:

$$(A, B, C, D) = \left(\begin{bmatrix} 0 & 1 \\ 5 & -1 \end{bmatrix}, \begin{bmatrix} 0 \\ 1 \end{bmatrix}, \begin{bmatrix} 1 & 0 \end{bmatrix}, 0 \right).$$

Plot the step responses for both the stable and unstable models and comment on the results.

2.12 Write a MATLAB routine which will check the stability and unstability of a LTI continuous time system (2.4). If the system is unstable, compute a feedback matrix K by solving CARE (using MATLAB command `care`) which will stabilize the system. Test the validity of your program applying to the system:

$$(A, B, C, D) = \left(\begin{bmatrix} 0 & 1 \\ 5 & -1 \end{bmatrix}, \begin{bmatrix} 0 \\ 1 \end{bmatrix}, \begin{bmatrix} 1 & 0 \end{bmatrix}, 0 \right).$$

Plot the step responses for both the stable and unstable models and comment on the results.

Chapter 3

Iterative Solution of Lyapunov Equations

3.1 Introduction

Lyapunov or Lyapunov-like equations play important roles in various disciplines of science and engineering such as system and control theory, optimization, linear algebra, differential equations, boundary value problems, signal processing, power systems, structural dynamics and so on (see, e.g., [13, 16, 60, 80, 119, 155, 182]). Although it was first introduced by Russian mathematician Alexander Mikhailovitch Lyapunov around a century ago (in 1992), it is now recognized for its wide range of applications. In control theory, besides the stability analysis and stabilization of systems, Lyapunov equations are also used to compute balancing transformation, Gramian-based model reduction, $\mathcal{H}_2$ optimal control and Riccati based optimal control. Due to its broad range of applications, solution of large-scale Lyapunov matrix equations has recently been the subject of attention for research in model order reduction (MOR), control theory, dynamical system and other fields. This chapter focuses on an efficient technique for solving Lyapunov matrix equations in the context of model reduction of large-scale sparse LTI dynamical systems. We consider two continuous-time algebraic Lyapunov equations the controllability Lyapunov equation (CLE)

$$APE^T + EPA^T = -BB^T, \tag{3.1}$$

and the observability Lyapunov equation (OLE)

$$A^TQE + E^TQA = -C^TC. \tag{3.2}$$

These Lyapunov equations arise from the LTI dynamical system

$$E\dot{x}(t) = Ax(t) + Bu(t); \quad x(t_0) = x_0, \, t \geq t_0, \tag{3.3a}$$
$$y(t) = Cx(t) + Du(t), \tag{3.3b}$$

where $E, A \in \mathbb{R}^{n\times n}$, $B \in \mathbb{R}^{n\times p}$ and $C \in \mathbb{R}^{n\times n}$. The solutions $P \in \mathbb{R}^{n\times n}$ and $Q \in \mathbb{R}^{n\times n}$ are called the controllability and observability Gramians, respectively.

There are numerous numerical methods that have been developed over the last fifty years to solve Lyapunov matrix equations. Broadly speaking, they can be divided into two types: the direct and the iterative methods. In 1972, Bartels and Stewart proposed a direct method [18] for solving dense Lyapunov equations. Their proposed algorithm, which is known as the *Bartels-Stewart algorithm*, is implemented in the MATLAB Control System Toolbox as the function `lyap`. Later on, this direct approach was generalized by Hammarling [91, 92] (see also [172]) to compute the Cholesky factors of the solutions of Lyapunov equations. This method can also be implemented by the SLICOT-based MATLAB functions `lyapchol`. All these methods are applicable for small- to medium-sized dense systems since the storage requirements are of the order $\mathcal{O}(n^2)$. Apart from the direct methods, there are also several iterative methods, for example: the Smith method [154], the alternating direction implicit (ADI) iteration method [175, 176, 177], the sign function iteration [24, 33], the Krylov subspace methods [94, 97] and the Smith(l) method [133]. These methods are particularly effective for large-scale sparse matrix equations but computing the solution in its dense form requires a quadratic amount of storage.

Over the last few decades, several methods have been developed [30, 114, 133, 152] that leverage the fact that often all coefficient matrices are sparse and the number of inputs and outputs is very small compared to the number of DoFs (degree of freedoms) to increase the efficiency of computation. The low-rank alternating direction implicit iteration (LR-ADI) [30, 114] is one such method. This prominent method is derived from the *ADI* (alternating direction implicit) *iteration* introduced in [117]. Details on the derivation of the LR-ADI iteration can be found in, e.g., [112].

In this chapter we first show the step-by-step procedure to derive the LR-ADI iteration from the ADI method. The ADI shift parameters play an important role to better the convergence of the method. We also discuss how to compute the ADI shift parameters efficiently. In the end, numerical results are discussed to show the efficiency of the method.

3.2 A brief history of alternating direction implicit method

Peaceman and Rachford [131] developed the alternating direction implicit (ADI) method for solving the linear system

$$\Gamma u = b \tag{3.4}$$

with $\Gamma \in \mathbb{R}^{n\times n}$ (symmetric positive definite) arising in the numerical solution of elliptic and parabolic differential equations. Considering $\Gamma = H + V$, where $H, V \in R^{n\times n}$ are symmetric and positive definite matrices, the above linear system can be equivalently stated as

$$Hu + Vu = b. \tag{3.5}$$

For instance, if A represents a centered finite difference discretization of a partial differential equation in two dimensions, then H and V can be chosen as centered finite difference discretization with respect to the x and y directions, respectively. The ADI iteration is for $i = 1, 2, \cdots$ defined in terms of double steps

$$\begin{aligned}(H + \mu_i I_n)u_{i-\frac{1}{2}} &= (\mu_i I - V)u_{i-1} + b,\\ (V + \mu_i I_n)u_i &= (\mu_j I - H)u_{i-\frac{1}{2}} + b,\end{aligned} \tag{3.6}$$

where μ_i are appropriately chosen parameters.

3.3 The ADI iteration for solving Lyapunov matrix-equations

In [117] Lu and Wachspress showed the ADI method can be applied for solving the matrix equation like

$$AX + XA^T = -FF^T, \tag{3.7}$$

where $A \in \mathbb{R}^{n\times n}$ and $F \in \mathbb{R}^{n\times r}$ (usually, $r \ll n$). This equation can be compared to the controllability Lyapunov equation (if $F = B$) and the observability Lyapunov equation (if $F = C^T$) for the system (3.5) when $E = I$. Following (3.6), the Lyapunov equation (3.7) can be solved using two steps

$$\begin{aligned}(A+\mu_i I_n)X_{i-\frac{1}{2}} &= -FF^T - X_{i-1}(A^T-\mu_i I_n),\\ (A+\mu_i I_n)X_i^* &= -FF^T - X_{i-\frac{1}{2}}^*(A^T-\mu_i I_n)\end{aligned} \tag{3.8}$$

for $i = 1, 2, \cdots, i_{\max}$, $\{\mu, \mu_2, \cdots, \mu_j\} \in \mathbb{C}^-$ (left complex half-plane) are called the *ADI shift parameters*. The iteration can be started with $i = 1$ and the initial guess $X_0 = 0_{n\times n}$. The two steps of the iterations can be combined with a single step as

$$\begin{aligned}X_i =&(A+\mu_i I)^{-1}(A-\bar{\mu}_i I)X_{i-1}(A-\bar{\mu}_i I)^*(A+\mu_i I)^{-*}\\ &-2\,\mathrm{Re}\,(\mu_i)(A+\mu_i I)^{-1}FF^T(A+\mu_i I)^{-*},\\ =&A_{\mu_i}X_{i-1}A_{\mu_i}^* - 2\,\mathrm{Re}\,(\mu_i)A_iFF^TA_i^*,\end{aligned} \tag{3.9}$$

where $A_{\mu_i} = (A+\mu_i I)^{-1}(A-\bar{\mu}_i I)$, $A_i = (A+\mu_i I)^{-1}$. The iterative scheme (3.9) is often referred to as the ADI iteration. If A is asymptotically stable, the iterative solution X_i converges to the exact solution X within a considerable tolerance. The convergence of the ADI iteration also depends on a choice of good shift parameters. Selection of ADI shift parameters will be discussed later in this chapter. The algorithm can be stopped if the normalized residual norm is sufficiently small, i.e.,

$$\frac{\|\mathcal{R}(X_i)\|}{\|FF^T\|} \le \tau, \tag{3.10}$$

where τ is the user-defined tolerance and

$$\mathcal{R}(X_i) = AX_i + X_iA^T + FF^T, \tag{3.11}$$

is known as *Lyapunov residual.* The whole procedure of ADI iteration for solving the Lyapunov equation (3.7) is presented in Algorithm 11.

3.4 Low-rank factor of the Lyapunov solutions

The ADI method discussed in the above section provides approximate solutions of the full Gramians. Generally, in the dynamical system (3.3) the input matrix has fewer columns and the output matrix has fewer rows than the dimension of the system, i.e., $p, m \ll n$. Hence P and Q, which are the

---

Algorithm 11: ADI iteration.

Input : A, F, $\{\mu_i\}_{i=1}^{J}$, $i_{\max}$, τ (tolerance).
Output: $X_i \approx X$.
$X_0 = 0, \quad i = 1.$
while $\frac{\|\mathcal{R}(X_i)\|}{\|FF^T\|} \geq \tau$ *or* $i \leq i_{max}$ **do**
  Construct $A_{\mu_i} = (A + \mu_i I)^{-1}(A - \bar{\mu}_i I)$ and $A_i = (A + \mu_i I)$
  Compute $X_i = A_{\mu_i} X_{i-1} A_{\mu_i}^* - 2\,\mathrm{Re}\,(\mu_i) A_i F F^T A_i^*$,
  $i = i + 1$

---

solutions of the Lyapunov equations (3.1) and (3.2), respectively, have rank deficiency. Therefore, instead of computing the full Gramians one can compute their low-rank factors. Computation of the low-rank factors not only saves memory but is also efficient in its computation. Note that for a large-scale dynamical system, computation of full Gramians is infeasible due to time constraints and memory limitations. The following example demonstrates this.

For the controllability Lyapunov equation (3.1), we consider

$$E = \begin{bmatrix} 1 & & & \\ & 2 & & \\ & & \ddots & \\ & & & 1000 \end{bmatrix}, \; A = \begin{bmatrix} -2 & 1 & & & \\ 1 & -2 & 1 & & \\ & \ddots & \ddots & \ddots & \\ & & 1 & -2 & 1 \\ & & & 1 & -2 \end{bmatrix} \in \mathbb{R}^{1000\times 1000},$$

and $B = A(:,2)$. Using the MATLAB function `lyap` to find the solution $P = P^T \in \mathbb{R}^{1000\times 1000}$ of the Lyapunov equation. Figure 3.1 shows the singular values (SVs) of the computed controllability Gramian P. We can see that most of the SVs are less than $\epsilon_{mach} = 2.2204\,.\,10^{-16}$ (i.e., machine precision). Truncating the smaller SVs we can compute the low-rank factor of the Gramian P by computing the singular value decomposition (SVD) of P as

$$P = U\Sigma V^T.$$

Since P is symmetric, $V = U$ holds true. Therefore,

$$P = U\Sigma U^T = \begin{bmatrix} U_r & U_{1000-r} \end{bmatrix} \begin{bmatrix} \Sigma_r & \\ & \Sigma_{1000-r} \end{bmatrix} \begin{bmatrix} U_r^T \\ U_{1000-r}^T \end{bmatrix},$$

where Σ_r consists of r large singular values of P. Truncating $(1000-r)$ smallest singular values of P we get

$$P \approx U_r \Sigma_r U_r^T = Z_r Z_r^T,$$

where $Z_r = U_r \Sigma_r^{\frac{1}{2}} \in \mathbb{R}^{n\times r}$. For the experiment, we consider $r = 21$ which leads to $\|X - Z_r Z_r^T\|_2 = 5.3372\,10^{-14}$. Figure 3.2 shows the sparsity pattern

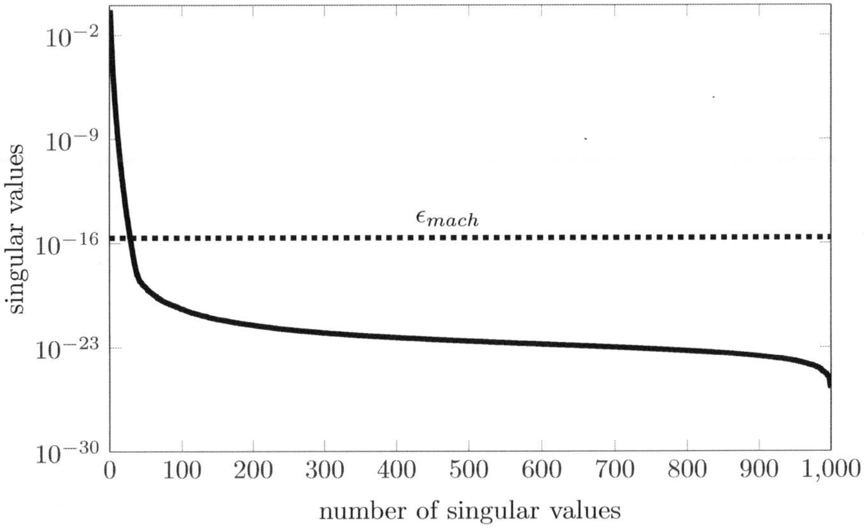

Figure 3.1: Singular value decay rate of the controllability Gramian.

of the full Gramian P and ranked 21 factor Z_{21}. The matrix P requires almost 50 times more memory than the matrix Z_{21}. Table 3.1 also shows the approximation errors of different dimensional low-rank factors of the Gramian P.

Computation of low-rank factors of the Lyapunov solutions is efficient in terms of computational complexity and memory requirements.

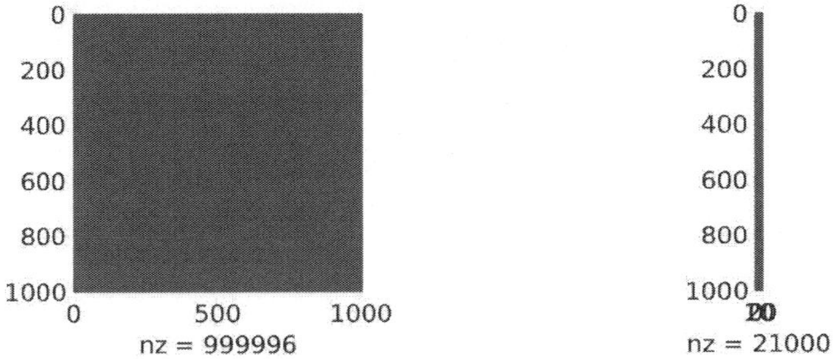

(a) Matrix P (b) Matrix Z_{21}

Figure 3.2: Sparsity pattern of the full Gramian and its low-rank factor.

$Z_r \in \mathbb{R}^{1000 \times r}$	$\|X - Z_r Z_r^T\|_2$
Z_{10}	$3.8391 \cdot 10^{-09}$
Z_{15}	$1.4005 \cdot 10^{-11}$
Z_{20}	$1.2668 \cdot 10^{-13}$
Z_{30}	$3.1739 \cdot 10^{-16}$

Table 3.1: Approximation of a Gramian by its low-rank factors.

3.5 Low-rank (LR-)ADI iteration

The above section motivates to compute low-rank factors of the controllability and observability Gramians. This section shows how to modify the ADI iteration (introduced in Section 3.3) for computing (approximate) low-rank factors of the Gramians.

3.5.1 Low-rank factors of the Gramian using ADI iteration

The ADI iteration as presented in Algorithm 11 saves X_i at i-th step which has a storage requirement of $\mathcal{O}(n^2)$. For a large-scale system, computing and storing a dense X_i is infeasible because of computational complexity and storage limitations. These can be considered as disadvantages of the ADI method. In order to minimize this drawback, a low-rank factored version of ADI iteration was realized. In this book, we call this the low-rank alternating direction implicit (LR-ADI) method.

We consider that the matrix F in the right-hand side of (3.7) is either the input or the output matrix of an LTI dynamical system. The number of inputs or outputs of the system is usually very small when compared to the number of DoFs. This implies that the matrix FF^T has rank deficiency. Therefore when using Algorithm 11, the final solution X_i of (3.7) has rank deficiency. Thus, instead of computing $X_i \in \mathbb{R}^{n \times n}$ it is possible to compute $Z_i \in \mathbb{R}^{n \times k}$ ($k \ll n$) which satisfies $X_i \approx Z_i Z_i^*$. Without a doubt, computing and saving the low-rank factor Z_i reduces computational complexity and saves computer memory. In the following subsection, we derive the LR-ADI iteration.

3.5.2 Derivation of LR-ADI iteration

Considering $X_i = Z_i Z_i^T$, (3.9) implies

$$Z_i Z_i^* = A_{\mu_i} Z_{i-1} Z_{i-1}^* A_{\mu_i}^* - 2\,\mathrm{Re}\,(\mu_i) A_i F F^T A_i^*. \tag{3.12}$$

If the initial guess in (3.9) is $X_0 = 0 \in \mathbb{R}^{n\times n}$, then $Z_0 Z_0^* = 0$. Hence, if $i = 1$, (3.12) becomes

$$\begin{aligned} Z_1 Z_1^* &= -2\,\mathrm{Re}\,(\mu_i) A_i F F^T A_i^* \\ &= \left(\sqrt{-2\,\mathrm{Re}\,(\mu_i)} A_i F\right) \left(\sqrt{-2\,\mathrm{Re}\,(\mu_i)} A_i F\right)^* . \end{aligned} \tag{3.13}$$

Note that the symbol "$*$" denotes the Hermitian matrix and in case of a real matrix it turns into "T" which denotes the transpose of a matrix. From (3.12), we obtain

$$Z_1 = \sqrt{-2\,\mathrm{Re}\,(\mu_1)} (A + \mu_1)^{-1} F. \tag{3.14}$$

If $i > 1$, (3.9) can be written as

$$Z_i Z_i^* = \begin{bmatrix} A_{\mu_i} Z_{i-1} & \sqrt{-2\,\mathrm{Re}\,(\mu_i)} A_i F \end{bmatrix} \begin{bmatrix} Z_{i-1}^* A_{\mu_i}^* \\ \sqrt{-2\,\mathrm{Re}\,(\mu_i)} F^T A_i^* \end{bmatrix}, \tag{3.15}$$

which implies

$$\begin{aligned} Z_i &= \begin{bmatrix} \sqrt{-2\,\mathrm{Re}\,(\mu_i)} A_i F & A_{\mu_i} Z_{i-1} \end{bmatrix} \\ &= \begin{bmatrix} Z_1 & A_{\mu_i} Z_{i-1} \end{bmatrix}. \end{aligned} \tag{3.16}$$

where Z_1 is defined in (3.14). This is called LR-ADI iteration. We observe that in each iteration, the number of columns in Z_i increased by p (i.e., the number of columns of F). At the ith step, we need to modify the previous factor Z_{i-1} by pre-multiplying it with A_{μ_i}. Thus for more iterations, the complexity increases since the number of columns which need to be modified at each iteration increases by p. To overcome this problem, the following step was taken to keep the number of modified columns constant. Let us consider

$$\gamma_i := \sqrt{-\,\mathrm{Re}\,(\mu_i)} F,\ S_i := A_\mu,\ T_i := A_{\mu_i} \text{ and } T_{i,j} := (A - \bar{\mu_i} I)(A + \mu_i I)^{-1}.$$

Now, at the ith step, the ip columns of Z_i can be rewritten as

$$Z_i = \begin{bmatrix} \gamma_i S_i & \gamma_{i-1} T_i S_{i-1} & \gamma_{i-2} T_i T_{i-1} S_{i-2} & \cdots & \gamma_1 T_i & \cdots & T_1 S_1 \end{bmatrix}. \tag{3.17}$$

One can observe that $T_j S_i = T_{j,i} S_j$ and $T_j T_i = T_{j,i} T_{i,j}$ for all i, j, and hence (3.17) becomes

$$Z_i = \begin{bmatrix} \gamma_i S_i & \gamma_{i-1} S_{i-1,i} T_i & \gamma_{i-2} T_{i-2,i-1} T_{i-1} S_{i-2} & \cdots & \gamma_1 T_{1,2} & \cdots & T_{i-1,i} S_i \end{bmatrix}. \tag{3.18}$$

At this point we can see that in each iteration step for $i > 1$, we need to modify only p columns in the augmented terms of Z_i. The order of the shift parameters has no significance in the LR-ADI iterations. Therefore, by reversing the order of the shift parameters, we obtain the LR-ADI iteration as presented in Algorithm 12 for solving the Lyapunov equations. In this algorithm, the main

---

Algorithm 12: LR-ADI iteration.

Input : A, F, $\{\mu_i\}_{i=1}^J$.
Output: $R = Z_i$, such that $X \approx RR^*$.
$V_1 = \sqrt{-2\,\mathrm{Re}\,(\mu_1)}(A + \mu_1 I)^{-1}F$ and $Z_1 = \sqrt{-2\,\mathrm{Re}\,(\mu_1)}V_1$
$i = 2$
while $\frac{\|\mathcal{R}(Z_i Z_i^*)\|}{\|FF^T\|} \geq \tau$ *or* $i \leq i_{max}$ **do**
    $V_i = \begin{bmatrix} V_{i-1} & (\mu_i + \bar{\mu}_{i-1})(A + \mu_i I)^{-1}V_{i-1} \end{bmatrix}$
    Update $Z_i = \begin{bmatrix} Z_{i-1} & \sqrt{-2\,\mathrm{Re}\,(\mu_i)}V_i \end{bmatrix}$.
    $i = i + 1$

---

computational effort comes from the solution of the associated shifted linear system with p right-hand sides. We can solve them efficiently by either direct [58, 63] or iterative [143, 170] solvers. If the shift parameters are complex, then the computed Gramian factor is also complex which is not desired. Complex Gramian factor will consume more memory and cause additional complexity in further applications. Another problem is to stop the algorithm, we need to compute the norm of an $n \times n$ matrix which is computationally expensive. These two issues are resolved in the following subsection.

3.5.3 Efficient handling of complex shift parameters

Although all of the input matrices A and F are real in Algorithm 12, due to the complex shift parameters in each iteration step, the updated Z_i stores complex data; this increases the overall complexity and memory requirements of the method. Moreover, the model reduction using these complex Gramian factors yields complex reduced systems by performing some complex arithmetic operations. By handling complex shift parameters, this problem is resolved in [27]. In this regard, the important assumption is that the selected ADI shift parameters should be proper.

Definition 33. *The ADI shift parameters $\{\mu_i\}_{i=1}^J \subset \mathbb{C}^-$ are called proper if μ_i and μ_{i+1} are complex conjugates of each other when one of them is complex.*

In [27] it is shown that at the $(i+1)$-th iteration of the LR-ADI iteration, V_{i+1} can be computed by

$$V_{i+1} = \overline{V}_i + 2\delta\,\mathrm{Im}\,(V_i), \tag{3.19}$$

where $\delta = \frac{\mathrm{Re}\,(\mu_i)}{\mathrm{Im}\,(\mu_i)}$. This identity states that, in Algorithm 12, corresponding to $\mu_{i+1} = \overline{\mu}_i$, V_{i+1} can be computed explicitly from V_i which frees us from solving a shifted linear system with $A + \overline{\mu}_i I$. This idea also results in the following theorem.

Algorithm 13: LR-ADI iteration (for a real low-rank Gramian factor).

Input : $A,\ F,\ \{\mu_i\}_{i=1}^{J}$.
Output: $R = Z_i$, such that $X \approx RR^T$.

$Z_1 = V_1 = \sqrt{-2\mu_1}(A + \mu_1 I)^{-1}B$
$i = 2$
while $\frac{\|\mathcal{R}(Z_i Z_i^*)\|}{\|FF^T\|} \geq \tau$ *or* $i \leq i_{max}$ **do**
    $V_i = \begin{bmatrix} V_{i-1} & (\mu_i + \bar{\mu}_{i-1})(A + \mu_i I)^{-1}V_{i-1} \end{bmatrix}$
    if $\operatorname{Im}(\mu_i) = 0$ **then**
        $Z_i = \begin{bmatrix} Z_{i-1} & \sqrt{-2\mu_i}V_i \end{bmatrix}$.
    else
        $\gamma = 2\sqrt{-\operatorname{Re}(\mu_i)}, \quad \delta = \frac{\operatorname{Re}(\mu_i)}{\operatorname{Im}(\mu_i)}$,
        $Z_{i+1} = \begin{bmatrix} Z_{i-1} & \sqrt{2\gamma}(\operatorname{Re}(V_i) + \delta \operatorname{Im}(V_i)) & \sqrt{2\gamma}\sqrt{(\delta^2+1)}.\operatorname{Im}(V_i) \end{bmatrix}$,
        $V_{i+1} = \overline{V} + \delta \operatorname{Im}(V_i)$.
        $i = i + 1$

Theorem 11. *Let us assume a set of proper ADI shift parameters. For a pair of complex conjugate shifts* $\{\mu_i, \mu_{i+1} := \overline{\mu_i}\}$, *the two subsequent block iterates* V_i *and* V_{i+1} *of Algorithm 12 satisfy*

$$[V_i, V_{i+1}] = \left[\sqrt{-2\operatorname{Re}(\mu_i)}(\operatorname{Re}(V_i) + \delta \operatorname{Im}(V_i)), \sqrt{-2\operatorname{Re}(\mu_i)}\sqrt{\delta^2+1}\operatorname{Im}(V_i)\right]. \tag{3.20}$$

This theorem reveals that for a pair of complex conjugate shifts at any iteration step in the LR-ADI iteration, Z_i can be updated by

$$Z_{i+1} = [Z_{i-1}, \sqrt{-2\operatorname{Re}(\mu_i)}(\operatorname{Re}(V_i) + \delta \operatorname{Im}(V_i)), \sqrt{-2\operatorname{Re}(\mu_i)}\sqrt{\delta^2+1}\operatorname{Im}(V_i)]. \tag{3.21}$$

A version of the LR-ADI algorithm is summarized in Algorithm 13, which computes low-rank real Gramian factors.

3.5.4 Low-rank Lyapunov residual factor based stopping technique

Algorithm 13 can be stopped whenever

$$\frac{\|\mathcal{R}(Z_i Z_i^*)\|}{\|FF^T\|} \leq \tau,$$

where τ is a user-defined tolerance and the Lyapunov residual is defined by

$$\mathcal{R}(Z_i Z_i^*) := AZ_i{Z_i}^T + Z_i{Z_i}^T A^T + FF^T. \tag{3.22}$$

Computing the Frobenius norm or 2-norm of the residual matrix is an expensive task since in each iteration, the formulated residual matrix is large (if n is large) and dense. This subsection will discuss how to compute the low-rank residual factor inside the LR-ADI iteration for terminating the Algorithm 13.

Low-rank factored form of the Lyapunov residual

For any $\mu \in \mathbb{R}^- := \{x \in \mathbb{R} | x < 0\}$, the equivalent form of the Lyapunov equation (3.7) is

$$(A^T + \mu I)X(A + \mu I) - (A^T - \mu I)X(A - \mu I) = -2\,\mathrm{Re}\,(\mu)F^T F, \qquad (3.23)$$

Since A is assumed to be stable, all eigenvalues of the matrix pencil $A + \mu I$ have negative real parts and hence for any $\mu \in \mathbb{R}^-$, $(A^T + \mu I)$ is invertible. Therefore, (3.23) yields

$$\begin{aligned} X =&(A^T + \mu I)^{-1}(A^T - \mu I)X(A - \mu I)(A + \mu I)^{-1} \\ &2\,\mathrm{Re}\,(\mu)(A^T + \mu I)^{-1}F^T F(A + \mu I)^{-1}. \end{aligned} \qquad (3.24)$$

On the other hand following (3.9), for a given ADI shift parameter μ at ith iteration, we obtain

$$\begin{aligned} X_i =&(A + \mu I)^{-1}(A - \bar{\mu} I)X_{i-1}(A - \bar{\mu} I)^*(A + \mu I)^{-*} \\ &- 2\,\mathrm{Re}\,(\mu)(A + \mu I)^{-1}FF^T(A + \mu I)^{-*}. \end{aligned} \qquad (3.25)$$

At the ith iteration of the ADI methods if X_i is the approximate solution of the Lyapunov equation (3.9), the error to the exact solution X is measured by

$$X - X_i = (A^T + \mu I)^{-1}(A^T - \mu I)(X - X_{i-1})(A - \mu I)(A + \mu I)^{-1}.$$

If $X_0 = 0_{0\times 0}$, then for $i = 1$,

$$\begin{aligned} X - X_1 =&(A^T + \mu_1 I)^{-1}(A^T - \mu_1 I)(X - X_0)(A - \mu_1 I)(A + \mu_1 I)^{-1}, \\ =&(A^T + \mu I)^{-1}(A^T - \mu I)X(A - \mu I)(A + \mu I)^{-1}. \end{aligned}$$

In the same way at $i = 2$,

$$\begin{aligned} X - X_2 =&(A^T + \mu_2 I)^{-1}(A^T - \mu_2 I)\Delta_1(A - \mu_2 I)(A + \mu_2 I)^{-1}, \\ =&(A^T + \mu_2 I)^{-1}(A^T - \mu_2 I)(A^T + \mu_1 I)^{-1}(A^T - \mu_1 I)X \\ &(A - \mu_1 I)(A + \mu_1 I)^{-1}(A - \mu_2 I)(A + \mu_2 I)^{-1}. \end{aligned}$$

Continuing the process at the ith ADI iteration, the error bound can be measured by

$$X - X_i = \prod_{k=0}^{i}(A^T + \mu_i I)^{-1}(A^T - \mu_i I)X$$
$$\times \prod_{k=0}^{i}(A - \mu_i I)(A + \mu_i I)^{-1}$$
$$=(\prod_{k=0}^{i} A_{\mu_i})X(\prod_{k=0}^{i} A_{\mu_i})^T$$

which measures the error of the Lyapunov solution at the ith iteration of the ADI methods. Again at the ith iteration of the ADI method, the Lyapunov residual as in (3.11) can be reformulated as

$$\mathcal{R}(X_i) = AX_i + X_i A^T + FF^T \tag{3.26}$$
$$= A(X - X_i) + (X - X_i)A^T \tag{3.27}$$
$$= A(\prod_{k=0}^{i} A_{\mu_i})X(\prod_{k=0}^{i} A_{\mu_i})^T + (\prod_{k=0}^{i} A_{\mu_i})X(\prod_{k=0}^{i} A_{\mu_i})^T A^T \tag{3.28}$$
$$= \prod_{k=0}^{i} A_{\mu_i}(AX + XA^T)\prod_{k=0}^{i} A_{\mu_i}^T \tag{3.29}$$
$$= \prod_{k=0}^{i} A_{\mu_i} FF^T \prod_{k=0}^{i} A_{\mu_i}^T \tag{3.30}$$
$$= W_i W_i^T, \tag{3.31}$$

where

$$W_i = \prod_{k=0}^{i} A_{\mu_i} = \prod_{k=0}^{i}(A^T + \mu_i I)^{-1}(A^T - \mu_i I)B. \tag{3.32}$$

This is known as Lyapunov residual factor. Note that the Frobenius norm or 2-norm of $W_i W_i^T$ is equal to the Frobenius norm or 2-norm of $W_i^T W_i$ which is cheaper to compute. Therefore, to terminate the LR-ADI iteration as presented in Algorithm 13, we can apply the following condition

$$\frac{\|W_i^T W_i|}{\|FF^T\|} \leq \tau.$$

We can update W_i at each iteration step without any cost inside the LR-ADI algorithm.

3.5.5 Reformulation of LR-ADI iteration using the low-rank factor based stopping criterion

In the LR-ADI iteration, the V_i can be expressed as

$$\begin{aligned}
V_i &= \left[I - (\mu_i + \overline{\mu_{i-1}})\,(A+\mu_i)^{-1}\right] V_{i-1} && (3.33)\\
&= (A - \overline{\mu_{i-1}})(A+\mu_i I)^{-1} V_{i-1}\\
&= (A - \overline{\mu_{i-1}} I)(A+\mu_i I)^{-1}(A - \overline{\mu_{i-2}} I)(A+\mu_{i-1} I)^{-1} V_{i-2}\\
&= \cdots = (A + \overline{\mu_i} I)^{-1}\left(\prod_{j=1}^{i-1}(A - \overline{\mu}_j I)(A+\mu_j I)^{-1}\right) F\\
&= (A+\mu_i I)^{-1} W_{i-1}. && (3.34)
\end{aligned}$$

From (3.32) we obtain

$$\begin{aligned}
W_i &= (A - \overline{\mu}_i I) V_i\\
&= (A - \overline{\mu_i} I)(A+\mu_i I)^{-1} W_{i-1}\\
&= \left[I - (\mu_i + \overline{\mu_{i-1}})(A+\mu_i I)^{-1}\right] W_{i-1}\\
&= W_{i-1} - 2\,\mathrm{Re}\,(\mu_i) V_i \quad (\text{using} \quad (3.33)). && (3.35)
\end{aligned}$$

In the case of real setting, when we consider $\mu_{i+1} := \overline{\mu_i}$, one must compute

$$\begin{aligned}
W_{i+1} &= W_i - 2\,\mathrm{Re}\,(\mu_i) V_{i+1}\\
&= W_{i-1} - 2\,\mathrm{Re}\,(\mu_i) V_i - 2\,\mathrm{Re}\,(\mu_i) V_{i+1}\\
&= W_{i-1} - 2\,\mathrm{Re}\,(\mu_i) E\left(V_i + \overline{V_i} + 2\delta\,\mathrm{Im}\,(V_i)\right) \quad (\text{using} \quad (3.19))\\
&= W_{i-1} - 4\,\mathrm{Re}\,(\mu_i)\left(\mathrm{Re}\,(V_i) + \delta\,\mathrm{Im}\,(V_i)\right), && (3.36)
\end{aligned}$$

where δ is defined in (3.19). The rank of W_i is at most m, i.e., the number of columns of F. Therefore, the computation of the Frobenius norm or 2-norm of $\|W_i W_i^T\| = \|W_i^T W_i\|$ in each iteration is extremely cheap. Applying these strategies (computation of real Gramian factors and low-rank residual based stopping techniques), the updated LR-ADI is rewritten in Algorithm 14.

3.5.6 LR-ADI for generalized system

This subsection mainly focuses on the solution of the Lyapunov equations (3.1) and (3.2) which are obtained from the generalized system (3.3). In the above subsection, we have presented algorithms for solving the Lyapunov equations for a standard system. This subsection contributes to generalize the algorithms for solving the Lyapunov equations of generalized systems.

First consider the controllability Lyapunov equation (3.1). If E^{-1} exists, multiplying left side by E^{-1} and right side by E^{-T} we obtain

$$\check{A}P + P\check{A}^T = -\check{B}\check{B}^T, \qquad (3.37)$$

Algorithm 14: LR-ADI iteration for standard system.

Input : A, F, $\{\mu_i\}_{i=1}^{J}$.
Output: $R = Z_i$, such that $X \approx RR^T$.

$W_0 = F, \quad Z_0 = [\ ], \quad i = 1.$
while $\|W_{i-1}^T W_{i-1}\| \geq \tau$ *or* $i \leq i_{max}$ **do**
  Compute $V_i = (A + \mu_i I)^{-1} W_{i-1}$.
  if $\mathrm{Im}\,(\mu_i) = 0$ **then**
    $Z_i = \begin{bmatrix} Z_{i-1} & \sqrt{-2\mu_i} V_i \end{bmatrix}$.
    $W_i = W_{i-1} - 2\mu_i V_i$.
  else
    $\gamma = -2\,\mathrm{Re}\,(\mu_i), \quad \delta = \frac{\mathrm{Re}\,(\mu_i)}{\mathrm{Im}\,(\mu_i)}$,
    $Z_{i+1} = \begin{bmatrix} Z_{i-1} & \sqrt{2\gamma}(\mathrm{Re}\,(V_i) + \delta\,\mathrm{Im}\,(V_i)) & \sqrt{2\gamma}\sqrt{(\delta^2+1)}\,\mathrm{Im}\,(V_i) \end{bmatrix}$,
    $W_{i+1} = W_{i-1} + 2\gamma\,(\mathrm{Re}\,(V_i) + \delta\,\mathrm{Im}\,(V_i))$.
    $i = i + 1$
  $i = i + 1$

---

where $\breve{A} = E^{-1}A$ and $\breve{B} = E^{-1}B$. This controllability Lyapunov equation can be compared with (3.7) in which $A = \breve{A}$ and $F = \breve{B}$. Therefore, using Algorithm 15, we can solve the Lyapunov equation (3.1) where the inputs are $\breve{A}, \breve{B}$ and the shift parameters. In this case, at the ith step, the Lyapunov residual can be written as

$$\begin{aligned} \breve{\mathcal{R}}(P_i) &= \breve{A}P_i + P_i\breve{A}^T + \breve{B}\breve{B}^T \\ &= \breve{A}(P_i - P) + (P_i - P)\breve{A}^T \\ &= \prod_{k=0}^{i} \breve{A}_{\mu_i} \breve{B}\breve{B}^T \prod_{k=0}^{i} \breve{A}_{\mu_i}^T \\ &= \breve{W}_i \breve{W}_i^H, \end{aligned} \tag{3.38}$$

with $\breve{W}_i = \prod_{k=0}^{i} \breve{A}_{\mu_i} \breve{B}$. To solve the (3.1), using Algorithm 15 at ith iteration we compute

$$V_i = (\breve{A} + \mu_i I)^{-1}(\breve{A} - \overline{\mu_{i-1}} I)V_{i-1} = (\breve{A} + \mu_i I)^{-1}\breve{W}_{i-1} = (A + \mu_i E)^{-1} E\breve{W}_{i-1}, \tag{3.39}$$

where the Lyapunov residual factor is

$$\breve{W}_i = \breve{W}_{i-1} - 2\,\mathrm{Re}\,(\mu_i)EV_i \tag{3.40}$$

Let us consider $E\breve{W}_i = W_i$. If $i = 1$ then $W_0 = E\breve{W}_0 = EE^{-1}B = B$ and

$$V_1 = (A + \mu_i E)^{-1}E\breve{W}_0 = (A + \mu_i E)^{-1}B. \tag{3.41}$$

Algorithm 15: LR-ADI iteration for generalized.

Input : E, A, B, $\{\mu_i\}_{i=1}^{J}$.

Output: $R = Z_i$, such that $P \approx RR^T$.

$W_0 = B, \quad Z_0 = [\,], \quad i = 1.$

while $\|W_{i-1}^T W_{i-1}\| \geq \tau$ *or* $i \leq i_{max}$ **do**

  Compute $V_i = (A + \mu_i E)^{-1} W_{i-1}$.

  if $\mathrm{Im}\,(\mu_i) = 0$ **then**

    $Z_i = \begin{bmatrix} Z_{i-1} & \sqrt{-2\mu_i} V_i \end{bmatrix}$.

    $W_i = W_{i-1} - 2\mu_i E V_i$.

  else

    $\gamma = -2\,\mathrm{Re}\,(\mu_i), \quad \delta = \frac{\mathrm{Re}\,(\mu_i)}{\mathrm{Im}\,(\mu_i)}$,

    $Z_{i+1} = \begin{bmatrix} Z_{i-1} & \sqrt{2\gamma}(\mathrm{Re}\,(V_i) + \delta\,\mathrm{Im}\,(V_i)) & \sqrt{2\gamma}\sqrt{(\delta^2+1)}\,\mathrm{Im}\,(V_i) \end{bmatrix}$,

    $W_{i+1} = W_{i-1} + 2\gamma E\,(\mathrm{Re}\,(V_i) + \delta\,\mathrm{Im}\,(V_i))$.

    $i = i + 1$

  $i = i + 1$

For $i > 1$

$$V_i = (A + \mu_i E)^{-1} W_i. \tag{3.42}$$

where

$$W_i = W_{i-1} - 2\,\mathrm{Re}\,(\mu_i) E V_i \tag{3.43}$$

In the case of real setting, when we consider $\mu_{i+1} := \overline{\mu_i}$, one must compute

$$W_{i+1} = W_{i-1} - 4\,\mathrm{Re}\,(\mu_i) E\,(\mathrm{Re}\,(V_i) + \delta\,\mathrm{Im}\,(V_i))\,, \tag{3.44}$$

where δ is defined in (3.19). Applying these strategies (computation of real Gramian factors and low-rank residual based stopping techniques), the updated LR-ADI is rewritten in Algorithm 15. Algorithm 15 can, however, solve the observability Lyapunov equation (3.2) by changing inputs $\{E, A, B\}$ to $\{E^T, A^T, C^T\}$.

3.6 ADI shift parameter selection

The convergence speed of the LR-ADI algorithms presented above heavily depends on a set of ADI shift parameters. The ADI shift parameters were originally introduced by Wachspress [175] to solve Lyapunov equations using

Algorithm 16: Computation of Penzl's heuristic shifts.

Input : E, A and k_+, k_- and J (number of shift parameters).
Output: $\{\mu_1, \mu_2, \cdots, \mu_J\}$.

Apply Algorithm 5 on to $E^{-1}A$ to compute k_+ ritzvalues $w_+ = \{\mu_1, \cdots \mu_{k_+}\}$.
Apply Algorithm 5 on to $A^{-1}E$ to compute k_- ritzvalues $w_- = \{\mu_{k_+ +1}, \cdots \mu_{k_-}\}$.
$w = (w_+ \cup \frac{1}{w_-}) = \{\mu_1, \cdots, \mu_{(k_+ + k_-)}\}$
if $w \not\subset \mathbb{C}^-$ **then**
  STOP
else
  By solving (3.45) using w compute $\{\mu_1, \mu_2, \cdots, \mu_J\}$

the ADI methods. The author shows that a set of optimal ADI shift parameters $\{\mu_i\}_{i=1}^{J}$ for the system (3.3) can be computed by solving the so-called *ADI min-max* problem [175, 177]

$$\min_{\mu_1, \cdots, \mu_j \subset \mathbb{C}^-} \left(\max_{1 \le l \le n} \left| \prod_{i=1}^{J} \frac{\overline{\mu_i} - \lambda_l}{\mu_i + \lambda_l} \right| \right), \quad \lambda_l \in \Lambda(A, E), \tag{3.45}$$

where $\Lambda(A, E)$ denotes the spectrum of the matrix pencil $A + \lambda E$. For a large-scale system, determining the entire spectrum of (A, E) is sometimes impossible. The literature proposes several techniques (see, e.g., [32, 112, 133, 145]) to solve the min-max problem (3.45) using a much smaller part of the spectrum. One such commonly used technique is Penzl's *heuristic approach* introduced in [133], where k_+ *Ritz values* (see, e.g., [78]) and k_- $(k_+, k_- \ll n)$ reciprocal ritzvalues with respect to $E^{-1}A$ and $A^{-1}E$ are employed respectively. A complete procedure of the heuristic approach can be found in Algorithm 16. Although computing the ritzvalues is computationally expensive, this approach is applicable for a large-scale standard or generalized (where E is invertible) state-space system. Another promising ADI shift selection criterion that we focus on is the *adaptive approach* introduced in [26, 41, 166]. This approach is reported to be superior to the heuristic approach, especially for the descriptor systems discussed regarding the computational issues. In this approach, the ADI shift parameters are generated and updated automatically by the LR-

ADI algorithm itself. There, the computed k shifts are the eigenvalues of the projected matrix pencil

$$\lambda U^T E U - U^T A U, \quad \lambda \in \mathbb{C}, \tag{3.46}$$

where $U \in \mathbb{R}^{n\times k}$ $(k \ll n)$. For a set of initial shifts, U in (3.46) is the span of W_0. In this case, whenever all shifts in the current set have been used, the matrix pencil is projected by using U as the span of the current V_i and the eigenvalues are used as the next set of shifts. In this procedure, especially for a SISO system or a system with few inputs and outputs, sometimes the projected pencil may become unstable. In this case, it is suggested in [26] to use the previous set of shift parameters for the next cycle of the iterations. In this procedure, the convergence rate of the LR-ADI iteration may not be as good as we want it to be. To resolve this problem, we propose [41, 166] slightly different initialization and also updating criterion for the adaptive ADI shift parameters approach. In this case, whenever all shifts in the current set have been used, the pencil is projected to the span of the current V_i and the eigenvalues are used as the next set of shifts. Here, we use the same initialization for the update step; however, we extend the subspace to all the V_i generated with the current set of shifts and then choose the next shifts following Penzl's heuristic with the ritzvalues replaced by the projected eigenvalues computed with respect to this larger subspace. Note that, due to the lack of the conditions for Bendixon's theorem, we cannot guarantee that the projected eigenvalues will be contained in $\mathbb{C}^-$. Should any of them end up in the wrong half-plane, we neglect them. Should the resulting set of shifts become empty due to this, we reuse the previous set as in [29].

3.7 Some useful MATLAB functions

Function name	Description
`ss`	construct state space model or convert model to state space
`rss`	generate randomized continuous-time state space models
`dss`	create generalized or descriptor state space models
`hsvd`	compute the Hankel singular values of linear systems
`hsvplot`	plot the Hankel singular values of an LTI model
`sigma`	singular value plot of dynamic systems
`gram`	controllability and observability Gramians
`lyap`	solve continuous-time Lyapunov equations
`chol`	Cholesky decomposition of a matrix
`qr`	QR factorization of a matrix
`schur`	Schur decomposition of a matrix
`norm`	matrix or vector norm
`normest`	compute the matrix 2-norm
`rank` `plot`	matrix rank computation linear plot

`loglog`
 	
`semilogx` `(semilogy)`
	log-log scale plot using logarithmic scaled for both the X-axis and Y-axis
	log-log scale plot using logarithmic scaled for X- (Y-) axis

`spy`	visualize the sparsity pattern
`nnz`	number of non-zero matrix elements

3.8 Numerical experiments

To assess the accuracy and efficiency of the proposed methods (LR-ADI iteration) in this section, we illustrate some numerical results. For the numerical experiments we consider the FOM model from the SLICOT Benchmark Examples [50]. See also Appendix A.2.2 in this book. Note that all the results are obtained by using MATLAB R2015a (8.5.0.197613) on a board with processor 4×Intel® Core™i5-4460s CPU with a 2.90 GHz clock speed and 16 GB RAM.

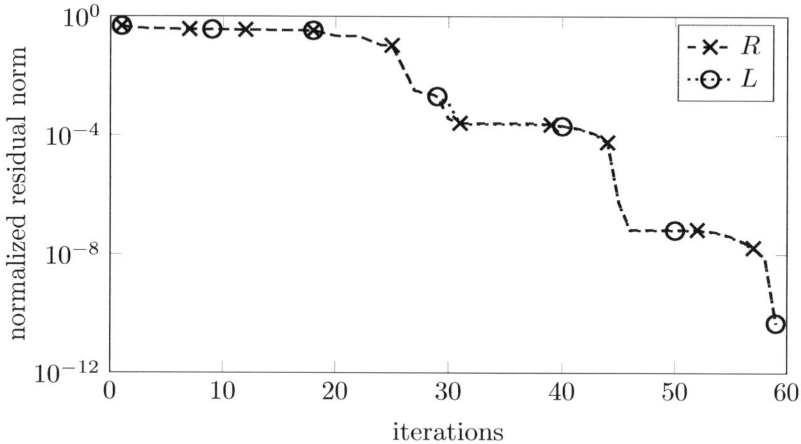

Figure 3.3: Convergence rate of Algorithm 15 to compute the low-rank controllability and observability Gramian factors.

tolerance (τ)	no. of iterations for R	CPU time (sec.)
10^{-5}	44	.058
10^{-10}	59	.066
10^{-15}	60	.067
10^{-20}	71	.072

Table 3.2: Number of iteration steps and computational time for computing the controllability Gramian factor with different tolerances.

We apply LR-ADI iteration (Algorithm 15) to the model for computing the low-rank factors of the controllability and observability Gramians using the adaptive shift computation approach. To compute both the Gramian factors, the algorithm converges by 59 iteration steps with the normalized residual tolerance $\tau = 10^{-10}$. The rate of convergence for both the Gramian factors are shown in Figure 3.3. If we decrease the residual tolerance τ, the iteration steps will increase; the resulting computational times are shown in Table 3.2. By increasing the iteration steps, we can minimize the approximation error to the computed Gramian factors. Figure 3.4 shows that taking more iteration steps yields more accurate HSVs of the system. We also investigated the performance of the heuristic and adaptive shift parameters. Figure 3.5 depicts that the adaptive shift approach gives faster convergence than the heuristic shift. We have selected 10 optimal heuristic shift parameters out of 25 large magnitude and 20 small magnitude ritzvalues using Algorithm 16. Perhaps the performance of the heuristic shift would be better if we can select better shift parameters. We will perform more investigations on these (heuristic and adaptive shift computation approaches) issues in exercise.

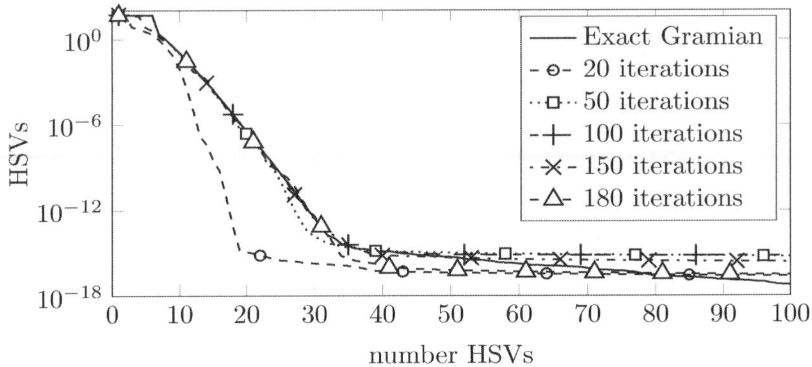

Figure 3.4: 100 large system HSVs computed by approximate Gramian factors and compared with the exact system HSVs.

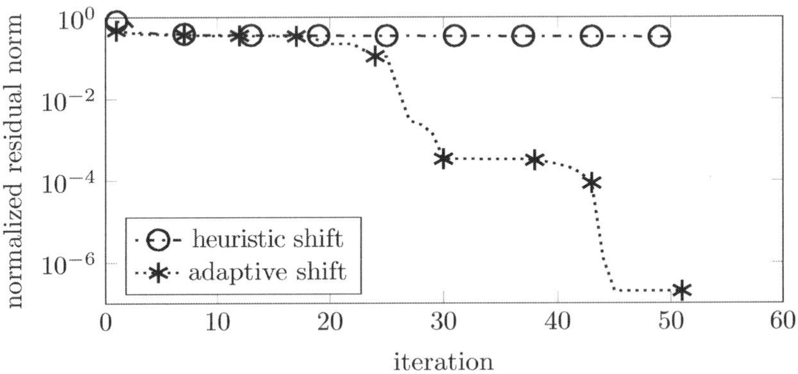

Figure 3.5: Convergence rate of low-rank controllability Gramian factor by using the heuristic and adaptive shifts.

Remark 4. *It is difficult to comment on which shift parameter selection approach is better. The performances are completely dependent on the data of the models and also the better choice of the shift parameters.*

■ Exercises:

3.1 Consider a LTI continuous-time system:

$$(E, A, B, C, D) = \left(\begin{bmatrix} 1 & 0 \\ 0 & 2 \end{bmatrix}, \begin{bmatrix} 0 & 1 \\ -5 & -1 \end{bmatrix}, \begin{bmatrix} 0 \\ 1 \end{bmatrix}, \begin{bmatrix} 1 & 0 \end{bmatrix}, 0 \right).$$

Show that $P = \begin{bmatrix} 0.1 & 0 \\ 0 & 0.1 \end{bmatrix}$ $Q = \begin{bmatrix} 1.1 & 0.1 \\ 0.1 & 0.1 \end{bmatrix}$ are the solutions of the Lyapunov equations: $APE^T + EPA^T + BB^T = 0$ and $A^TQE + E^TQA + C^TC = 0$, respectively.

3.2 For the LTI continuous-time system, as given in Exercise 3.1, find the controllability and observability Gramians analytically by solving the corresponding Lyapunov equations and also compute the controllability and observability Gramians using MATLAB command (`lyap`). Are they symmetric and positive definite?

3.3 In a LTI continuous-time system (E, A, B, C, D), if $A = A^T$, $B = C^T$ and $E = E^T$ then both the Lyapunov equations are the same and the solutions are coincided with each other. Verify these results with the model

$$(E, A, B, C, D) = \left(\begin{bmatrix} 1 & 0 \\ 0 & 2 \end{bmatrix}, \begin{bmatrix} 0 & -5 \\ -5 & -1 \end{bmatrix}, \begin{bmatrix} 1 \\ 0 \end{bmatrix}, \begin{bmatrix} 1 & 0 \end{bmatrix}, 0 \right).$$

3.4 Load the data of FOM model. Using MATLAB commands `lyap` and `lyapchol` find the controllability Gramian (P) and its Cholesky factor R such that $P \approx RR^T$. Show the sparsity pattern of P and R to compare the number of nonzero elements and also the memory requirements to store them.

3.5 We know that system HSVs is $h_s = svd(L^TR)$, where R and L are the Cholesky factor of the controllability and observability Gramian factors, respectively. Using MATLAB function `lyapchol` find R and L of FOM. Using these Gramian factors, compute the HSVs of the system and compare them with the HSVs of the system computed by `hsvd`. You can plot them in a single figure using the command `semilogy` to show the comparisons.

3.6 Using MATLAB function `lyap` compute the controllability Gramian ($P \in \mathbb{R}^{n \times n}$; n is the dimension of the model) of transmission line model (tline) and find the rank of P. Do you think P can be approximated by its low-rank Gramian factor? If so, using SVD as discussed in Section 3.4 find $Z_k \in \mathbb{R}^{n \times k}$ $(k \ll n)$ such that $P \approx Z_k Z_k^T$. Compute Z_k

for considering $k = 5, 10, 15, 20, 25$ and 30. For each k find the approximation error $err_k = \|P - Z_k Z_k^T\|_2$ and plot a bar graph to compare the approximation error of various ranking Gramian factors by scaling horizontal axis with k and vertical axis with err_k.

3.7 One of main advantages of computing the low-rank Gramian factor is to save the memory of CPU. Using the results from Exercise 3.6, you can discuss this issue. Compute the memory requirement for various Z_k then plot a graph showing approximation error against the consumed storages of Z_k.

3.8 Apply Algorithm 15 to compute the low-rank controllability and observability Gramian factors of the model CD player (CDP) (Appendix A.2.1). Investigate both heuristic and adaptive shift to execute this algorithm. Show the comparisons of the performance of both the shift parameters in term of convergence rate and computational time. You may use either a figure or table to show this comparison.

3.9 Efficient handling of complex shift parameters dramatically reduces the computational time to compute a low-rank Gramian factor using the LR-ADI iteration as discussed in Subsection 3.5.3. Investigate this by computing the low-rank controllability or observability Gramian factor by applying Algorithms 12 and 13 using the data of FOM.

3.10 As defined in Subsection 3.5.4 at i-th step in the LR-ADI iteration the Lyapunov residual ($\mathcal{R}(Z_i Z_i^*)$) norm can be computed by

$$\|AZ_i{Z_i}^T + Z_i{Z_i}^T A^T + FF^T\|_2.$$

The residual norm can also be computed efficiently by $W_i W_i^T$ where W_i is the low-rank factor of the Gramians at i-th iteration. Show this observation using numerical experiments by using the data of FOM.

3.11 Update Algorithm 12 for generalized state space model. Then compare the performances in terms of time computation of the updated and Algorithm 15 using the data of FOM.

Chapter 4

Model Reduction of Generalized State Space Systems

4.1 Introduction

In Chapter 2, we have introduced a linear time invariant continuous-time system of the form

$$E\dot{x}(t) = Ax(t) + Bu(t); \quad x(t_0) = x_0, \, t \geq t_0, \tag{4.1a}$$

$$y(t) = Cx(t) + Du(t) \tag{4.1b}$$

and discussed some important properties of the system. In system and control theory, such systems arise when solving many real-life applications such as controller design, design optimization and so on. Simulation and analysis of the system are also needed to make crucial decision about the physical system. Either of the tasks becomes challenging if the mathematical model is very large. Practically speaking, a large-scale model may appear for many reasons. The mathematical models are generated in many different ways; most commonly, they are obtained by finite element method (FEM) or finite difference method (FDM) discretization. In order to model a system accurately, a sufficient number of grid points must be generated because many geometrical details must be resolved. Sometimes physical systems consist of several bodies and each body is composed of a large number of disparate devices. Therefore, the mathematical models become more detailed and different coupling effects must be included. In either case, the resulting systems are typically very *large* and *sparse*. Moreover, often they might be *well-structured.*

A *large*[1]*-scale* system leads to additional memory requirements and enormous computational efforts. They also prevent frequent simulations which is often required in many applications. Sometimes, the generated systems are too large for storage due to limitations of computer memory. To circumvent these complexities, reducing the size of the systems is unavoidable. The method to reduce a higher dimensional to a lower one is called *model order reduction* (MOR). See, e.g., [6, 7, 25, 31, 135, 149] for motivations, applications, restrictions and techniques of MOR.

4.2 Goal of model order reduction

The aim of model reduction is to replace the system (4.1) by a substantially lower dimensional system

$$\begin{aligned} \hat{E}\dot{\hat{x}}(t) &= \hat{A}\hat{x}(t) + \hat{B}u(t), \\ \hat{y}(t) &= \hat{C}\hat{x}(t) + \hat{D}_a u(t), \end{aligned} \tag{4.2}$$

where $\hat{E}$, $\hat{A} \in \mathbb{R}^{r\times r}$, $\hat{B} \in \mathbb{R}^{r\times m}$, $\hat{C} \in \mathbb{R}^{m\times p}$, $\hat{D}_a := D_a$. Here the goal is to ensure that the approximation error which can be measured by

$$\|y - \hat{y}\|, \tag{4.3}$$

($\|.\|$ denotes a suitable norm) must be sufficiently small. In Section 2.4 we have already shown that in the complex domain, the input-output relation of the system (4.1) is

$$Y(s) = \mathrm{G}(s)U(s), \tag{4.4}$$

where

$$\mathrm{G}(s) = \mathrm{C}(sE - A)^{-1}B + D_a. \tag{4.5}$$

is called the *transfer function* of the system. Its analogue for the ROM (4.2) can be written as

$$\hat{Y}(s) = \hat{G}(s)U(s), \tag{4.6}$$

---

[1]The notion of what is considered large is constantly changing with increasing capabilities of computer hardware.

where

$$\hat{G}(s) = \hat{C}(s\hat{E} - \hat{A})^{-1}\hat{B} + \hat{D}_a, \tag{4.7}$$

is the transfer function for the reduced model. Subtracting (4.6) from (4.4) we obtain

$$Y(s) - \hat{Y}(s) = (\mathrm{G}(s) - \hat{\mathrm{G}}(s))U(s).$$

This relation indicates that the error between original and reduced systems can also be measured from

$$\|Y - \hat{Y}\|_{\mathcal{H}_2} = \|\mathrm{G}U - \hat{\mathrm{G}}U\|_{\mathcal{H}_2} \leq \|\mathrm{G} - \hat{\mathrm{G}}\|_{\mathcal{H}_\infty}\|U\|_{\mathcal{H}_2}. \tag{4.8}$$

Hence, in the frequency domain, for the same input, the difference between two output responses can be bounded by

$$\|\mathrm{G} - \hat{\mathrm{G}}\|_{\mathcal{H}_\infty}.$$

Therefore, by minimizing $\|\mathrm{G} - \hat{\mathrm{G}}\|_{\mathcal{H}_\infty}$ we can guarantee that $\|Y - \hat{Y}\|_{\mathcal{H}_2}$ is minimized. Hence in model reduction, the approximation error between the original and reduced models can be obtained by computing the $\mathcal{H}_\infty$ norm of the difference of two transfer functions, i.e., $\|\mathrm{G} - \hat{\mathrm{G}}\|_{\mathcal{H}_\infty}$ in a certain range of the frequency domain.

Besides minimizing the approximation error, in some cases, features such as *stability, passivity, definiteness, symmetry* and so forth of the original system must be preserved in the reduced systems. That is to say if the original model consists of all these properties, the ROM should preserve them. Moreover, the ROM should be more robust than the original model in practical applications.

4.3 Model order reduction methods

In principle, the model reduction of LTI system can be carried out by projection methods. That is the system (4.1) is projected onto the lower dimensional subspaces to obtain the ROM (4.2). The idea of projection-based MOR is riffed as follows.

Definition 34. *Let the matrix* $\Pi \in \mathbb{R}^{n\times n}$ *be a projector and* S_1 *is a subspace with dimension* r*. If* $S_1 = \mathrm{Range}(\Pi)$*, then* Π *is a projector onto* S_1*. Let* $V = \begin{bmatrix} v_1 & v_2 & \cdots & v_r \end{bmatrix}$ *and* $S_1 = \mathrm{Range}(V)$*, then* $\Pi = V(V^TV)V^T$ *is the projector onto* S_1*.*

Lemma 3. *Suppose* Π *is a projector. The following are then true for* Π*:*

1. *The matrix* $I - \Pi$ *is also a projector, called complementary projector.*

2. *The projector* Π *is orthogonal if* $\Pi = \Pi^T$*, otherwise it is an oblique projector.*

3. *Let* S_2 *be another* r *dimensional subspace and* $S_2 = \text{Range}\,(W)$*, where* $W = \begin{bmatrix} w_1 & w_2 & \cdots & w_r \end{bmatrix}$*, then* $\Pi = V(W^TV)^{-1}W^T$ *is called an oblique projector.*

Recalling the system (4.1), let us assume that the state vector $x(t)$ is contained in a lower dimensional subspace S_1. Thus, we can project $x(t)$ onto S_1 along S_2 by applying an orthogonal or an oblique projector. To achieve this we construct

$$V = \begin{bmatrix} v_1 & v_2 & \cdots & v_r \end{bmatrix}$$

and

$$W = \begin{bmatrix} w_1 & w_2 & \cdots & w_r \end{bmatrix}$$

such that

$$S_1 = \text{Range}\,(V) \quad \text{and} \quad S_2 = \text{Range}\,(W). \tag{4.9}$$

Now approximating $x(t)$ by $VW^Tx(t)$ in (4.1) we obtain

$$EVW^T\dot{x}(t) \approx AVW^Tx(t) + Bu(t), \tag{4.10a}$$

$$y(t) \approx CVW^Tx(t) + D_au(t). \tag{4.10b}$$

By considering $\hat{x}(t) = W^Tx(t)$ we define an error $e = EV\dot{\hat{x}}(t) - AVx(t) - Bu(t)$ in the state equation. Because of this error here we write "$\approx$" instead of "$=$". This error is called residual. By construction, each column of W is perpendicular to e, i.e., $W^Te = 0$. Thus (4.10) becomes

$$\begin{aligned} W^TEV\dot{\hat{x}}(t) &= W^TAV\hat{x}(t) + W^TBu(t), \\ \hat{y}(t) &= CV\hat{x}(t) + D_au(t), \end{aligned} \tag{4.11}$$

which is exactly the reduced model as in (4.2) with

$$\hat{E} = W^TEV \quad \hat{A} = W^TAV, \quad \hat{B} = W^TB \;\text{ and }\; \hat{C} = CV. \tag{4.12}$$

To compute the reduced models (4.2) by using the projection-based model reduction methods, one needs to compute the reduced coefficient matrices (4.12) by applying thin rectangular matrices V and W; these are called the right and left transformation matrices, respectively. In this method the basic task is to construct the transformations by using the bases vectors of the subspaces S_1 and S_2. However, the choice of the basis for S_1 and S_2 is not unique. Therefore, different types of model reduction methods are available in the literature based on the different choices of the basis for these subspaces. In the following, we discuss some prominent MOR methods which compute the transformation matrices V and W in different ways.

The techniques to reduce the state space dimension of an LTI continuous-time system (4.1) are well established. See, e.g., [6, 7, 31, 83] for an overview. In a broad sense, there are two techniques: the *Gramian*-based methods and the *moment matching*-based methods. The Gramian-based methods include *optimal Hankel norm approximation* [77], *singular perturbation approximation* [36, 70, 116], *dominant subspaces projection* [113, 134], *frequency weighted balanced truncation* [67, 193], *dominant pole algorithm* and *balanced truncation* (BT) [125, 146, 163]. On the other hand, moment matching can be implemented efficiently via *rational Krylov* methods discussed in [5, 66, 69, 72, 74, 180]. The concept of projection for rational interpolation of the transfer function was first proposed in [173]. In [81] Grimme showed how to obtain the required projection using the rational Krylov method of Ruhe [140]. Later on, the authors in [5, 85] generalized Grimme's idea to generate a reduced model which is an optimal $\mathcal{H}_2$ approximation to the original system in the sense that it minimizes the $\mathcal{H}_2$ norm. The implementing algorithm is called the Iterative Rational Krylov Algorithm (IRKA).

Among all the aforementioned methods, the balanced truncation (BT) and the interpolatory method via IRKA are currently the most commonly used techniques for large-scale dynamical systems. Since this book focuses on these two methods, in the following sections we briefly introduce them.

4.4 Gramian-based model reduction

The fundamental idea of balanced truncation is to truncate the *less-important states* from the systems. A less-important state is any state that is difficult to control and observe. These states essentially correspond to the smallest HSVs. In reality, the states which are difficult to control may not be difficult to observe and vice versa. This implicates that if we eliminate the states that are hard to control directly from the original system, then we may also eliminate some states that are easy to observe. However, in an application, the easily observable states are essential to be preserved. The same contradiction might appear for those states that are difficult to observe but are easily controlled. This problem can be resolved by transforming the system into a balanced form. In a balanced system, the degree of controllability and the degree of observability of each state are the same.

4.4.1 Balancing criterion

Definition 35 (Balanced system). *A stable and minimal LTI system is called balanced if the controllability Gramian and the observability Gramian of the system are equal and diagonal. The diagonal elements are the system's HSVs.*

Now if we eliminate those states of the balanced system that are hard to control, we have eliminated the states that are hard to observe at the same time. The systems can be balanced via a *balancing transformation.* A balancing transformation that creates the controllability and observability Gramian equal and diagonal can be computed several ways.

Definition 36 (Balancing transformation). *A state space transformation T as defined in (2.33) is called a balancing transformation if it causes*

$$T^{-T}PT^{-1} = TQT^{T} = \Sigma = \begin{bmatrix} \Sigma_1 & \\ & \Sigma_2 \end{bmatrix}, \tag{4.13}$$

where P and Q are the controllability and observability Gramians, respectively, $\Sigma_1 = \operatorname{diag}(\sigma_1, \cdots, \sigma_r)$, $\Sigma_2 = \operatorname{diag}(\sigma_{r+1}, \cdots, \sigma_n)$, and $\{\sigma_i\}_{i=1}^{n}$ are the system's HSVs.

In Chapter 2 we have shown that the controllability Gramian (P) and the observability Gramian (Q) of the system (4.1) are the solutions of the two continuous-time algebraic Lyapunov equations:

$$APE^T + EPA^T = -BB^T \quad \text{and} \tag{4.14}$$

$$A^TQE + E^TQA = -C^TC, \tag{4.15}$$

respectively. Keeping Definition 36 in mind, in the text that follows, we discuss one of the criteria to compute a balancing transformation of the system (4.1). We know that P and Q are symmetric and positive definite matrices. Therefore, they have the Cholesky factorizations:

$$P = RR^T \qquad \text{and} \qquad Q = LL^T. \tag{4.16}$$

Computing the singular value decomposition of L^TR as

$$L^TR = U\Sigma V^T, \tag{4.17}$$

form the transformations

$$V := RV\Sigma^{-\frac{1}{2}} \quad \text{and} \quad W := LU\Sigma^{-\frac{1}{2}}, \tag{4.18}$$

which satisfy the properties

$$W^TV = \Sigma^{-\frac{1}{2}}U^T(L^TR)V\Sigma^{-\frac{1}{2}} = \Sigma^{-\frac{1}{2}}U^T(U\Sigma V^T)V\Sigma^{-\frac{1}{2}} = I, \tag{4.19}$$

$$VW^T = RV\Sigma^{-\frac{1}{2}}\Sigma^{-\frac{1}{2}}U^TL^T = R(V\Sigma^{-1}U^T)L^T = R(L^TR)^{-1}L^T = I. \tag{4.20}$$

Now applying the transformations onto the system (4.1) we obtain

$$\begin{aligned} W^TEVW^T\dot{x}(t) &= W^TAVW^Tx(t) + W^TBu(t), \\ y(t) &= CVW^Tx(t) + D_au(t). \end{aligned}$$

Algorithm 17: Balanced system.

Input : E, A, B, C, D_a.
Output: $\tilde{E}$, $\tilde{A}$, $\tilde{B}$, $\tilde{C}$, D_a.

1 Compute R and L as defined in (4.16) by solving the Lyapunov equations:

$$APE^T + EPA^T = -BB^T,$$
$$A^TQE + E^TQA = -C^TC.$$

2 Compute the singular value decomposition (SVD)

$$L^TR = U\Sigma V^T$$

3 Construct $V := RV\Sigma^{-\frac{1}{2}}$ and $W := LU\Sigma^{-\frac{1}{2}}$,

4 Form $\tilde{E} = W^TEV$, $\tilde{A} = W^TAV$, $\tilde{B} = W^TB$ and $\tilde{C} = CV$.

Considering $W^Tx = \tilde{x}$, this will yield

$$\begin{aligned} \tilde{E}\dot{\tilde{x}}(t) &= \tilde{A}\tilde{x}(t) + \tilde{B}u(t); \\ \tilde{y}(t) &= \tilde{C}x(t) + D_au(t), \end{aligned} \tag{4.21}$$

where $(\tilde{E}, \tilde{A}, \tilde{B}, \tilde{C}, D_a) = (W^TEV,\ W^TAV,\ W^TB,\ CV,\ D_a)$. Now we want to check whether the system (4.21) is balanced or not. The corresponding Lyapunov equations of this system are

$$\tilde{A}\tilde{P}\tilde{E}^T + \tilde{E}\tilde{P}\tilde{A}^T = -\tilde{B}\tilde{B}^T, \tag{4.22}$$
$$\tilde{A}^T\tilde{Q}\tilde{E} + \tilde{E}^T\tilde{Q}\tilde{A} = -\tilde{C}^T\tilde{C}, \tag{4.23}$$

where $\tilde{P} = W^TPW$ and $\tilde{Q} = V^TQV$. It can be shown that

$$\begin{aligned} \tilde{P} &= W^TPW \\ &= \Sigma^{-\frac{1}{2}}U^T(L^TR)(R^TL)U\Sigma^{-\frac{1}{2}} \\ &= \Sigma^{-\frac{1}{2}}U^T(U\Sigma V^T)(V\Sigma U^T)U\Sigma^{-\frac{1}{2}} \\ &= \Sigma^{-\frac{1}{2}}\Sigma^2\Sigma^{-\frac{1}{2}} = \Sigma. \end{aligned}$$

Analogously, we can show that $\tilde{Q} = V^TQV = \Sigma$. Since the transformed Gramians $\tilde{P} = \tilde{Q} = \Sigma$, by Definition 35, the transformed system in (4.21) is balanced. The above procedure to obtain the balanced system (4.21) from (4.1) is summarized in Algorithm 17. The following proposition shows that the balanced system preserves all HSVs of the system.

Proposition 1. *Let us consider that the system in (4.21) is a balanced form of the system (4.1) under the balancing transformation T defined in (4.18). Then the balanced system preserves all HSVs of the original system.*

Proof.

$$\tilde{P}\tilde{Q} = T^{-1}PT^{-T}T^{T}QT = T^{-1}PQT.$$

This proves $\tilde{P}\tilde{Q}$ and PQ are similar and, therefore, $\sqrt{\lambda_i(\tilde{P}\tilde{Q})} = \sqrt{\lambda_i(PQ)}$. That is the transformed system preserves all the system HSVs of the original one. □

Remark 5. *Similar matrices have the same eigenvalues.*

4.4.2 Truncation of balanced system

In the balanced system, the Gramian can be partitioned as

$$\tilde{P} = \tilde{Q} = \Sigma = \begin{bmatrix} \Sigma_1 & \\ & \Sigma_2 \end{bmatrix} = \begin{bmatrix} \sigma_1 & & & & & \\ & \ddots & & & & \\ & & \sigma_r & & & \\ & & & \sigma_{r+1} & & \\ & & & & \ddots & \\ & & & & & \sigma_n \end{bmatrix}, \tag{4.24}$$

where $\sigma_1, \sigma_2, \cdots, \sigma_n$ are the system's HSVs. In terms of the Gramians, the balanced system (4.21) can also be partitioned as

$$\begin{aligned} \begin{bmatrix} E_{11} & E_{12} \\ E_{21} & E_{22} \end{bmatrix} \begin{bmatrix} \dot{\tilde{x}}_1(t) \\ \dot{\tilde{x}}_2(t) \end{bmatrix} &= \begin{bmatrix} A_{11} & A_{12} \\ A_{21} & A_{22} \end{bmatrix} \begin{bmatrix} \tilde{x}_1(t) \\ \tilde{x}_2(t) \end{bmatrix} + \begin{bmatrix} \tilde{B}_1 \\ \tilde{B}_2 \end{bmatrix} u(t), \\ \tilde{y}(t) &= \begin{bmatrix} \tilde{C}_1 & \tilde{C}_1 \end{bmatrix} \begin{bmatrix} \tilde{x}_1(t) \\ \tilde{x}_2(t) \end{bmatrix} + D_a u(t), \end{aligned} \tag{4.25}$$

where $\tilde{x}_1 \in \mathbb{R}^r$ and $\tilde{x}_2 \in \mathbb{R}^{n-r}$. In the transformed system, the system Hankel singular values are arranged in descending order. Therefore, we can easily identify which states are most important and which are less important. We know that, in the balanced system, the first state is associated with the largest HSVs of the system and accordingly the other states. If we select a system by picking up the block matrices E_{11}, A_{11}, B_1, C_1 we can construct r dimensional ROM matrices

$$\begin{aligned} E_{11}\dot{\tilde{x}}_1(t) &= A_{11}\tilde{x}_1(t) + \tilde{B}_1 u(t); \\ \tilde{y}(t) &= \tilde{C}_1 + D_a u(t), \end{aligned} \tag{4.26}$$

which preserves r larger system HSVs $\sigma_1, \sigma_2, \cdots, \sigma_r$. The ROM is obtained by truncating the states $x_2(t)$ which are related to the singular values $\sigma_{r+1}, \cdots, \sigma_n$.

This concludes the process of balanced truncation. The ROM (4.26) is balanced. Moreover, the following two theorems characterize the properties of the ROM and the balanced truncation-based model reduction.

Theorem 12. *Let us consider that system (4.1) is asymptotically stable and* Σ_1 *and* Σ_2 *defined in (4.24) have no entries in common. Then the truncated balanced system (4.26) is asymptotically stable.*

Theorem 13. *Let us consider that* $G(s)$ *and* $\hat{G}(s)$ *are the transfer functions of the original and reduced systems (4.24) and (4.26), respectively. Then the error system satisfies the following condition.*

$$\left\|G - \hat{G}\right\|_{\mathcal{H}_\infty} \leq 2(\sigma_{r+1} + \cdots, \sigma_n) \tag{4.27}$$

where $\sigma_{r+1}, \cdots, \sigma_n$ *are the truncated singular values from the given system.*

The relation (4.27) is an *a priori* error bound. Thus, for a given error bound (tolerance) one can use it to fix the required dimension of the reduced system.

4.4.3 Balancing and truncating transformations

From the above discussions, we can observe that in the balancing-based model reduction one must first compute the balancing transformation to convert the system into its balanced form. Then the truncation is performed on the balanced system. For a large-scale system, balancing the entire system before truncation is infeasible. Hence, for such systems, the balancing and truncation are usually carried out simultaneously, by using the so-called *balancing and truncating* transformations.

To compute the balancing and truncating transformations, we rewrite the relation (4.17) as

$$L^T R = \begin{bmatrix} U_1 & U_2 \end{bmatrix} \begin{bmatrix} \Sigma_1 & \\ & \Sigma_2 \end{bmatrix} \begin{bmatrix} V_1^T \\ V_2^T \end{bmatrix},$$

and define

$$V := RV_1\Sigma_1^{-\frac{1}{2}}, \quad W := LU_1\Sigma_1^{-\frac{1}{2}}, \tag{4.28}$$

where U_1 and V_1 are composed of the leading r columns of U and V, respectively; Σ_1 is the first $r \times r$ block of the matrix $\Sigma = \text{diag}(\sigma_1, \sigma_2, \ldots, \sigma_r, \ldots, \sigma_n)$. Finally, by applying the balancing transformations (4.28) to the system (4.1), one can derive the ROM (4.2) where the reduced matrices are obtained as shown in (4.12). This procedure is the so-called *square root method* (SRM), originally defined in [163]. The whole procedure of the SRM is summarized in Algorithm 18.

Algorithm 18: SRM for balanced truncation.

Input : E, A, B, C, D_a from (4.1).
Output: $\hat{E}$, $\hat{A}$, $\hat{B}$, $\hat{C}$, $\hat{D}_a := D_a$ in (4.2).

1 Compute R and L as defined in (4.16) by solving the Lyapunov equations:

$$APE^T + EPA^T = -BB^T,$$
$$A^TQE + E^TQA = -C^TC.$$

2 Compute SVD

$$L^TR = U\Sigma V^T = \begin{bmatrix} U_1 & U_2 \end{bmatrix} \begin{bmatrix} \Sigma_1 & \\ & \Sigma_2 \end{bmatrix} \begin{bmatrix} V_1^T \\ V_2^T \end{bmatrix}.$$

3 Construct $V := LV_1\Sigma_1^{-\frac{1}{2}}, \quad W := RU_1\Sigma_1^{-\frac{1}{2}}$.

4 Form $\hat{E} = W^TEV, \quad \hat{A} = W^TAV, \quad \hat{B} = W^TB$ and $\hat{C} = CV$.

4.4.4 Balanced truncation by low-rank Gramian factors

The main components needed to implement Algorithm 18 are the two Gramians factors R and L. These Gramian factors are in fact the Cholesky factors of the Gramians P and Q, respectively. Note that since both the Gramians are symmetric positive definite (SPD) matrices, they must have Cholesky decompositions. To compute P and Q we need to solve the Lyapunov equations in (4.14). Solution of the Lyapunov equations is one of the challenging tasks in the balancing-based model order reduction. Some researchers consider it as a drawback of the BT-based model reduction.

To compute the Gramians or their Cholesky factors by solving the corresponding Lyapunov equations, there exist direct methods. See, e.g., [18, 91]. The Lyapunov equations can also be solved iteratively [33, 97, 175] to compute the Gramians or the Cholesky factors of the Gramians. Unfortunately, all these methods cited here are applicable for a small dense system. If the number of inputs and outputs is much smaller than the dimension of the system, then the Gramians P and Q can usually be approximated by low-rank factors, that is

$$P \approx RR^T \quad \text{and} \quad Q \approx LL^T. \tag{4.29}$$

Here R and L are thin rectangular matrices. Therefore, instead of computing the full Gramian factors, one can compute low-rank factors of the Gramians. In the last few decades, several iterative methods were proposed, e.g., LRCF-ADI (low-rank Cholesky factor alternating direction implicit) iterations [30, 114], cyclic low-rank Smith methods [88, 133], projection methods [62, 97, 99, 141,

Algorithm 19: LR-SRM for balanced truncation.

Input : E, A, B, C, D_a from (4.1).
Output: $\hat{E}$, $\hat{A}$, $\hat{B}$, $\hat{C}$, $\hat{D}_a := D_a$ in (4.2).

1 Compute R and L as defined in (4.29) by solving the Lyapunov equations:

$$APE^T + EPA^T = -BB^T,$$
$$A^TQE + E^TQA = -C^TC.$$

2 Compute SVD $L^TER = U\Sigma V^T = \begin{bmatrix} U_1 & U_2 \end{bmatrix} \begin{bmatrix} \Sigma_1 & \\ & \Sigma_2 \end{bmatrix} \begin{bmatrix} V_1^T \\ V_2^T \end{bmatrix}$.

3 Construct $V := LV_1\Sigma_1^{-\frac{1}{2}}$, $\quad W := RU_1\Sigma_1^{-\frac{1}{2}}$.

4 Form $\hat{E} = W^TEV$, $\quad \hat{A} = W^TAV$, $\quad \hat{B} = W^TB \quad$ and $\quad \hat{C} = CV$.

152], and sign function methods [19, 34, 35]. Although most of the methods are shown to be applicable for large-scale dynamical systems, the LR-ADI iteration is more attractive in the context of Gramian-based model reduction for large sparse systems with few inputs and outputs. A motivation of this prominent method can be found in [38]. Note that in Chapter 3 we have shown how to compute the low-rank Gramian factors by solving the Lyapunov equations (4.14) using the LR-ADI iteration.

Chapter 3 discusses the solution of the Lyapunov equations (4.14) using the LR-ADI iteration to compute the low-rank Gramian factors.

Using the low-rank Gramian factors, R and L, the square root method is summarized in Algorithm 19.

4.5 Rational Krylov subspace-based model reduction

This section discusses the Rational Krylov subspace-based model reduction method. In particular, we focus on the interpolatory projection method. Like balanced truncation, interpolatory projection method is currently another prominent method for the model reduction of large-scale LTI dynamical system as in (4.1). This idea begets from moment matching-based model reduction [65]. The concept of projection for interpolatory model reduction was initially introduced by Skelton et al. in [1, 173, 190]. Later in [81], Grimme modified the approach by utilizing the rational Krylov method as in [140]. We emphasize on the Krylov-based methods, since they are able to match moments without ever computing them explicitly. This issue is important since the computation of moments is ill-conditioned. This is the fundamental motivation behind the Krylov-based methods [68]. The Krylov-based interpolation method also generates the projectors V and W to compute the reduce-order model (4.2). Unlike the balanced truncation, the Krylov-based methods requires only matrix-vector multiplications, some sparse linear solvers, and can be iteratively implemented; therefore, it is computationally effective.

One of the most crucial tasks of applying the rational Krylov subspace-based interpolation method is the selection of interpolation points. The quality of the reduced model is highly dependent on the choice of interpolation points. Various techniques [173] have been proposed in the literature for the selection of interpolation points. Recently, in [85], the issue of selecting good interpolation points has been linked to the problem of $\mathcal{H}_2$-optimal model reduction. The optimal $\mathcal{H}_2$ model reduction is of great importance in the area of dynamical systems and simulation. The optimal $\mathcal{H}_2$ model reduction was discussed in quite a few research papers; see, e.g., [46, 96, 111, 157, 189]. The optimal $\mathcal{H}_2$ model reduction computes a ROM which minimizes the $\mathcal{H}_2$ norm of the error; this is an optimization problem. In general, there exist two approaches for solving this problem - the Lyapunov-based optimal $\mathcal{H}_2$ methods presented in [90, 96, 157, 183, 189, 196] and the interpolation-based optimal $\mathcal{H}_2$ methods considered in [20, 21, 47, 85, 171]. While the Lyapunov-based approaches require solving a series of Lyapunov equations which becomes costly and sometimes intractable in large-scale settings, the interpolatory approaches only require solving a series of sparse linear systems and have proved to be numerically very effective. This is the motivation behind using the interpolatory model reduction techniques for the optimal $\mathcal{H}_2$ model reduction.

For SISO systems, interpolation-based $\mathcal{H}_2$ optimality conditions were originally developed by Meier and Luenberger [123]. Based on this condition, an effective interpolatory optimal $\mathcal{H}_2$ approximation algorithm was introduced by Gugercin, Antoulas and Beattie [82, 84]. Their proposed algorithm is called the Iterative Rational Krylov Algorithm (IRKA); the details of IRKA for the SISO system can be found in [85]. However, the algorithm has been generalized for MIMO systems using the tangential interpolation framework; see, e.g., [47, 171] for more details. This book emphasizes the IRKA-based interpolation (for the SISO case) and tangential interpolation (for the MIMO case) frameworks for the model reduction of large-scale dynamical systems.

4.5.1 Interpolatory projections for SISO systems

Let us consider (4.1) is a SISO system. Interpolatory projection methods seek a ROM (4.2) by constructing the matrices V and W in such a way that the reduced transfer function (4.7) interpolates the original transfer function

(4.5) at a predefined set of interpolation points. That is to say, find $\hat{G}(s)$ such that

$$\begin{aligned} G(\alpha_i) &= \hat{G}(\alpha_i), \\ C(\alpha_i E - A)^{-1} B &= \hat{C}(\alpha_i \hat{E} - \hat{A})^{-1} \hat{B}, \qquad \text{for } i = 1, \cdots, r, \end{aligned} \tag{4.30}$$

where $\alpha_i \in \mathbb{C}$ are the interpolation points. Often, in addition to the above conditions, we are interested in matching more quantities, that is

$$\begin{gathered} \mathrm{G}^{(j)}(\alpha_i) = \hat{\mathrm{G}}^{(j)}(\alpha_i), \\ C[(\alpha_i E - A)^{-1} E]^j (\alpha_i E - A)^{-1} B = \hat{C}[(\alpha_i \hat{E} - \hat{A})^{-1} \hat{E}]^j (\alpha_i \hat{E} - \hat{A})^{-1} \hat{B}, \end{gathered} \tag{4.31}$$

for $j = 0, 1, \cdots, q$, where $C[-(\alpha_i E - A)^{-1} E]^j (\alpha_i E - A)^{-1} B$ is called the j-th moment of $\mathrm{G}(s)$ at α_i and represents the j-th derivative of $\mathrm{G}(s)$ evaluated at σ_i. Note that for $j = 0$, these conditions reduce to (4.30). In this book, we restrict ourselves to simple Hermite interpolation, where $j = 0$ and $j = 1$. In the text that follows, we discuss how projection can ensure a reduced interpolating approximation by carefully selecting the matrices V and W.

The concept of projection for interpolatory model reduction was initially introduced in [173]; later, Grimme [81] modified the approach by utilizing the rational Krylov method [140]. Since Krylov-based methods can achieve moment matching without explicitly computing moments (explicit computation of moments is known to be ill-conditioned [69]), they are extremely useful for model reduction of large-scale systems.

The following result suggests a choice of V and W that ensure Hermite interpolation with the use of a rational Krylov subspace.

Lemma 4. *Consider two sets of distinct interpolation points, $\{\alpha_i\}_{i=1}^{r} \subset \mathbb{C}$ and $\{\beta_i\}_{i=1}^{r} \subset \mathbb{C}$, which are closed under conjugation (i.e., the points are either real or appear in conjugate pairs). Suppose V and W satisfy*

$$\mathrm{Range}\,(V) = \mathrm{span}\Big\{(\alpha_1 E - A)^{-1} B, \cdots, (\alpha_r E - A)^{-1} B\Big\}, \tag{4.32a}$$

$$\mathrm{Range}\,(W) = \mathrm{span}\Big\{(\beta_1 E^T - A^T)^{-1} C^T, \cdots, (\beta_r E^T - A^T)^{-1} C^T\Big\}. \tag{4.32b}$$

Then V and W can be real and $\hat{G}(s) = \hat{C}(s\hat{E} - \hat{A})^{-1}\hat{C}$, where $\hat{E}$, $\hat{A}$, $\hat{B}$ and $\hat{C}$ define the ROM (4.2). The ROM satisfies the interpolation conditions

$$\begin{aligned} G(\alpha_i) &= \hat{G}(\alpha_i), \quad G(\beta_i) = \hat{G}(\beta_i), \quad \textit{and} \\ G'(\alpha_i) &= \hat{G}'(\alpha_i) \quad \textit{when} \quad \alpha_i = \beta_i, \end{aligned} \tag{4.33}$$

for $i = 1, \cdots, r$.

The subspace in (4.32a), i.e., the span of the column vectors $(\alpha_i E - A)^{-1} B$ for $i = 1, \cdots, r$, can be considered as the union of shifted rational Krylov subspaces. For a given shift frequency $\alpha \in \mathbb{C}$, the rational Krylov subspace $\mathcal{K}_q((\alpha E - A)^{-1}, (\alpha E - A)^{-1}B)$ is defined as

$$\mathcal{K}_q((\alpha E - A)^{-1}, (\alpha E - A)^{-1}B) := \text{span}\Big\{(\alpha E - A)^{-1}B, \cdots, (\alpha E - A)^{-q}B\Big\}. \tag{4.34}$$

If $q = 1$ for each $\alpha_i, i = 1, \cdots, r$, then the union of such shifted rational Krylov subspaces is equivalent to the subspace in (4.32a). Analogously, the subspace in (4.32b) can also be defined as the union of shifted rational Krylov subspaces given above. To summarize, rational Krylov-based model reduction requires a suitable choice of interpolation points, the construction of V and W as in Lemma 4 and the use of Petrov-Galerkin conditions.

The quality of the reduced model is highly dependent on the choice of the interpolation points; therefore, various techniques [173] have been developed for the selection of interpolation points. The issue of choosing good interpolation points is interpreted in [85]. The proposed method in [85] is called IRKA. In this algorithm the interpolation points are not predefined. In each iteration, the interpolation points are updated by $\sigma_i = -\hat{\lambda}_i(\hat{A}, \hat{E})$ where $\hat{\lambda}_i(A)$ are the eigenvalues of the ROM. This process is continued until the interpolation points converge to a fixed set of interpolation points (i.e., the set of Ritz values of (A, E)). The authors of [85] show that IRKA is linked to the problem of $\mathcal{H}_2$-optimal model reduction.

Definition 37. *A ROM (4.2) is called $\mathcal{H}_2$ optimal if it satisfies*

$$\|G\|_{\mathcal{H}_2} = \min_{\dim(\hat{G})=r} \|G - \hat{G}\|_{\mathcal{H}_2}. \tag{4.35}$$

Note that the $\mathcal{H}_2$-optimal model reduction is one of the most important classes of model reduction techniques of LTI dynamical systems. See, e.g., [46, 96, 111, 157, 189]. A complete procedure of IRKA for a SISO system is given in Algorithm 20.

4.5.2 Interpolatory projections for MIMO systems

For the model reduction of MIMO dynamical systems, rational tangential interpolation has been developed by Gallivan et al. [75]. The problem of rational tangential interpolation is to construct V and W such that the reduced transfer function $\hat{\mathrm{G}}(s)$ tangentially interpolates the original transfer function $\mathrm{G}(s)$ at a predefined set of interpolation points and some fixed tangent directions. That is

$$\begin{aligned} &\mathrm{G}(\alpha_i)b_i = \hat{\mathrm{G}}(\alpha_i)b_i, \quad c_i^T\mathrm{G}(\alpha_i) = c_i^T\hat{\mathrm{G}}(\alpha_i), \quad \text{and} \\ &c_i^T\mathrm{G}(\alpha_i)b_i = c_i^T\hat{\mathrm{G}}(\alpha_i)b_i, \quad \text{for } i = 1, \cdots, r, \end{aligned}$$

where $b_i \in \mathbb{C}^m$ and $c_i \in \mathbb{C}^p$ are the right and left tangential directions, respectively, and correspond to the interpolation points α_i. With these quantities,

Algorithm 20: IRKA for SISO systems.

Input : E, A, B, C, D_a.

Output: $\hat{E}$, $\hat{A}$, $\hat{B}$, $\hat{C}$, $\hat{D}_a := D_a$.

Make an initial selection of the interpolation points $\{\alpha_i\}_{i=1}^r$.

$V = \left[(\alpha_1 E - A)^{-1}B, \cdots, (\alpha_r E - A)^{-1}B\right]$,
$W = \left[(\alpha_1 E^T - A^T)^{-1}C^T, \cdots, (\alpha_r E^T - A^T)^{-1}C^T\right]$.

while *(not converged)* **do**

  $\hat{E} = W^T EV$, $\hat{A} = W^T AV$, $\hat{B} = W^T B$, $\hat{C} = CV$.

  Compute $\alpha_i \leftarrow -\lambda_i$, where $\{\lambda_i\}_{i=1}^r$ are the eigenvalues of $(\hat{A}, \hat{E})$.

  $V = \left[(\alpha_1 E - A)^{-1}B, \cdots, (\alpha_r E - A)^{-1}B\right]$,
  $W = \left[(\alpha_1 E^T - A^T)^{-1}C^T, \cdots, (\alpha_r E^T - A^T)^{-1}C\right]$.

  $i = i + 1$

$\hat{E} = W^T EV$, $\hat{A} = W^T AV$, $\hat{B} = W^T B$ and $\hat{C} = \mathcal{C}V$.

the rational tangential interpolation can be performed. The IRKA-based interpolatory projection methods for MIMO systems have been discussed in [47, 85], where the algorithm updates interpolation points as well as tangential directions until the reduced system satisfies the necessary condition for $\mathcal{H}_2$-optimality. We have summarized a complete procedure for a MIMO system in Algorithm 21.

Algorithm 21: IRKA for MIMO systems.

Input : E, A, B, C, D_a.
Output: $\hat{E}$, $\hat{A}$, $\hat{B}$, $\hat{C}$, $\hat{D}_a := D_a$.

Make an initial selection of the interpolation points $\{\alpha_i\}_{i=1}^r$ and the tangential directions $\{b_i\}_{i=1}^r$ and $\{c_i\}_{i=1}^r$.

$V = \left[(\alpha_1 E - A)^{-1} B b_1, \cdots, (\alpha_r E - A)^{-1} B b_r\right]$,
$W = \left[(\alpha_1 E^T - A^T)^{-1} C^T c_1, \cdots, (\alpha_r E^T - A^T)^{-1} C^T c_r\right]$.

while *(not converged)* **do**

  $\hat{E} = W^T E V$, $\hat{A} = W^T A V$, $\hat{B} = W^T B$, $\hat{C} = CV$.

  Compute $\hat{A} z_i = \hat{\lambda}_i \hat{E} z_i$ and $y_i^* \hat{A} = \hat{\lambda}_i y_i^* \hat{E}$.

  $\alpha_i \leftarrow -\lambda_i$, $b_i^* \leftarrow -y_i^* \hat{B}$ and $c_i \leftarrow \hat{C} z_i$, for $i = 1, \cdots, r$.

  $V = \left[(\alpha_1 E - A)^{-1} B b_1, \cdots, (\alpha_r E - A)^{-1} B b_r\right]$,
  $W = \left[(\alpha_1 E^T - A^T)^{-1} C^T c_1, \cdots, (\alpha_r E^T - A^T)^{-1} C c_r\right]$.

  $i = i + 1$

$\hat{E} = W^T E V$, $\hat{A} = W^T A V$, $\hat{B} = W^T B$ and $\hat{C} = \mathcal{C} V$.

4.6 Some useful MATLAB functions

Function name	Description
ss	construct state space model or convert model to state space
rss	generate randomized continuous-time state space models
dss	create generalized or descriptor state space models
balreal	compute a balanced realization for the linear system
balred	compute a reduced-order approximation of the LTI system
hsvd	compute the Hankel singular values of linear systems
step	step response of dynamic systems
stepplot	plot step response of linear systems
freqresp	frequency response of dynamic systems
sigma	singular value plot of the frequency response of the dynamic system
lyapchol	Cholesky factor of solution of Lyapunov equation
norm	compute $\mathcal{H}_2$ norm and $\mathcal{H}_\infty$ norm of LTI systems
loglog	log-log scale plot using logarithmic scaled for both the X-axis and Y-axis
semilogx (semilogy)	log-log scale plot using logarithmic scaled for X- (Y-) axis

n	p/m	truncation tolerance	r
1006	1/1	10^{-3}	15
		10^{-4}	17
		10^{-5}	19
		10^{-6}	21
		10^{-7}	22

Table 4.1: Dimension of the original model (n), number of input(s)/output(s)(p/m), and the dimension of reduced models (r) with different truncation tolerances.

4.7 Numerical experiments

To assess the accuracy and efficiency of the (model reduction) methods proposed in this chapter, we illustrate some numerical results. For the numerical experiments, we consider two model examples: the FOM model [134] and the CD player [61, 187]. See also Appendix A.2.1 and A.2.2 for details. The data of the models are available from [50] SLICOT Benchmark Examples[2].

All the results are obtained by using MATLAB R2015a (8.5.0.197613) on a board with 4×Intel® Core™i5-4460 CPUs with a 2.90 GHz clock speed and 16 GB RAM.

Balancing based results. Considering a truncation tolerance of 10^{-5}, we applied the balanced truncation method (i.e., Algorithm 19) to the FOM model. The algorithm provides different dimensional reduced models for different truncation tolerances that are listed in Table 4.1. Figure 4.1 shows the comparison of the original and the different dimensional reduced models. In all cases, sigma-plots of the original and the reduced models match very well and the error between the original and the reduced models obey the global error bound.

Interpolation based results. Applying the IRKA (Algorithm 20) to the same model, we obtain different dimensional ROMs. Figure 4.2 shows the absolute and the relative errors between the original and the reduced models. To implement this algorithm, we consider 100 iteration steps for computing 30 dimensional reduced models.

Comparisons between balanced truncation and interpolatory techniques. In order to show the comparisons between BT and IRKA, both methods are applied to the model CDP to compute 30 dimensional reduced models.

[2]`http://slicot.org/20-site/126-benchmark-examples-for-model-reduction`

Figure 4.3 shows the comparison of both methods. From the absolute and relative errors shown in Figure 4.3(b) and 4.3(c), respectively, we observe that the IRKA performs better than the balanced truncation method in the lower-frequency. If the frequency level gets higher, balanced truncation shows better performance.

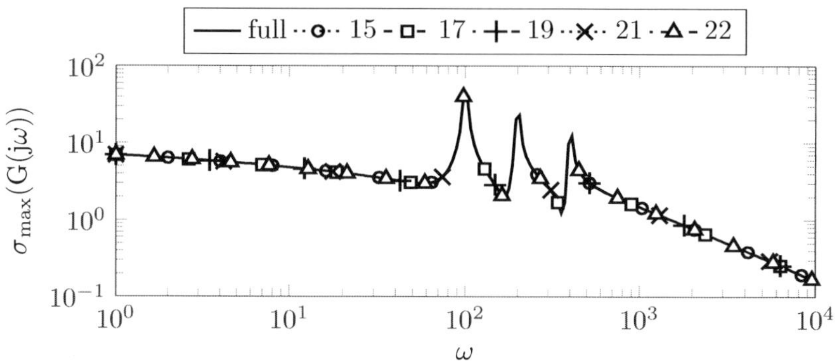

(a) Sigma plot.

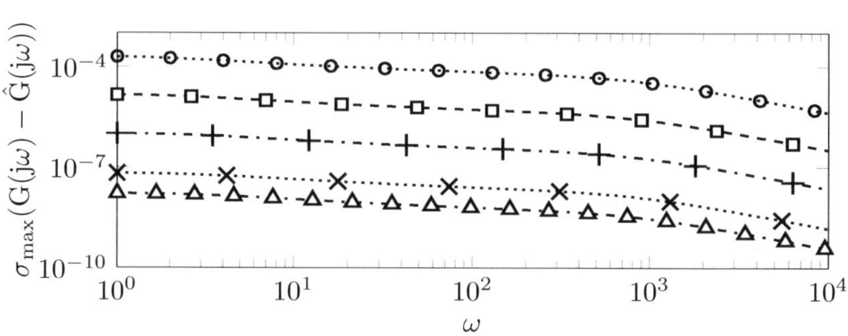

(b) Absolute error.

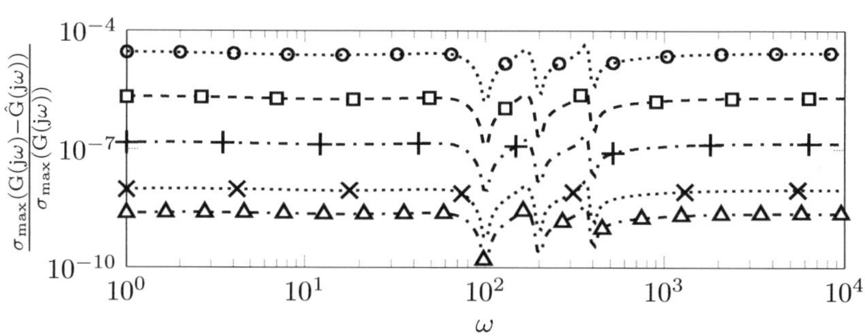

(c) Relative error.

Figure 4.1: Comparison between the original and reduced models obtained by balanced truncation using the FOM model.

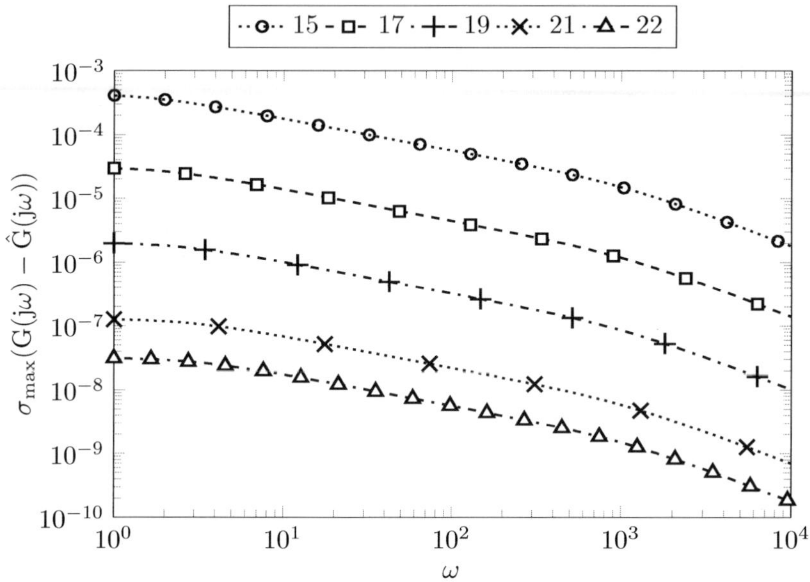

(a) Absolute error.

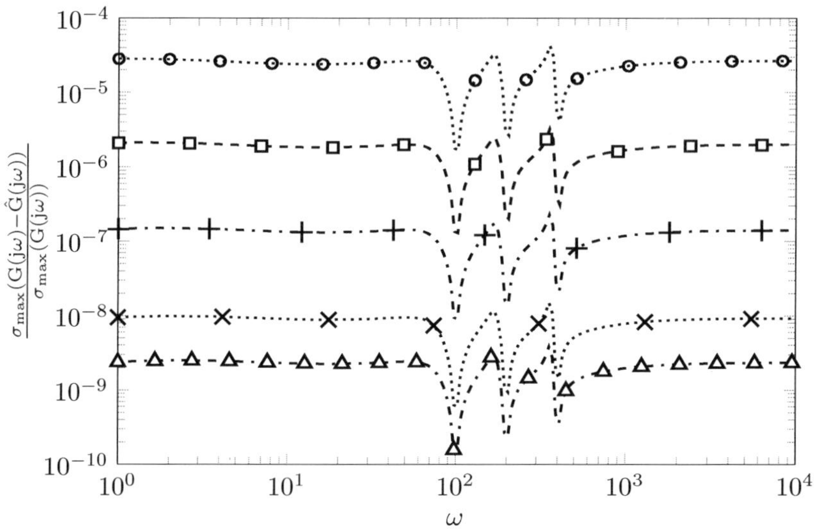

(b) Relative error.

Figure 4.2: Error between original and reduced models obtained by IRKA using the FOM model.

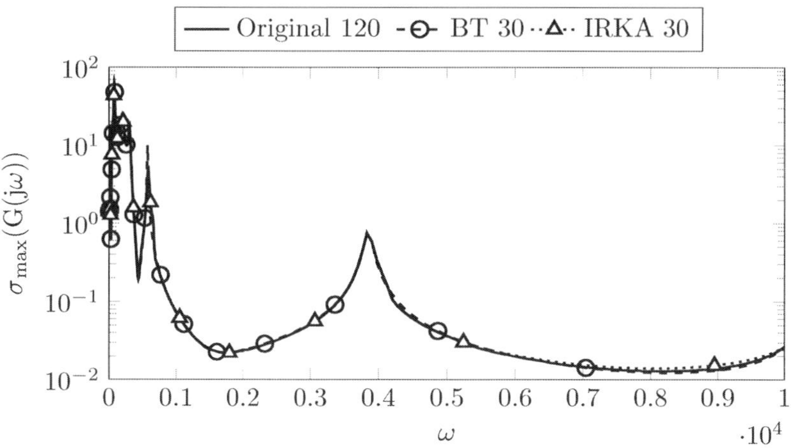

(a) Sigma plot.

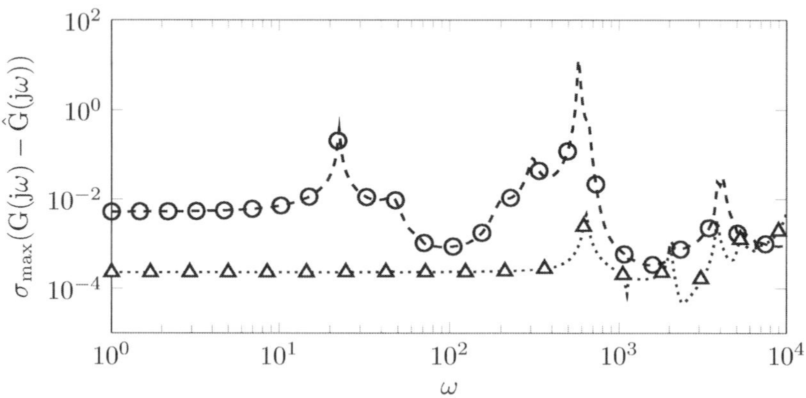

(b) Absolute error.

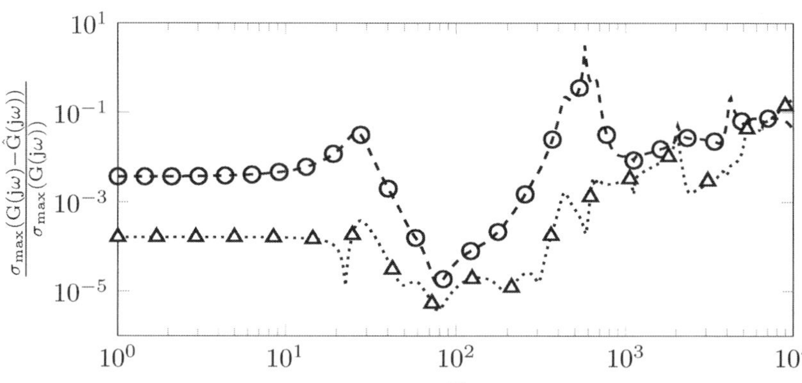

(c) Relative error.

Figure 4.3: Comparison of the performances between the BT and IRKA based MOR applying the CDP model.

■ Exercises:

4.1 Consider a generalized state space system:

$$(E, A, B, C, D) = \left(\begin{bmatrix} 1 & 0 & 0 \\ 0 & 3 & 0 \\ 0 & 0 & 4 \end{bmatrix}, \begin{bmatrix} -1 & 2 & 3 \\ 0 & -6 & 1 \\ 0 & 0 & -12 \end{bmatrix}, \begin{bmatrix} 0 \\ 1 \\ 1 \end{bmatrix}, \begin{bmatrix} 1 & 0 & 0 \end{bmatrix}, 0 \right).$$

Compute the controllability and the observability Gramians of the system and hence comment whether the system is balanced or not. If the system is not balanced, compute balancing transformations following (4.18) and find a balanced system.

4.2 Using MATLAB function `balreal` convert a non-balanced system into a balanced system. Consider the generalized state space system as in Exercise 2.1 to convert it into a balanced system. (Hint: You can compute controllability and observability Gramians of a LTI system using MATLAB function `gram` to test whether the system is balanced or not).

4.3 For the LTI system in Exercise 4.1, find a balanced system using MATLAB function `balreal`. Then compute Hankel singular values (HSVs) using the command `hsvd`, the controllability and observability Gramians by `gram` of the balanced system. Are the HSVs equal to the diagonal elements of both Gramians?

4.4 For the LTI system in Exercise 4.1, we can compute HSVs of the system by `hsvd`. Now truncating the smallest HSV of the system and compute a 2-dimensional reduced model. Use `stepplot` to plot the step responses on a figure to compare the original and reduced models.

4.5 MATLAB function `balred` can compute a balanced reduced order model. Use this function to compute different dimensional, e.g., 50, 100, 150 and 200 dimensional reduced models of the FOM model and compare them with the original model. For comparison, you can analyze the (absolute and relative) errors between the original and reduced models in both time and frequency domains.

4.6 Algorithm 19 can find a balanced reduced model. For the FOM model compute 50 dimensional reduced model applying this algorithm. Also compute the same dimensional reduced model by MATLAB function `balred`. Measure the computational time for both cases and comment about their efficiency.

4.7 Test the stability of the FOM model. Apply Algorithm 19 to find a reduced model with a truncation tolerance 10^{-5}. Is the reduced model also stable? Plot the frequency responses and the absolute error (in $\mathcal{H}_\infty$-norm) of the original and reduced models over the interval $[10^1, 10^3]$ by using 500 logarithmically distributed sample points. Does the reduced model satisfy the error bound?

4.8 Write some advantages and disadvantages of balanced truncation and interpolatory projection techniques via IRKA. Apply both methods (i.e., Algorithms 19 and 20) to the FOM model for finding a suitable ROM. Now discuss the results to compare the efficiency of the methods.

4.9 Compute, for example 50, 100, 150 and 200 dimensional reduced models by applying the IRKA to the CD player (CDP)(in Appendix A.2.1) model. Using MATLAB function `hsvplot`, plot the singular values of the original and reduced models on a figure and discuss the results.

4.10 Applying Algorithms 19 and 20, compute 50 dimensional reduced models of the model CDP. Using MATLAB function `norm` find the $\mathcal{H}_2$ and $\mathcal{H}_\infty$ norm of the error systems for both cases and show them in a table for a comparison.

Chapter 5

Model Reduction of Second-Order Systems

5.1 Introduction

We consider second-order LTI continuous-time systems

$$M\ddot{\xi}(t) + D\dot{\xi}(t) + K\xi(t) = Hu(t), \tag{5.1a}$$

$$y(t) = L_1\xi(t), \tag{5.1b}$$

where $M, D, K \in \mathbb{R}^{n_\xi \times n_\xi}$, input matrix $H \in \mathbb{R}^{n_\xi \times p}$ and output matrix $L_1 \in \mathbb{R}^{m \times n_\xi}$. Such systems usually appear in mechanics [8] or structural and multi-body dynamics [54, 64] where the velocity is taken into account in the modeling and thus the acceleration becomes part of the system. In mechanics, usually, the matrices M, D and K are known as mass, damping and stiffness matrices, respectively, while the vector $\xi(t)$ is known as mechanical displacement. Such systems also appear in electrical engineering when RLC circuits are designed for nodal analysis [188] here, the matrices M, D and K are called the conductance, capacitance and susceptance matrices while the vector $\xi(t)$ is denoted as electric charge.

If the size of the model is very large, performing the simulation, optimization and controller design is difficult or infeasible. Therefore, reducing the size of the model is a crucial task. A classical approach to find a reduced-order model of second-order systems is to first rewrite the systems into first-order form, then we can apply model order reduction techniques as discussed in Chapter 4. The reduced-order model would be in state space [28, 37, 48, 156] form. Having the state space form is often significant because some softwares, for example, MATLAB simulink® is designed for the state space model.

Now the question arises, can we return to the second-order form from the reduced systems? In general, the answer is no, since the structure of the original model is destroyed in the reduced form. Sometimes, the preservation of second-order structure in the reduced systems is essential to perform the simulation, optimization and controller design if the software tools are specially designed for second-order systems. Moreover, structure preserving reduced models allow meaningful physical interpretation and provide more accurate approximation which we will see later. In recent time, structure preserving model reduction of second-order systems has been receiving much attention. See, e.g., [12, 22, 28, 37, 87, 107, 137, 147, 148] and the references therein.

This chapter contributes in finding the reduced-order model of second-order systems of the form (5.1). We will discuss both *second-order-to-first-order* and *second-order-to-second-order* reduction methods with several model reduction methods. In this book, second-order-to-second-order is called *structure preserving model order reduction* (SPMOR). We also discuss equivalent various first-order representations of second-order Gramians and transfer function of the second-order system. To perform balanced truncation and PDEG methods, we need to solve Lyapunov equations. Moreover, this chapter presents low-rank ADI-based algorithms to solve the Lyapunov equations arising from the second-order systems. The efficiency of the proposed algorithms are tested by applying them to benchmark model examples and the experimental results are discussed at the end of the chapter.

5.2 Preliminaries

This section discusses some preliminary results and introduces important definitions and notations of the underlying second-order systems which will be exploited in the remaining parts of this and upcoming chapters.

5.2.1 Equivalent first-order representations

A second-order form of system (5.1) can be converted into a first-order form. There are several transformations to convert the second-order system into a first-order form; all of which can be proved to be equivalent. The text that follows will show several ways of first-order representations of the second-order system (5.1).

From Equation (5.1a) we obtain

$$M\ddot{\xi}(t) = -D\dot{\xi}(t) - K\xi(t) + Hu(t) \tag{5.2}$$

Suppose

$$x_1(t) = \xi(t) \quad \text{and} \tag{5.3}$$

$$x_2(t) = \dot{\xi}(t). \tag{5.4}$$

Differentiating Equation (5.3) with respect to t yields

$$\begin{aligned} \dot{x}_1(t) &= \dot{\xi}(t) \\ &= x_2(t) \qquad \text{(using (5.4))} \end{aligned} \tag{5.5}$$

Now Equation (5.4) results in

$$\begin{aligned} \dot{x}_2(t) &= \ddot{\xi}(t) \\ M\dot{x}_2(t) &= M\ddot{\xi}(t) \\ &= -Dx_2(t) - Kx_1(t) + Hu(t). \end{aligned} \tag{5.6}$$

At the same time, the output equation in (5.1) becomes

$$y(t) = L_1 x_1(t) \tag{5.7}$$

Combining Equations (5.5), (5.6) and (5.7) we obtain the first-order form of the second-order system (5.1) as

$$\begin{aligned} \underbrace{\begin{bmatrix} I & 0 \\ 0 & M \end{bmatrix}}_{E} \underbrace{\begin{bmatrix} \dot{x}_1(t) \\ \dot{x}_2(t) \end{bmatrix}}_{\dot{x}(t)} &= \underbrace{\begin{bmatrix} 0 & I \\ -K & -D \end{bmatrix}}_{A} \underbrace{\begin{bmatrix} x_1(t) \\ x_2(t) \end{bmatrix}}_{x(t)} + \underbrace{\begin{bmatrix} 0 \\ H \end{bmatrix}}_{B} u(t), \\ y(t) &= \underbrace{\begin{bmatrix} L_1 & 0 \end{bmatrix}}_{C} \begin{bmatrix} x_1(t) \\ x_2(t) \end{bmatrix}, \end{aligned} \tag{5.8}$$

where I is an $n \times n$ identity matrix, $x(t) \in \mathbb{R}^{2n}$, and the matrices E, A, B and C have appropriate sizes. In many fields of engineering, the mass, damping and stiffness matrices are symmetric and even positive definite. In that case, it is recommended to transform model (5.8) as,

$$\underbrace{\begin{bmatrix} -K & 0 \\ 0 & M \end{bmatrix}}_{E} \underbrace{\begin{bmatrix} \dot{x}_1(t) \\ \dot{x}_2(t) \end{bmatrix}}_{\dot{x}(t)} = \underbrace{\begin{bmatrix} 0 & -K \\ -K & -D \end{bmatrix}}_{A} \underbrace{\begin{bmatrix} x_1(t) \\ x_2(t) \end{bmatrix}}_{x(t)} + \underbrace{\begin{bmatrix} 0 \\ H \end{bmatrix}}_{B} u(t),$$
$$y(t) = \underbrace{\begin{bmatrix} L_1 & 0 \end{bmatrix}}_{C} \begin{bmatrix} x_1(t) \\ x_2(t) \end{bmatrix}. \tag{5.9}$$

Here, the symmetry and definiteness of M and K are automatically transferred into E and symmetry of K and D are automatically transferred into A; in other words, if M, K are symmetric and positive (semi-)definite and D is symmetric, then A is symmetric and E is symmetric and positive (semi-)definite.

Reordering the state vectors $[x_1(t)^T, x_2(t)^T]^T$ in (5.8), a second-order model can also be transformed into first-order form as,

$$\underbrace{\begin{bmatrix} 0 & I \\ M & D \end{bmatrix}}_{E} \underbrace{\begin{bmatrix} \dot{x}_2(t) \\ \dot{x}_1(t) \end{bmatrix}}_{\dot{x}(t)} = \underbrace{\begin{bmatrix} I & 0 \\ 0 & -K \end{bmatrix}}_{A} \underbrace{\begin{bmatrix} x_2(t) \\ x_1(t) \end{bmatrix}}_{z(t)} + \underbrace{\begin{bmatrix} 0 \\ H \end{bmatrix}}_{B} u(t),$$
$$y(t) = \underbrace{\begin{bmatrix} 0 & L_1 \end{bmatrix}}_{C} \begin{bmatrix} x_2(t) \\ x_1(t) \end{bmatrix}, \tag{5.10}$$

where all the matrices and vectors can be defined as in (5.8). In this formulation the system matrices E and A become symmetric if $I = M$ and that changes the realization into,

$$\underbrace{\begin{bmatrix} 0 & M \\ M & D \end{bmatrix}}_{E} \underbrace{\begin{bmatrix} \dot{x}_2(t) \\ \dot{x}_1(t) \end{bmatrix}}_{\dot{x}(t)} = \underbrace{\begin{bmatrix} M & 0 \\ 0 & -K \end{bmatrix}}_{A} \underbrace{\begin{bmatrix} x_2(t) \\ x_1(t) \end{bmatrix}}_{z(t)} + \underbrace{\begin{bmatrix} 0 \\ H \end{bmatrix}}_{B} u(t),$$
$$y(t) = \underbrace{\begin{bmatrix} 0 & L_1 \end{bmatrix}}_{C} \begin{bmatrix} x_2(t) \\ x_1(t) \end{bmatrix}. \tag{5.11}$$

This formulation is particularly important in some practical implementations for a special class of systems. For instance, if the matrices M, D and K are symmetric and also L_1 and H are the transpose of each other, then in (5.11) the matrices E and A are symmetric and the input and output matrices are also the same. Therefore, the corresponding Lyapunov equations arising from this system coincide. We will discuss this issue later in this and the upcoming chapters as well. The first-order representations of second-order systems are essentially the different realizations of the same transfer function. The following section will show this result.

5.2.2 Transfer function of second-order systems

To find the transfer function of the second-order system (5.1), we apply the Laplace Transform (see Definition 21) to the system and obtain the following transformed form.

$$s^2 M\Xi(s) + sD\Xi(s) + K\Xi(s) = HU(s), \tag{5.12a}$$

$$Y(s) = L_1\Xi(s), \tag{5.12b}$$

where $\Xi(s)$, $Y(s)$ and $U(s)$ are the Laplace transformation of $\xi(t)$, $y(t)$ and $u(t)$, respectively. Equation (5.12a) implies

$$\Xi(s) = (s^2 M + sD + K)^{-1} HU(s). \tag{5.13}$$

Inserting this identity into (5.12b), we obtain

$$Y(s) = L_1(s^2 M + sD + K)^{-1} HU(s). \tag{5.14}$$

This input-output relation can be represented as

$$Y(s) = \mathcal{G}(s)U(s), \tag{5.15}$$

where

$$\mathcal{G}(s) = L_1(s^2 M + sD + K)^{-1} H. \tag{5.16}$$

is called *transfer function* of the second-order system (5.1). On the other hand, both representations in (5.8) and (5.10) of the second-order system can be written as

$$G(s) = C(sE - A)^{-1} B, \tag{5.17}$$

where E, A, B and C are defined either in (5.8) or (5.10). The following observation shows that transfer functions of a second-order system and its first-order form are the same.

Theorem 14. *Let $\mathcal{G}(s)$ and $G(s)$, as defined in (5.16) and (5.17), be the transfer functions of system (5.1) and its first-order representation (5.8) or (5.10), respectively. Then it can be shown that $G(s) \Leftrightarrow \mathcal{G}(s)$.*

Proof. We want to prove Theorem 14 for the first-order representation as in (5.8). The same procedure can be followed for the representation in (5.10) also. From (5.17) we obtain,

$$\begin{aligned}
G(s) &= C(sE - A)^{-1} B \\
&= \begin{bmatrix} L_1 & 0 \end{bmatrix} \left(s \begin{bmatrix} I & 0 \\ 0 & M \end{bmatrix} - \begin{bmatrix} 0 & I \\ -K & -D \end{bmatrix} \right)^{-1} \begin{bmatrix} 0 \\ H \end{bmatrix} \\
&= \begin{bmatrix} L_1 & 0 \end{bmatrix} \left(\begin{bmatrix} s & -I \\ K & sM + D \end{bmatrix} \right)^{-1} \begin{bmatrix} 0 \\ H \end{bmatrix} \\
&= \begin{bmatrix} L_1 & 0 \end{bmatrix} \begin{bmatrix} \mathbf{x}_1 \\ \mathbf{x}_2 \end{bmatrix},
\end{aligned} \tag{5.18}$$

where,

$$\begin{bmatrix} \mathbf{x}_1 \\ \mathbf{x}_2 \end{bmatrix} = \left(\begin{bmatrix} s & -I \\ K & sM + D \end{bmatrix} \right)^{-1} \begin{bmatrix} 0 \\ H \end{bmatrix}.$$

This implies

$$\begin{bmatrix} s & -I \\ K & sM + D \end{bmatrix} \begin{bmatrix} \mathbf{x}_1 \\ \mathbf{x}_2 \end{bmatrix} = \begin{bmatrix} 0 \\ H \end{bmatrix},$$

and the equivalent linear equations,

$$s\mathbf{x}_1 - \mathbf{x}_2 = 0 \tag{5.19}$$

$$K\mathbf{x}_1 + (sM + D)\mathbf{x}_2 = H. \tag{5.20}$$

Equation (5.19) gives $\mathbf{x}_2 = s\mathbf{x}_1$. Inserting this into (5.20), we obtain

$$\begin{aligned} K\mathbf{x}_1 + (sM + D)s\mathbf{x}_1 &= H \\ (s^2M + sD + K)\mathbf{x}_1 &= H \\ \mathbf{x}_1 &= (s^2M + sD + K)^{-1}H. \end{aligned}$$

Therefore, $\mathbf{x}_2 = s(s^2M + sD + K)^{-1}H$. Plugging $\mathbf{x}_1$ and $\mathbf{x}_2$ into (5.18), we obtain

$$\begin{aligned} G(s) &= \begin{bmatrix} L_1 & 0 \end{bmatrix} \begin{bmatrix} (s^2M + sD + K)^{-1}H \\ s(s^2M + sD + K)^{-1}H \end{bmatrix} \\ &= L_1(s^2M + sD + K)^{-1}H + 0 \\ &= L_1(s^2M + sD + K)^{-1}H \\ &= \mathcal{G}(s). \end{aligned}$$

Conversely, it can be shown that $\mathcal{G}(s) = G(s)$. □

5.2.3 Gramians of the second-order system

In Chapter 2, the Gramians of the first-order systems have been introduced from a physical point of view. Analogously, the Gramians of the second-order systems can be defined from an energy interpretation perspective. Second-order Gramians are defined in [49, 124]. For the sake of convenience, they are briefly reviewed as follows.

Recall the Gramians and the corresponding Lyapunov equations of first-order systems as presented in Chapter 2. The above section shows that the second-order system (5.1) can be converted into its first-order form (5.8). Let us consider that P is the controllability Gramian of the system (5.8), which satisfies the Lyapunov equation

$$APE^T + EPA^T = -BB^T, \tag{5.21}$$

where the matrices E, A and B are defined in (5.8). By defining

$$J(u) = \int_{-\infty}^{0} u^*(t)u(t)dt,$$

it can be shown that

$$x_0^* P^{-1} x_0 \tag{5.22}$$

is the solution of the problem

$$\begin{aligned} &\min_u \; J(u) \\ &\text{s. t.} \quad E\dot{x}(t) = Ax(t) + Bu(t), \quad x(0) = x_0, \end{aligned} \tag{5.23}$$

with $x_0 = \begin{bmatrix} \xi_0 \\ \dot{\xi}_0 \end{bmatrix}$. Equation (5.22) in fact represents the required minimal energy to reach to the state x_0 from $t = -\infty$ at time $t = 0$. Now consider the two optimization problems

$$\begin{aligned} &\min_{\xi_0} \min_u \; J(u) \\ &\text{s. t.} \quad M\ddot{\xi}(t) + D\dot{\xi}(t) + K\xi(t) = Hu(t), \quad \xi(0) = \xi_0, \end{aligned} \tag{5.24}$$

and

$$\begin{aligned} &\min_{\dot{\xi}_0} \min_u \; J(u) \\ &\text{s. t.} \quad M\ddot{\xi}(t) + D\dot{\xi}(t) + K\xi(t) = Hu(t), \quad \dot{\xi}(0) = \dot{\xi}_0. \end{aligned} \tag{5.25}$$

Due to the structure, the controllability Gramian P of the system (5.8) can be compatibly partitioned as

$$P = \begin{bmatrix} P_p & P_o \\ P_o^T & P_v \end{bmatrix}. \tag{5.26}$$

It can be shown that (see [124]) the optimal solution to problem (5.24) is $\xi_0 P_p^{-1} \xi_0$ which is the minimal energy required to reach the given position ξ_0 over all past inputs and initial values. The optimal solution to problem (5.25) is $\dot{\xi}_0 P_v^{-1} \dot{\xi}_0$ which is the minimal energy required to reach the given velocity $\dot{\xi}_0$ over all past inputs and initial values. Therefore, P_p and P_v are called the second-order controllability position Gramian and the velocity Gramian, respectively.

Similarly, consider Q to be the observability Gramian of system (5.8) which satisfies the observability Lyapunov equation

$$A^T QE + E^T QA = -C^T C, \tag{5.27}$$

where E, A and C are defined in (5.8). Like the controllability Gramian (P) we can also partition Q as

$$Q = \begin{bmatrix} Q_p & Q_o \\ Q_o^T & Q_v \end{bmatrix}, \tag{5.28}$$

in which Q_p and Q_v can be defined as the second-order *observability velocity* and *position Gramians*, respectively.

> The second-order system (5.1) has four types of Gramians: controllability position Gramian, controllability velocity Gramian, observability position Gramian and observability velocity Gramian.

Here we have defined the Gramians of the second-order system (5.1) by considering the first-order representation (5.8). However, they can also be defined for the first-order representation (5.9). This task remains for the Exercises. Note that in later parts of this chapter we will discuss how to find the low-rank Gramian factors of the second-order system by solving corresponding Lyapunov equations using the low-rank ADI iteration.

5.3 Second-order-to-first-order reduction

This section shows how to find the first-order reduced system from the second-order system (5.1). We consider the balanced truncation and interpolation technique via IRKA. Note that both methods are introduced in the previous chapter for the model reduction of first-order systems. To apply them for the second-order-to-first-order reduction (SOFOR) of second-order system (5.1), we convert the system into its first-order form as in either (5.8) or (5.9). The converted system looks similar to a generalized system (4.1). Therefore, we can apply the model reduction methods introduced in Chapter 4. Here we directly present the algorithm for SOFOR. Readers can see Chapter 4 for further elaborations. The goal of this section is to find a considerably small-sized generalized state space system

$$\begin{aligned} \hat{E}\dot{\hat{x}}(t) &= \hat{A}\hat{x}(t) + \hat{B}u(t), \\ \hat{y}(t) &= \hat{C}\hat{x}(t), \end{aligned} \tag{5.29}$$

from the second-order system (5.1).

5.3.1 Balancing-based algorithm

Recall Algorithm 19 (low-rank square root method [LR-SRM] for balanced truncation). Once we have the first-order form (5.8) of the second-order system (5.1), we can apply this algorithm to get the reduced-order model (5.29). The whole procedure of SOFOR of second-order system (5.1) is discussed in Algorithm 22.

Algorithm 22: LR-SRM for second-order system.

Input : M, D, K, H and L_1 (from 5.1).
Output: $\hat{E}$, $\hat{A}$, $\hat{B}$ and $\hat{C}$ (as in (5.29).

1 Form

$$E = \begin{bmatrix} I & 0 \\ 0 & M \end{bmatrix}, \; A = \begin{bmatrix} 0 & I \\ -K & -D \end{bmatrix}, \; B = \begin{bmatrix} 0 \\ H \end{bmatrix} \text{ and } C = \begin{bmatrix} L_1 & 0 \end{bmatrix}$$

2 Solve the Lyapunov equations

$$APE^T + EPA^T = -BB^T \text{ and } A^TQE + E^TQA = -C^TC,$$

to compute low-rank Gramian factors R and L, respectively, which satisfy $P \approx RR^T$ and $Q \approx LL^T$.

3 Compute the singular value decomposition (SVD)

$$L^TR = U\Sigma V^T = \begin{bmatrix} U_1 & U_2 \end{bmatrix} \begin{bmatrix} \Sigma_1 & \\ & \Sigma_2 \end{bmatrix} \begin{bmatrix} V_1^T \\ V_2^T \end{bmatrix}.$$

4 Construct $V := LV_1\Sigma_1^{-\frac{1}{2}}, \quad W := RU_1\Sigma_1^{-\frac{1}{2}}$.

5 Form $\hat{E} = W^TEV, \quad \hat{A} = W^TAV, \quad \hat{B} = W^TB \quad \text{and} \quad \hat{C} = CV$.

5.3.2 Interpolatory projection via IRKA

Interpolatory projection method via IRKA for the generalized (first-order) systems was detailed in Chapter 4. If second-order system (5.1) is converted into first-order system (5.8), we can directly apply the procedures as discussed in Section 4.5. Following the discussion in Section 4.5, IRKA is summarized in Algorithm 23 for the model reduction of second-order system (5.1). In this algorithm, the main computational task is to construct V and W at each iteration. Because at each iteration step, we require to simultaneously solve two shifted linear systems with large-scale matrices. To accelerate the computation, we need to modify the linear systems. Each column of V can be computed by solving a shifted linear system like

$$(\alpha E - A)\chi = Bb,$$

which implies

$$\left(\alpha \begin{bmatrix} I & 0 \\ 0 & M \end{bmatrix} - \begin{bmatrix} 0 & I \\ -K & -D \end{bmatrix}\right) \begin{bmatrix} \chi_1 \\ \chi_2 \end{bmatrix} = \begin{bmatrix} 0 \\ Hb \end{bmatrix},$$

or

$$\begin{bmatrix} \alpha I & -I \\ K & \alpha M + D \end{bmatrix} \begin{bmatrix} \chi_1 \\ \chi_2 \end{bmatrix} = \begin{bmatrix} 0 \\ Hb \end{bmatrix}.$$

Algorithm 23: IRKA for second-order MIMO systems.

Input : M, D, K, H and L_1 (from 5.1).
Output: $\hat{E}$, $\hat{A}$, $\hat{B}$, $\hat{C}$ as in (5.29).

Form

$$E = \begin{bmatrix} I & 0 \\ 0 & M \end{bmatrix}, A = \begin{bmatrix} 0 & I \\ -K & -D \end{bmatrix}, B = \begin{bmatrix} 0 \\ H \end{bmatrix} \quad \text{and} \quad C = \begin{bmatrix} L_1 & 0 \end{bmatrix}$$

Make an initial selection of the interpolation points $\{\alpha_i\}_{i=1}^r$ and the tangential directions $\{b_i\}_{i=1}^r$ and $\{c_i\}_{i=1}^r$.

Construct

$$V = \left[(\alpha_1 E - A)^{-1} B b_1, \cdots, (\alpha_r E - A)^{-1} B b_r\right],$$
$$W = \left[(\alpha_1 E^T - A^T)^{-1} C^T c_1, \cdots, (\alpha_r E^T - A^T)^{-1} C^T c_r\right].$$

while *(not converged)* **do**

Form
$$\hat{E} = W^T E V,\ \hat{A} = W^T A V,\ \hat{B} = W^T B,\ \hat{C} = C V.$$

Compute $\hat{A} z_i = \hat{\lambda}_i \hat{E} z_i$ and $y_i^* \hat{A} = \hat{\lambda}_i y_i^* \hat{E}$.

$\alpha_i \leftarrow -\lambda_i$, $b_i^* \leftarrow -y_i^* \hat{B}$ and $c_i \leftarrow \hat{C} z_i$, for $i = 1, \cdots, r$.

Construct

$$V = \left[(\alpha_1 E - A)^{-1} B b_1, \cdots, (\alpha_r E - A)^{-1} B b_r\right],$$
$$W = \left[(\alpha_1 E^T - A^T)^{-1} C^T c_1, \cdots, (\alpha_r E^T - A^T)^{-1} C^T c_r\right].$$

$i = i + 1$

Form $\hat{E} = W^T E V$, $\quad \hat{A} = W^T A V$, $\quad \hat{B} = W^T B$ and $\hat{C} = C V$.

Split this linear system as

$$\alpha \chi_1 - \chi_2 = 0, \tag{5.30a}$$
$$K\chi_1 + (\alpha M + D)\chi_2 = Hb, \tag{5.30b}$$

From (5.30a) we obtain,

$$\chi_2 = \alpha \chi_1. \tag{5.31}$$

Now inserting χ_2 into (5.30b) yields

$$K\chi_1 + (\alpha M + D)\alpha\chi_1 = Hb,$$

which is again equivalent to the solution of the linear system

$$(\alpha^2 M + \alpha D + K)\chi_1 = Hb. \tag{5.32}$$

Again each column of W can be computed by solving a shifted linear system like

$$(\alpha E^T - A^T)\chi = C^T c,$$

which implies

$$\left(\alpha \begin{bmatrix} I & 0 \\ 0 & M^T \end{bmatrix} - \begin{bmatrix} 0 & -K^T \\ I & -D^T \end{bmatrix}\right) \begin{bmatrix} \chi_1 \\ \chi_2 \end{bmatrix} = \begin{bmatrix} L_1^T c \\ 0 \end{bmatrix},$$

or

$$\begin{bmatrix} \alpha I & K^T \\ -I & \alpha M^T + D^T \end{bmatrix} \begin{bmatrix} \chi_1 \\ \chi_2 \end{bmatrix} = \begin{bmatrix} L_1^T c \\ 0 \end{bmatrix}.$$

Splitting this linear system as

$$\alpha\chi_1 - K^T\chi_2 = L_1^T c, \tag{5.33a}$$

$$-\chi_1 + (\alpha M^T + D^T)\chi_2 = 0. \tag{5.33b}$$

From (5.33b) we obtain,

$$\chi_1 = (\alpha M^T + D^T)\chi_2. \tag{5.34}$$

Now inserting χ_1 into (5.33a) yields

$$\alpha(\alpha M^T + D^T)\chi_2 - K^T\chi_2 = L_1^T c,$$

which is again equivalent to the solution of the linear system

$$(\alpha^2 M^T + \alpha D^T - K^T)\chi_2 = L_1^T c. \tag{5.35}$$

Applying the idea of splitting to the systems, instead of solving an $2n_\xi$ dimensional linear system, we can solve only an n_ξ dimensional linear system, which ensures faster computation.

Note that Algorithm 23 yields ROM for second-order system (5.1) when its first-order form is (5.8).

5.4 Second-order-to-second-order reduction

This section considers second-order-to-second-order model order reduction which is also known as structure preserving model order reduction (SPMOR).

We have already mentioned earlier that such SPMOR is particularly important when we are required to preserve the structure of the systems. Note that in the previous section, the computed reduced models do not preserve the structure of the second-order system. Here our goal is to apply the model reduction onto the system (5.1); we want to find a sufficiently reduced-order model

$$\begin{aligned} \hat{M}\ddot{\hat{\xi}}(t) + \hat{D}\dot{\hat{\xi}}(t) + \hat{K}\hat{\xi}(t) &= \hat{H}u(t), \\ \hat{y}(t) &= \hat{L}_1\hat{\xi}(t). \end{aligned} \tag{5.36}$$

Mainly, two important methods such as balanced truncation (BT) and projection onto the dominant eigenspace of the Gramian (PDEG) will be discussed here. Both techniques are based on the Gramians of the system. We have introduced the Gramians of the second-order system in Subsection (5.2.3).

5.4.1 Balancing-based methods

We consider R as a low-rank factor of controllability Gramian (P) such that $P \approx RR^T$. The structure of first-order system allows us to split R as

$$R = \begin{bmatrix} R_p^T & R_v^T \end{bmatrix}^T, \tag{5.37}$$

where R_p and R_v are respectively defined as low-rank factors of the controllability position and velocity Gramians. Therefore, the controllability Gramian defined in (5.26) can be written [37] as

$$P = \begin{bmatrix} P_p & P_o \\ P_o^T & P_v \end{bmatrix} \approx RR^T = \begin{bmatrix} R_p \\ R_v \end{bmatrix} \begin{bmatrix} R_p^T & R_v^T \end{bmatrix} = \begin{bmatrix} R_pR_p^T & R_pR_v^T \\ R_vR_p^T & R_vR_v^T \end{bmatrix}.$$

Hence we have

$$P_p \approx R_pR_p^T \qquad \text{and} \qquad P_v \approx R_vR_v^T.$$

Similarly, if L is the low-rank factor of Q as defined in (5.28) satisfying $Q \approx LL^T$ we get

$$Q_p \approx L_pL_p^T \qquad \text{and} \qquad Q_v \approx L_vL_v^T,$$

where $L = \begin{bmatrix} L_p^T & L_v^T \end{bmatrix}^T$. We will discuss how to compute the low-rank Gramian factors of second-order Gramians by solving the Lyapunov equations in the coming sections.

Once we have R_α and L_β ($\alpha \in \{p, v\}$, $\beta \in \{p, v\}$), the balancing and truncation transformation can be formed [28, 137] using the SVD

$$R_\alpha^T M L_\beta = U_{\alpha\beta}\Sigma_{\alpha\beta}V_{\alpha\beta}^T = \begin{bmatrix} U_{\alpha\beta,1} & U_{\alpha\beta,2} \end{bmatrix} \begin{bmatrix} \Sigma_{\alpha\beta,1} & \\ & \Sigma_{\alpha\beta,2} \end{bmatrix} \begin{bmatrix} V_{\alpha\beta,1}^T \\ V_{\alpha\beta,2}^T \end{bmatrix}, \tag{5.38}$$

type	SVD	left proj. W_s	right proj. V_s
PP	$L_p^T M R_p = U_{pp}\Sigma_{pp}V_{pp}^T$	$L_p U_{pp,1}\Sigma_{pp,1}^{-\frac{1}{2}}$	$R_p V_{pp,1}\Sigma_{pp,1}^{-\frac{1}{2}}$
VV	$L_v^T M R_v = U_{vv}\Sigma_{vv}V_{vv}^T$	$L_v U_{vv,1}\Sigma_{vv,1}^{-\frac{1}{2}}$	$R_v V_{vv,1}\Sigma_{vv,1}^{-\frac{1}{2}}$
PV	$L_p^T M R_v = U_{pv}\Sigma_{pv}V_{pv}^T$	$L_p U_{pv,1}\Sigma_{pv,1}^{-\frac{1}{2}}$	$R_v V_{pv,1}\Sigma_{pv,1}^{-\frac{1}{2}}$
VP	$L_v^T M R_p = U_{vp}\Sigma_{vp}V_{vp}^T$	$L_v U_{vp,1}\Sigma_{vp,1}^{-\frac{1}{2}}$	$R_p V_{vp,1}\Sigma_{vp,1}^{-\frac{1}{2}}$

Table 5.1: Balancing transformations for the second-order systems.

and defining

$$W_s := L_\beta U_{\alpha\beta,1}\Sigma_{\alpha\beta,1}^{-\frac{1}{2}}, \quad V_s := R_\alpha V_{\alpha\beta,1}\Sigma_{\alpha\beta,1}^{-\frac{1}{2}}, \tag{5.39}$$

where $U_{\alpha\beta,1}$ and $V_{\alpha\beta,1}$ are composed of the leading k columns of $U_{\alpha\beta}$ and $V_{\alpha\beta}$, respectively, and $\Sigma_{\alpha\beta,1}$ is the first $k \times k$ block of the matrix $\Sigma_{\alpha\beta}$. Applying W_s, $V_s \in \mathbb{R}^{n_\xi \times k}$ with $k \ll n_\xi$ in (5.1), we obtain the reduced models (5.36) where the reduced matrices are constructed as

$$\begin{aligned} \hat{M} &= W_s^T M V_s, \quad \hat{D} = W_s^T D V_s, \quad \hat{K} = W_s^T K V_s, \\ \hat{H} &= W_s^T H, \quad \hat{L}_1 = L_1 V_s. \end{aligned} \tag{5.40}$$

When $\alpha = \beta = p$, the balancing technique by the above procedure is called position-position (PP) balancing. Likewise velocity-velocity (VV) balancing is obtained if $\alpha = \beta = v$, position-velocity (PV) balancing is obtained if $\alpha = p, \beta = v$ and velocity-position (VP) balancing is obtained if $\alpha = v, \beta = p$. Now, using these projectors, we obtain four types of reduced models. Construction of balancing and truncating transformations for different balancing levels are summarized in Table 5.1. Algorithm 24 briefly presents how to construct the reduced second-order system from large-scale second-order system via balanced truncation model order reduction (BT-MOR).

Algorithm 24: BT-MOR for second-order system (LR-SRM).

Input : M, D, K, H, L_1 from (5.1).
Output: $\hat{M}$, $\hat{D}$, $\hat{K}$, $\hat{H}$ and $\hat{L}_1$ as in (5.36).

1. Solve Lyapunov equation (5.21) and (5.27) to compute R_p, R_v, L_p and L_v.
2. Compute four types of transformations following Table 5.1.
3. Construct $\hat{M}$, $\hat{D}$, $\hat{K}$, $\hat{H}$ and $\hat{L}_1$ following (5.40).

5.4.2 Projection onto dominant eigenspaces of the Gramian

In general, second-order-to-second-order reduced models via balanced truncation discussed above do not preserve the stability of the system [137]. However, for a special case, i.e., when M, D, K are symmetric and input-output matrices are transpose of each other, a particular first-order representation can construct a stable reduced model (see, e.g., [42] for details). Note that herein we do not consider such a special model. In this subsection, we propose a new projection technique for a structure preserving model reduction of second-order system. The method is based on projecting the systems onto the dominant eigenspace of the second-order Gramians. In short we call this technique the PDEG (projection onto dominant eigenspace of the Gramian) method. The technique is superior to BT since it preserves important properties such as stability, symmetry and definiteness of the original system. The following text discusses how to implement the PDEG method by constructing the projectors cheaply from the low-rank factors of the second-order Gramians.

Gramians of the second-order systems are already defined in Subsection 5.2.3. We first consider the controllability position Gramian P_p. Since P_p is symmetric positive definite (SPD), it has a symmetric decomposition, i.e.,

$$P_p = R_p R_p^T. \tag{5.41}$$

The *thin* SVD of R_p is

$$R_p = U_p \Sigma_p V_p^T, \tag{5.42}$$

where the diagonal matrix Σ_p consists of the decreasingly ordered singular values σ_{v_i}, $i = 1, 2, \ldots, n_1$ of R_p. Using this SVD, we have

$$P_p = (U_p \Sigma_p V_p^T)(V_p \Sigma_p U_p^T) = U_p \Sigma_p^2 U_p^T. \tag{5.43}$$

This is also an eigenvalue decomposition where Σ_p^2 is a diagonal matrix whose entries are the decreasingly ordered eigenvalues of P_p, and U_p is the orthogonal matrix consisting of the eigenvectors corresponding to the eigenvalues. We observe that U_p is the left singular vector matrix of R_p. Hence U_p is obtained by the SVD of R_p. To identify the k largest eigenvalues of P_p, i.e., the k largest singular values of R_p, we construct

$$U_k = \left[u_1, u_2, \ldots, u_k\right], \tag{5.44}$$

where u_i, $i = 1, 2, \ldots, k$ are the eigenvectors corresponding to the eigenvalues σ_i^2. Then we construct the k dimensional reduced-order model (5.36), by forming the matrices as

$$\begin{aligned} \hat{M} &= U_k^T M U_k, \quad \hat{D} = U_k^T D U_k, \quad \hat{K} = U_k^T K U_k, \\ \hat{H} &= U_k^T H, \quad \hat{L}_1 = L_1 U_k. \end{aligned} \tag{5.45}$$

Note that U_k in (5.44) can be constructed cheaply from the low-rank factor of the controllability position Gramian. If we consider R_p is a low-rank Gramian factor of the controllability position Gramian P_p such that $P_p = \begin{bmatrix} R_p^T & R_p^T \end{bmatrix}^T$, then we can compute U_k in (5.44) by identifying the k largest left singular vectors of the SVD of R_p as in Equation (5.43).

The above procedure that constructs a k dimensional ROM (5.36) via projecting the system onto the dominant eigenspaces of the controllability position Gramian is summarized in Algorithm 25. However, this algorithm

---

Algorithm 25: PDEG method for SPMOR.

Input : M, D, K, H, L_1 from (5.1) and k (dimension of ROM).
Output: $\hat{M}$, $\hat{D}$, $\hat{K}$, $\hat{H}$ and $\hat{L}_1$ as in (5.36).

1. Compute the low-rank factor (R_p) of the controllability position Gramian by solving Lyapunov equation (5.21).
2. Construct U_k as in (5.44) using SVD of R_p.
3. Compute the reduced matrices as

$$\hat{M} = U_k^T M U_k, \quad \hat{D} = U_k^T D U_k, \quad \hat{K} = U_k^T K U_k, \quad \hat{H} = U_k^T H, \quad \hat{L}_1 = L_1 U_k$$

---

can also be used to obtain a k dimensional ROM via projecting the system onto the dominant eigenspace of the controllability velocity Gramian P_v, observability position Gramian Q_p and the observability velocity Gramian Q_v. In these cases, we are required to compute the low-rank factor of the respective Gramians by solving the corresponding Lyapunov equations. In the next section, we will discuss how to find these low-rank Gramian factors by solving Lyapunov equations using low-rank ADI iteration.

The transformation U_k is called *contra-gradient transformation* [109]; this is because by using this transformation we can show that

$$\begin{aligned}
U_k^T R_p U_k &= U_k^T U_v \Sigma_v^2 U_v^T U_k \\
&= U_k^T \begin{bmatrix} U_k & U_{n_1-k} \end{bmatrix} \begin{bmatrix} \Sigma_k^2 & 0 \\ 0 & \Sigma_{n_1-k}^2 \end{bmatrix} \begin{bmatrix} U_k^T \\ U_{n_1-k}^T \end{bmatrix} U_k \\
&= \begin{bmatrix} I_k & 0 \end{bmatrix} \begin{bmatrix} \Sigma_k^2 & 0 \\ 0 & \Sigma_{n_1-k}^2 \end{bmatrix} \begin{bmatrix} I_k \\ 0 \end{bmatrix} \\
&= \Sigma_k^2,
\end{aligned}$$

that is to say the Gramian of the reduced model is diagonal. Hence, U_k is the balancing transformation. It can easily be shown that $\hat{M}$, $\hat{D}$ and $\hat{K}$ are all symmetric. Moreover, the ROM preserves the definiteness of the original system. Therefore, the stability is also preserved.

> ROM that is constructed by the PDEG method is balanced and preserves the stability, symmetry and definiteness of the original model.

5.5 LR-ADI iteration for solving second-order Lyapunov equation

To implement the model reduction (using the BT and PDEG methods) of second-order systems, we need to compute the low-rank Gramian factors. The low-rank Gramian factors can be computed by solving the Lyapunov equations (5.21) and (5.27). Chapter 3 discussed the computation of the low-rank Gramin factors by solving the Lyapunov of generalized state space system by using low-rank alternating direction implicit (LR-ADI) iteration. This section generalizes the ideas of LR-ADI iteration for solving the Lyapunov equations arising from the second-order systems. In the following subsection, we discuss the solution of controllability Lyapunov equation in detail. Following the same procedure, we will also briefly show how to solve observability Lyapunov equation.

5.5.1 Solution of second-order controllability Lyapunov equation

One can apply Algorithm 15 to solve the second-order controllability Lyapunov equation (5.21) for computing the low-rank factor of the second-order controllability Gramian. In this case, input matrices E, A and B are coming from one of the first-order representations of the second-order system (5.1) as discussed is Subsection 5.2.1. Considering the matrices E, A and B defined in (5.9), we discuss how to modify Algorithm 15 to accelerate the solution of Lyapunov equation (5.21).

The initial guess of the Lyapunov residual factor is $W_0 = B$, which is equivalent to

$$\begin{bmatrix} W_0^1 \\ W_0^2 \end{bmatrix} = \begin{bmatrix} 0 \\ H \end{bmatrix}.$$

Thus,

$$W_0^1 = 0 \quad \text{and} \quad W_0^2 = H. \tag{5.46}$$

We know that the most expensive part in the LR-ADI iteration is to solve a linear system in each iteration step. In the $i-$th step (see Step 3 in Algorithm 15), we need to compute $V_i = (A + \mu_i E)^{-1} W_{i-1}$, which is equivalent to solve the linear system

$$(A + \mu_i E) V_i = W_{i-1}. \tag{5.47}$$

This linear system can be solved directly or iteratively. In either case, we can split the linear system as follows. Insert A and E from (5.1) and considering

$W_i = \begin{bmatrix} W_i^1 \\ W_i^2 \end{bmatrix}$ and $V_i = \begin{bmatrix} V_i^1 \\ V_i^2 \end{bmatrix}$, (5.53) yields

$$\left(\begin{bmatrix} 0 & I \\ -K & -D \end{bmatrix} + \mu_i \begin{bmatrix} I & 0 \\ 0 & M \end{bmatrix} \right) \begin{bmatrix} V_i^1 \\ V_i^2 \end{bmatrix} = \begin{bmatrix} W_{i-1}^1 \\ W_{i-1}^2 \end{bmatrix}. \tag{5.48}$$

Which is equivalent to

$$\begin{bmatrix} \mu_i I & I \\ -K & \mu_i M - D \end{bmatrix} \begin{bmatrix} V_i^1 \\ V_i^2 \end{bmatrix} = \begin{bmatrix} W_{i-1}^1 \\ W_{i-1}^2 \end{bmatrix}. \tag{5.49}$$

From the first line of (5.49) we obtain

$$V_i^2 = W_{i-1}^1 - \mu_i V_i^1. \tag{5.50}$$

Inserting this identity into the second line of (5.49) and algebraic manipulation will give

$$(\mu_i^2 M - \mu_i D + K)V_i^1 = \mu M W_{i-1}^1 - D W_{i-1}^1 - W_{i-1}^2. \tag{5.51}$$

To compute V_i in (5.47), we first solve (5.51) for V_i^1 and plugging this into (5.50) computes V_i^2. Following this strategy, we just need to solve a linear system with order n_ξ instead of $2n_\xi$. Following Step 6 in Algorithm 15 at each iteration, the Lyapunov residual can be updated as follows

$$\begin{bmatrix} W_i^1 \\ W_i^2 \end{bmatrix} = \begin{bmatrix} W_i^1 \\ W_i^2 \end{bmatrix} - 2\mu_i \begin{bmatrix} I & 0 \\ 0 & M \end{bmatrix} \begin{bmatrix} V_i^1 \\ V_i^2 \end{bmatrix},$$

which gives

$$\begin{aligned} W_i^1 &= W_{i-1}^1 - 2\mu_i V_i^1, \\ W_i^2 &= W_{i-1}^2 - 2\mu_i M V_i^2. \end{aligned}$$

In the case of complex shift parameters, the residual factor is updated by

$$\begin{aligned} W_i^1 &= W_{i-1}^1 - 4\mu_i V_i^1, \\ W_i^2 &= W_{i-1}^2 - 4\mu_i M V_i^2. \end{aligned}$$

The algorithm can be stopped if the $\|W_i^T W_i\| = \| \begin{bmatrix} W_i^{1T} & W_i^{2T} \end{bmatrix} \begin{bmatrix} W_i^1 \\ W_i^2 \end{bmatrix} \| = \|{W_{i-1}^{(1)}}^T W_{i-1}^{(1)} + {W_{i-1}^{(2)}}^T W_{i-1}^{(2)}\|$ is sufficiently small. The whole procedure for solving the second-order controllability Lyapunov equation (5.21) is summarized in Algorithm 26. Note that this algorithm is developed following the first-order representation as in (5.8) of the second-order system. However, one can modify this algorithm following other first-order formulations as discussed above. We leave this for the Exercises.

> Algorithm 26 is developed by following the first-order representation (5.8) of the second-order system (5.1).

Algorithm 26: LR-ADI for second-order controllability Gramian factor.

Input : M, D, K, H and shift parameters $\{\mu_i\}_{i=1}^{J}$.

Output: $R = Z_i$ such that $P \approx RR^T$.

Set $Z_0 = []$, $i = 1$, $W_0^1 = 0$ and $W_0^2 = H$.

while $\|{W_{i-1}^{(1)}}^T W_{i-1}^{(1)} + {W_{i-1}^{(2)}}^T W_{i-1}^{(2)}\| \geq tol$ *or* $i \leq i_{max}$ **do**

  Solve (5.51) for V_i^1, then compute $V_i^2 = W_{i-1}^1 - \mu_i V_i^1$.

  Form $V_i = \begin{bmatrix} V_i^{1^T} & V_i^{2^T} \end{bmatrix}^T$.

  if $\mathrm{Im}\,(\mu_i) = 0$ **then**

    $Z_i = \begin{bmatrix} Z_{i-1} & \sqrt{\gamma}\,\mathrm{Re}\,(V_i) \end{bmatrix}$, where $\gamma = -2\,\mathrm{Re}\,(\mu_i)$,

    $W_i^1 = W_{i-1}^1 + \gamma M V_i^1, \quad W_i^2 = W_{i-1}^2 + \gamma M V_i^2$.

  else

    $\delta = \frac{\mathrm{Re}\,(\mu_i)}{\mathrm{Im}\,(\mu_i)}, \quad V_{i+1} = \overline{V}_i + 2\delta\,\mathrm{Im}\,(V_i)$.

    Update low-rank solution factor

    $Z_{i+1} = \begin{bmatrix} Z_{i-1} & \sqrt{2\gamma}(\mathrm{Re}\,(V_i) + \delta\,\mathrm{Im}\,(V_i)) & \sqrt{2\gamma}\sqrt{(\delta^2+1)}.\,\mathrm{Im}\,(V_i) \end{bmatrix}$.

    Compute $V_i^1 = \mathrm{Re}\,(V_i^1) + \delta\,\mathrm{Im}\,(V_i^1)$, $V_i^2 = \mathrm{Re}\,(V_i^2) + \delta\,\mathrm{Im}\,(V_i^2)$,

    and $W_{i+1}^1 = W_{i-1}^1 + 2\gamma V_i^2, \quad W_{i+1}^2 = W_{i-1}^2 + 2\gamma M V_i^2$.

    $i = i + 1$

  $i = i + 1$

5.5.2 Solution of second-order observability Lyapunov equation

Algorithm 26 can also be applied for solving the second-order observability Lyapunov equation (5.27) by changing a few steps. These changes are as follows.

- **Initialization of Lyapunov residual factor:** Line (5.46) becomes

$$W_0^1 = L_1 \quad \text{and} \quad W_0^2 = 0. \tag{5.52}$$

- **Solution of linear system:** At the ith iteration, we solve the linear system

$$(A^T + \mu_i E^T)V_i = W_{i-1}. \tag{5.53}$$

Insert A and E from (5.8) and considering $W_i = \begin{bmatrix} W_i^1 \\ W_i^2 \end{bmatrix}$ and $V_i = \begin{bmatrix} V_i^1 \\ V_i^2 \end{bmatrix}$

$$\left(\begin{bmatrix} 0 & -K^T \\ I & -D^T \end{bmatrix} + \mu_i \begin{bmatrix} I & 0 \\ 0 & M^T \end{bmatrix} \right) \begin{bmatrix} V_i^1 \\ V_i^2 \end{bmatrix} = \begin{bmatrix} W_{i-1}^1 \\ W_{i-1}^2 \end{bmatrix}. \tag{5.54}$$

Which is equivalent to

$$\begin{bmatrix} \mu_i I & -K^T \\ I & \mu_i M^T - D^T \end{bmatrix} \begin{bmatrix} V_i^1 \\ V_i^2 \end{bmatrix} = \begin{bmatrix} W_{i-1}^1 \\ W_{i-1}^2 \end{bmatrix}. \tag{5.55}$$

From the first line of (5.55), we obtain

$$V_i^1 = \frac{1}{\mu_i}(W_{i-1}^1 + K^T V_i^2) \tag{5.56}$$

Inserting this identity into the second line of (5.55) with some algebraic manipulation will give

$$(\mu_i^2 M^T - \mu_i D^T + K^T)V_i^2 = \mu W_{i-1}^2 - W_{i-1}^1. \tag{5.57}$$

Therefore, to find V_i, instead of solving (5.53) we solve (5.57) for V_i^2 and then compute V_i^1 from (5.56).

- **Update the Lyapunov residual factor:** Following Step 6 in Algorithm 15 at each iteration, the Lyapunov residual can be updated as follows

$$\begin{bmatrix} W_i^1 \\ W_i^2 \end{bmatrix} = \begin{bmatrix} W_i^1 \\ W_i^2 \end{bmatrix} - 2\mu_i \begin{bmatrix} I & 0 \\ 0 & M^T \end{bmatrix} \begin{bmatrix} V_i^1 \\ V_i^2 \end{bmatrix},$$

which gives

$$\begin{aligned} W_i^1 &= W_{i-1}^1 - 2\mu_i V_i^1, \\ W_i^2 &= W_{i-1}^2 - 2\mu_i M^T V_i^2. \end{aligned}$$

In the case of complex shift parameters, the residual factor is updated by

$$\begin{aligned} W_i^1 &= W_{i-1}^1 - 4\mu_i V_i^1, \\ W_i^2 &= W_{i-1}^2 - 4\mu_i M^T V_i^2. \end{aligned}$$

Applying these changes, we can modify Algorithm 26 to solve the observability Lyapunov equation (5.27) for computing low-rank controllability Gramian factor $L = Z_i$.

Once we have computed the low-rank controllability Gramian factor R, then the upper n_ξ rows and lower n_ξ rows of R define the low-rank controllability position and velocity Gramian factors, respectively. The upper n_ξ rows and lower n_ξ rows of L can be used to compute the low-rank observability position (L_p) and velocity (L_v) Gramian factors, respectively.

5.6 MOR of symmetric second-order systems

This section focuses on the model reduction of symmetric second-order system. We consider the second-order system

$$M\ddot{\xi}(t) + D\dot{\xi}(t) + K\xi(t) = Hu(t), \tag{5.58a}$$

$$y(t) = H^T\xi(t), \tag{5.58b}$$

where

$$M = M^T, \quad D = D^T, \quad K = K^T.$$

Under these conditions, there are some advantages in implementing the reduced-order model.

In this case, we choose first-order representation:

$$\begin{aligned}\underbrace{\begin{bmatrix} 0 & M \\ M & D \end{bmatrix}}_{E}\underbrace{\begin{bmatrix} \dot{x}_2(t) \\ \dot{x}_1(t) \end{bmatrix}}_{\dot{x}(t)} &= \underbrace{\begin{bmatrix} M & 0 \\ 0 & -K \end{bmatrix}}_{A}\underbrace{\begin{bmatrix} x_2(t) \\ x_1(t) \end{bmatrix}}_{z(t)} + \underbrace{\begin{bmatrix} 0 \\ H \end{bmatrix}}_{B} u(t), \\ y(t) &= \underbrace{\begin{bmatrix} 0 & H^T \end{bmatrix}}_{B^T}\begin{bmatrix} x_2(t) \\ x_1(t) \end{bmatrix}.\end{aligned} \tag{5.59}$$

This representation is particularly interesting since it yields a first-order symmetric system (since $A^T = A$, $E^T = E$) and the input-output matrices are transposes of each other. In this case, the controllability Gramian and the observability Gramian are the same and, therefore, we need only solve one Lyapunov equation, i.e.,

$$APE + EPA = -BB^T, \tag{5.60}$$

where $P \in \mathbb{R}^{n_\xi \times n_\xi}$ is either the controllability Gramian or the observability Gramian of the system. In the Exercise problems, we will practice how to solve this Lyapunov equation by modifying the LR-ADI iteration as presented in Algorithm 26.

Based on the discussion in Sections 5.3 and 5.4, we can find the second-order-to-first-order and second-order-to-second-order reductions of the second-order symmetric system (5.58), respectively. We will discuss more on this topic in the Exercise problems.

5.7 Some useful MATLAB functions

Function name	Description
`ss`	construct state-space model or convert model to state space
`rss`	generate randomized continuous-time state space models
`dss`	create generalized or descriptor state space models
`balreal`	compute a balanced realization for the linear system
`balred`	compute a reduced-order approximation of the LTI system
`hsvd`	compute the Hankel singular values of linear systems
`step`	step response of dynamic systems
`impulse`	impulse response of dynamic systems
`sigma`	singular value plot of the frequency response of the dynamic system
`lyap`	solve continuous-time Lyapunov equations
`norm`	compute $\mathcal{H}_2$-norm and $\mathcal{H}_\infty$-norm of LTI systems
`loglog`	log-log scale plot using logarithmic scaled for both the X-axis and Y-axis
`semilogx` (`semilogy`)	log-log scale plot using logarithmic scaled for X- (Y-) axis

5.8 Numerical results

In this section we present some numerical results for assessing the accuracy and efficiency of the model reduction methods discussed above. For the numerical experiments, we consider two model examples: the ISS (International Space Station) model [134], see Appendix A.3.1 for details. This data was collected from [50] SLICOT Benchmark Examples[1] which was in first-order form; for our purposes, we have converted it into a second-order structured form.

All the results were obtained by using MATLAB R2015a (8.5.0.197613) on a board with 4×Intel® Core™i5-4460 CPUs with a 2.90 GHz clock speed and 16 GB RAM.

[1] `http://slicot.org/20-site/126-benchmark-examples-for-model-reduction.`

Second-order-to-first-order reduction. Considering a truncating tolerance of 10^{-3}, we apply the balanced truncation method (i.e., Algorithm 22) to the ISS model. The algorithm computes a 27 dimensional reduced model. Note that, to implement this algorithm, we have to compute the controllability and observability Gramians by applying the procedure discussed in Section 5.5. We have also computed a 27 dimensional reduced model using IRKA (i.e., Algorithm 23). The comparisons between the original and reduced models are depicted in Figure 5.1; showing the transfer functions (in Figure 5.1(a)) and their absolute and relative errors. Referring to both the absolute and relative errors (in Figure 5.1(b) and 5.1(c)), we see that the accuracy of balancing-based model is slightly better than the IRKA.

Second-order-to-second-order reduction. To asses the structure preserving model reduction methods, i.e., Algorithms 24 and 25, we apply both the algorithms to the ISS model. Algorithm 24 produces different reduced models for different balancing levels with a truncation tolerance of 10^{-3}. The comparison between the original and reduced models are shown in Figure 5.2. The transfer functions, as shown in Figure 5.2(a), are matching accurately with the original transfer function for all balancing labels. Applying both the absolute and relative errors are satisfactory at all levels. We have also computed 20 dimensional reduced models by projecting the system onto the dominant eigenspace of the system Gramians using Algorithm 25. Analyzing the absolute and relative errors from Figure 5.3, one can see that the reduced models show good accuracy in all cases. However, the reduced model that is obtained by projecting onto the controllability velocity and observability position Gramians provide more accuracy than the other two cases.

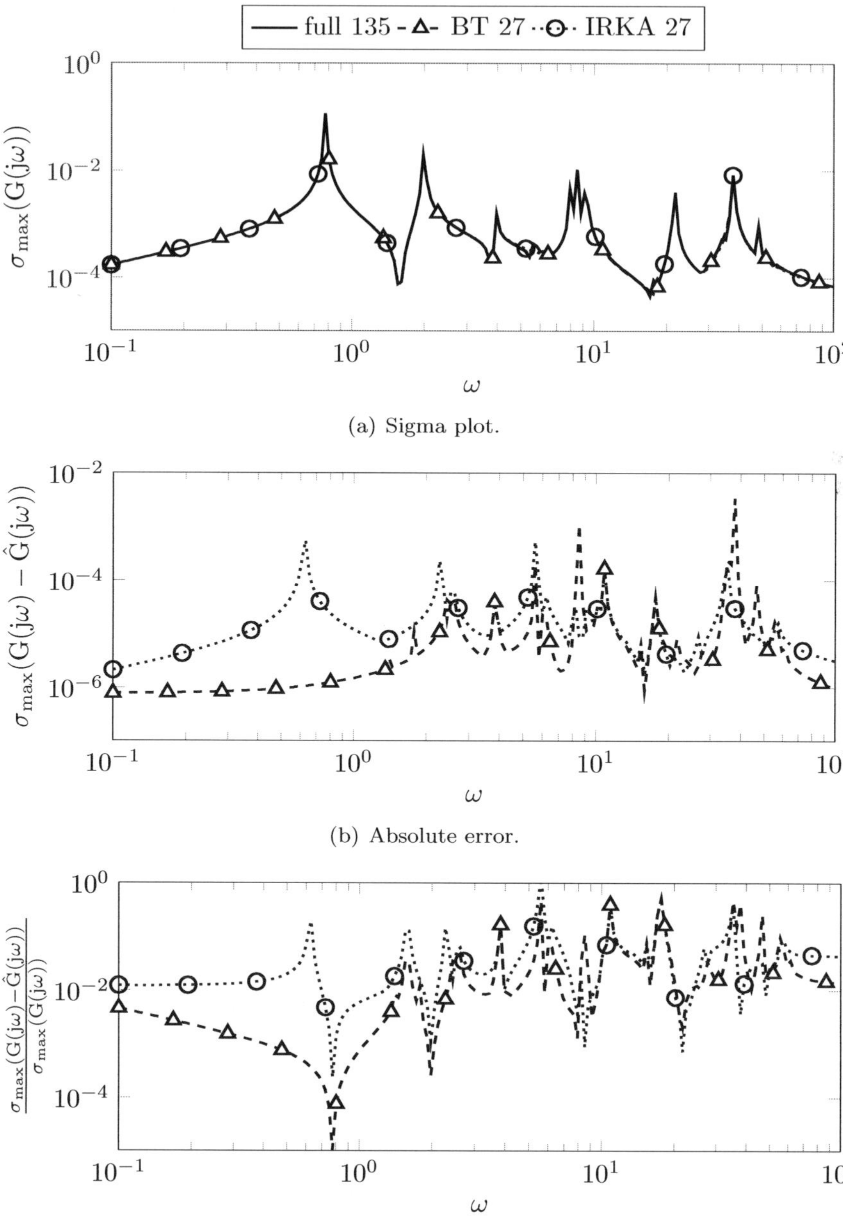

(c) Relative error.

Figure 5.1: Comparison between the original and reduced models obtained by balanced truncation and IRKA using the ISS model.

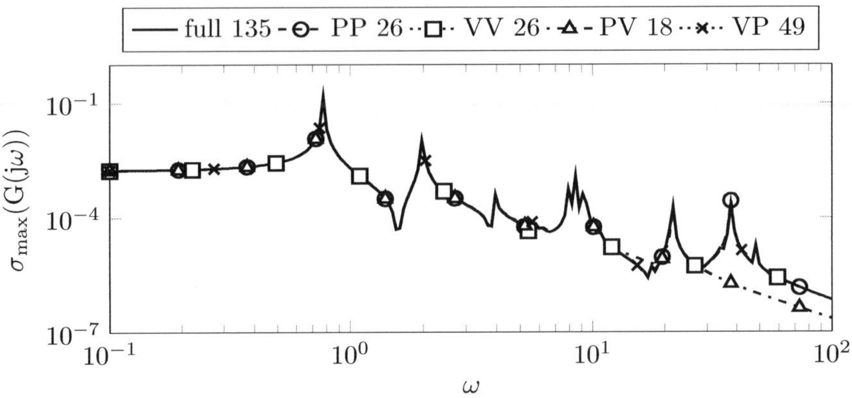

(a) Sigma plot.

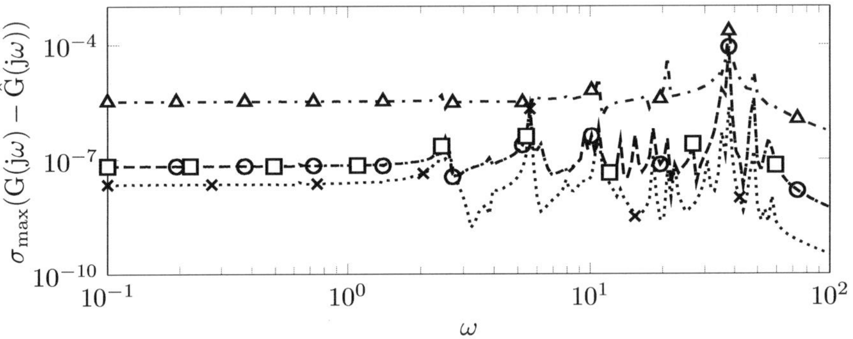

(b) Absolute error.

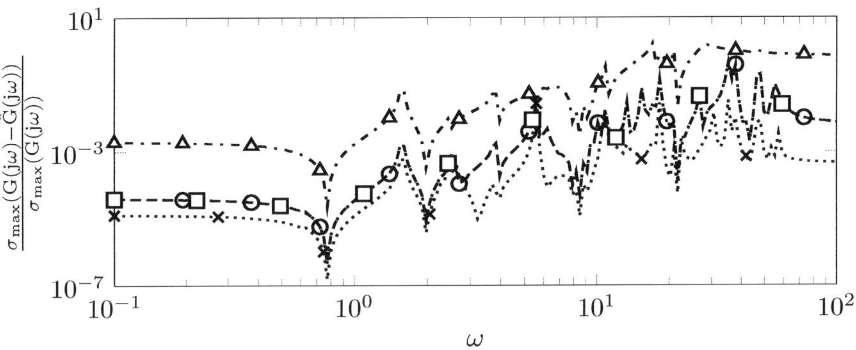

(c) Relative error.

Figure 5.2: Comparison between the original and reduced models obtained by balanced truncation using ISS model.

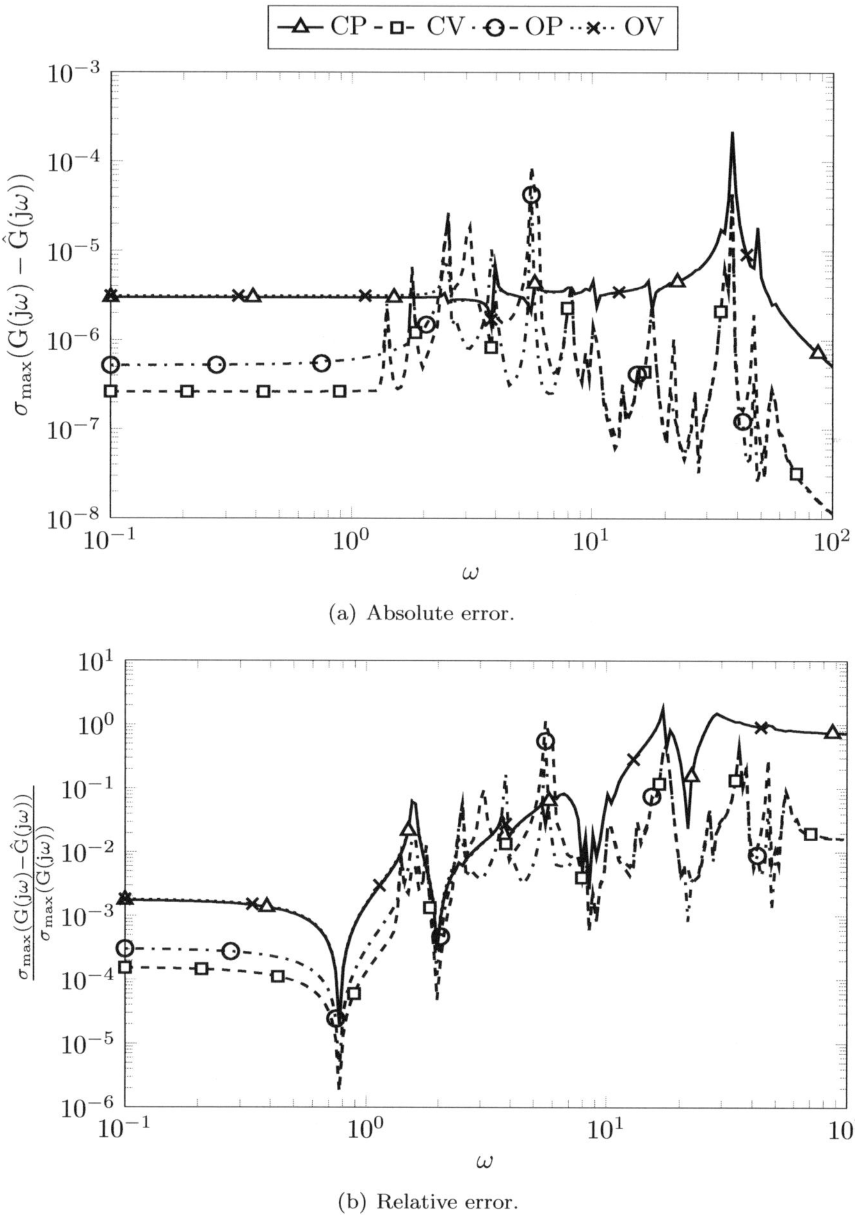

Figure 5.3: Comparison between the original and reduced models obtained by the PDEG using ISS model.

■ Exercises:

5.1 Find the transfer function of the second-order system:

$$\begin{aligned} M\ddot{\xi}(t) + D\dot{\xi}(t) + K\xi(t) &= Hu(t), \\ y(t) &= L_1\xi(t) + L_2\dot{\xi}(t). \end{aligned}$$

If

$$(E, A, B, C, D) = \left(\begin{bmatrix} 0 & I \\ D & M \end{bmatrix}, \begin{bmatrix} I & 0 \\ 0 & -K \end{bmatrix}, \begin{bmatrix} 0 \\ H \end{bmatrix}, \begin{bmatrix} L_1 & L_2 \end{bmatrix}, 0 \right).$$

is the first-order representation of the second-order system, then show that the transfer functions of the second- and first-order systems are the same.

5.2 Consider the symmetric second-order system (5.58). Convert this into its first-order form in a way that will give a symmetric system and the input-output matrices will be the transpose of each other. Then write the corresponding (controllability and observability) Lyapunov equations for the first-order system. Do the Lyapunov equations coincide?

5.3 Write a MATLAB routine that will plot the frequency responses of the second-order system (5.1) and its various first-order representations in a single figure using the frequency range $[10^1, 10^3]$ by using 500 logarithmically distributed sample points. To verify your code, use the data of the ISS model. Also show the absolute deviation of the frequency responses between the second- and first-order forms of the model.

5.4 Apply Algorithms 22 and 23 to the clamped beam model (CBM) (Appendix A.3.2) for computing various size ROMs and analyze the errors of the original and reduced models in both frequency and time domains.

5.5 Using the MATLAB function `lyap`, compute a second-order Gramian (e.g., controllability position Gramian) for the CBM. Now compute the eigenvalues of this Gramian. Also compute the singular values of the Cholesky factor of the Gramian. How many singular values are matching with the positive square root of the eigenvalues of the Gramian?

5.6 Apply Algorithm 24 to compute the reduced model in different balancing labels for the triple chain oscillation model (TCOM) (Appendix A.3.3), and compare the transfer functions of the original and reduced models.

5.7 Update Algorithm 26 for solving the Lyapunov equation (5.60). Then verify the efficiency of the algorithm by applying it to the triple chain oscillation model (TCOM) (Appendix A.3.3).

5.8 Write some advantages and disadvantages of balanced truncation and interpolatory projection techniques via IRKA. Apply both methods (i.e., Algorithms 22 and 23) to the Butterfly Gyro (in Appendix A.3.4) to find the ROM. Discuss the results and also mention the challenges that you encounter.

5.9 By applying the PDEG method, balanced truncation and IRKA as discussed in this chapter, find the second-order-to-first-order reduced models of the clamped beam model (CBM).

5.10 Apply the PDEG method (Algorithm 25) to find the structure preserving ROMs of CBM via projection onto the dominant eigenspace of the Gramians. Show the reduced models preserve the stability of the original system.

5.11 Develop efficient algorithms for the model reduction of the second-order symmetric system (5.58) based on the following methods:

- Second-order-to-first-order reduction
  - (a) Balanced truncation
  - (b) IRKA
- Second-order-to-second-order reduction
  - (a) Balanced truncation
  - (b) PDEG

Test the algorithms by applying to the TCOM model in Appendix A.3.3.

5.12 Develop an efficient algorithm based on LR-ADI iteration for solving the Lyapunov equations arising from the second-order system (5.1) with the first-order formulation (5.11).

Part II

MODEL REDUCTION OF DESCRIPTOR SYSTEMS

Chapter 6

Introduction to Descriptor Systems

6.1 Introduction

The main purpose of this book is to describe and discuss the model reduction of different types of descriptor systems. This chapter briefly introduces some special types of descriptor systems, which we will be studying in the coming chapters. We advise readers to refer to [9, 76, 106, 138, 158, 174] for more information on descriptor systems including their properties, solution techniques and so on. A descriptor system is a special form of a generalized state space model. Systems of the form in (2.4) with singular matrix E, i.e., $\det(E) = 0$, are often referred to as *descriptor systems.* In some references, they are also known as *singular systems* or *differential-algebraic equation (DAE) systems* (see, e.g., [106]). Such a system appears in various disciplines including electrical and mechanical engineering, fluid dynamics, mechatronics, system biology and multibody dynamics. In the coming chapters, we will consider some real-world examples of descriptor systems that arise in some of the aforementioned disciplines. The behavior of a descriptor system is different from a standard state-space system. In the text that follows, we briefly discuss some important properties of descriptor systems.

6.2 Solvability

Let us consider the following descriptor system

$$\begin{aligned} E\dot{x}(t) &= Ax(t) + Bu(t), \qquad x(t_0) = x_0, \\ y(t) &= Cx(t), \end{aligned} \tag{6.1}$$

where $x(t) \in \mathbb{R}^n$ is the vector of states, $u(t) \in \mathbb{R}^p$ is the vector of inputs and $y(t) \in \mathbb{R}^m$ is the outputs vector. The solution characteristics of this DAE system are determined by the corresponding matrix pencil

$$\mathcal{P} = (\mu E - A). \tag{6.2}$$

More precisely, the system has a solution if $\mathcal{P}$ is regular (see Definition 13). Due to the singularity of E, some of the eigenvalues of the system are infinite [102]. The subspace that is generated by solutions $x(t)$ is separated into two subspaces. One is formed by the eigenspaces associated with corresponding finite eigenvalues and the other one is formed by infinite eigenvalues of the system. Since E is singular, $\text{rank}(E) = r < n$ and $\det(\mathcal{P})$ is a (non-zero) polynomial of degree k $(0 \leq k \leq r)$; therefore, there exist nonsingular matrices $\Psi, \upsilon \in \mathbb{R}^{n \times n}$ such that premultiplying (6.1) with Ψ and employing a coordinate change

$$\upsilon x(t) = \begin{pmatrix} z_1(t) \\ z_2(t) \end{pmatrix} \tag{6.3}$$

would yield the following standard canonical form representation (see [76, 105, 122]):

$$\dot{\bar{z}}_1(t) = A_1 \bar{z}_1(t) + B_1 u(t), \tag{6.4a}$$

$$N\dot{\bar{z}}_2(t) = \bar{z}_2(t) + B_2 u(t), \tag{6.4b}$$

$$y(t) = C_1 \bar{z}_1(t) + C_2 \bar{z}_2(t), \tag{6.4c}$$

where $\bar{z}_1 \in \mathbb{R}^k$ and $\bar{z}_2 \in \mathbb{R}^{n-k}$ while A_1, B_1, B_2, C_1 and C_2 are constant matrices of appropriate dimensions and $\mathcal{N}$ is nilpotent with nil-potency ν, i.e., $N^{\nu-1} \neq 0$ but $N^{\nu} = 0$. Usually, the nilpotency ν indicates the index of the descriptor system. In the literature, it is known as the *algebraic index*. However, there are other types of indices for descriptor systems such as the *differentiation index* [9] the *tractability index* [121], and so on. The most commonly used one is the differentiation index which is defined by the number of derivatives that are required to take place in a DAE system to convert it into an equivalent ODE system.

Definition 38. *The differentiation index of a DAE system is the minimum number of times that all or part of the system must be differentiated with respect to t in order to find explicit ODE systems.*

Note that for an LTI system, the algebraic index and the differential index coincide.

For any initial condition $\bar{z}_1(0)$ and any input $u(t)$, (6.4) has the following unique solution

$$\bar{z}_1(t) = e^{A_1 t}\bar{z}_1(0) + \int_0^t e^{A_1(t-\tau)} B_1 u(\tau)\, d\tau, \qquad t \geq 0, \tag{6.5}$$

while the solution of (6.5), including initial condition $\bar{z}_2(0)$, is uniquely determined by the forcing inputs $u(t)$:

$$\bar{z}_2(t) = -\sum_{i=0}^{\nu-1} N^i B_2 u^{(i)}(t), \qquad t \geq 0, \tag{6.6}$$

where $u^{(i)}(t)$ denotes the ith derivatives of the input. Thus, unlike ODE systems, DAE systems do not have smooth solutions for arbitrary initial conditions. Only initial conditions $x(0)$ that are consistent, i.e., $x(0)$ for which $\bar{z}_2(0)$ satisfies (6.6) at $t = 0$, yield smooth solutions. Furthermore, unlike ODE systems, if the index ν for the DAE system exceeds one, then the solution of the DAE system may depend on the derivative of the forcing input $u(t)$ which must be smooth accordingly.

6.3 Transfer function

The *Laplace transformation* of a function $f(t)$, defined by

$$F(s) = \mathcal{L}[f(t)] = \int_0^\infty f(t)e^{-st}\, dt.$$

where s is a complex number, $s = a + ib$ with a and b being real numbers. Applying the Laplace transformation to the DAEs in (6.1), the following relation can be found

$$Y(S) = C(sE - A)^{-1}BU(S) + C(sE - A)^{-1}EX(0), \tag{6.7}$$

where $X(s)$, $Y(s)$ and $U(s)$ are Laplace transformations of $x(t)$, $y(t)$ and $u(t)$, respectively. The rational matrix valued function

$$G(s) = C(sE - A)^{-1}B$$

in (6.7) is called *transfer function* (TF) of the system in (6.7). It was already mentioned in Chapter 2 that the TF is one of the most powerful tools for simulation and stability analysis of LTI continuous-time systems. The TF $G(s)$ is called *proper* if $\lim_{s\to\infty} G(s) < \infty$ and *strictly proper* if $\lim_{s\to\infty} G(s) = 0$. Otherwise, $G(s)$ is called *improper*. For a descriptor system such as the one in (6.1), the TF can be divided into two parts (see [122]): $G(s) = G_{sp}(s) + G_p(s)$ where $G_{sp}(s)$ is strictly proper and $G_p(s)$ is the polynomial part of $G(s)$. We should remember that if the index of the descriptor system is at most one, then $G(s)$ is proper.

6.4 Stability

The descriptor system (6.1) is called asymptotically stable if $\lim_{t\to\infty} x(t) = 0$ for all solutions $x(t)$ of $E\dot{x}(t) = Ax(t)$. We know that the corresponding matrix pencil of the descriptor system (6.1) consists of finite and infinite eigenvalues. If all the finite eigenvalues lie in $\mathbb{C}^-$, then the system is asymptotically stable. If at least one of the finite eigenvalues has a positive real part, then the system is said to be unstable.

6.5 Structured DAE system

We have discussed general form of descriptor systems and their properties above briefly. However, this book is concerned with special structured descriptor systems considering their applications in different fields. The descriptor systems that we focus on have the following form

$$\begin{bmatrix} E_{11} & 0 \\ 0 & 0 \end{bmatrix} \begin{bmatrix} \dot{x}_1(t) \\ \dot{x}_2(t) \end{bmatrix} = \begin{bmatrix} A_{11} & A_{12} \\ A_{21} & A_{22} \end{bmatrix} \begin{bmatrix} x_1(t) \\ x_2(t) \end{bmatrix} + \begin{bmatrix} B_1 \\ B_2 \end{bmatrix} u(t), \tag{6.8a}$$

$$y(t) = \begin{bmatrix} C_1 & C_2 \end{bmatrix} \begin{bmatrix} x_1(t) \\ x_2(t) \end{bmatrix}, \tag{6.8b}$$

where $x_1(t) \in \mathbb{R}^{n_1}$ and $x_2(t) \in \mathbb{R}^{n_2}$. The block matrices E_{11} and A_{11} have full rank. The system can be divided into different indices as follows:

- index 1 if $\det(A_{22}) \neq 0$,
- index 2 if $A_{22} = 0$ and $\det(A_{21}A_{12}) \neq 0$, and
- index 3 if $A_{22} = 0$ and $\det(A_{21}A_{12}) = 0$.

In the coming chapters, we will study these structured DAE systems including their numerical properties, applications and model order reduction techniques.

6.6 Some useful MATLAB functions

Function name	Description
`dss`	create a descriptor state-space models
`isproper`	return true for proper dynamic systems
`ssdata`	quick access to state-space data or eliminate the algebraic variables and compute an explicit realization
`step`	step response of dynamic systems
`stepplot`	plot step response of linear systems
`impulse`	impulse response of dynamic systems
`impulseplot`	plot impulse response of linear systems
`tf`	construct transfer function or convert to transfer function of LTI dynamical system
`freqresp`	frequency response of dynamic systems
`sigma`	singular value plot of the frequency response of the dynamic system
`bode`	plot frequency response of dynamic system
`hsvd`	compute the Hankel singular values of linear systems
`hsvplot`	plot the Hankel singular values of an LTI model
`sigma`	singular value plot of dynamic systems

Exercises:

6.1 Define a DAE system including its indices.

6.2 Consider the LTI continuous-time system:

$$(E, A, B, C, D) = \left(\begin{bmatrix} 1 & 0 \\ 0 & 0 \end{bmatrix}, \begin{bmatrix} 3 & -2 \\ 12 & -7 \end{bmatrix}, \begin{bmatrix} 1 \\ 1 \end{bmatrix}, \begin{bmatrix} 1 & 0 \end{bmatrix}, 0 \right).$$

Is it a DAE system? If so, can you find the index of the DAE system manually?

6.3 Consider the LTI continuous-time system:

$$(E, A, B, C, D) = \left(\begin{bmatrix} 1 & 0 \\ 0 & 0 \end{bmatrix}, \begin{bmatrix} 3 & -2 \\ 12 & 0 \end{bmatrix}, \begin{bmatrix} 1 \\ 1 \end{bmatrix}, \begin{bmatrix} 1 & 0 \end{bmatrix}, 0 \right).$$

Is it a DAE system? If so, can you find the index of the DAE system manually?

6.4 By using an appropriate MATLAB command, determine whether the system in Exercise 6.2 is proper or improper.

6.5 Write the relations between finite and infinite eigenvalues of index 1, index 2 and index 3 DAE systems. Verify your answers with the data given for the different types of DAE system in Appendix A.

6.6 Using the MATLAB command `ssdata`, convert the DAE system in Exercise 6.2 into an ODE system. Now the DAE and ODE interms of the finite eigenvalues and transfer functions. Are both the systems equivalent?

Chapter 7

Model Reduction of First-Order Index 1 Descriptor Systems

7.1 Introduction

This chapter presents model reduction techniques of first-order index 1 descriptor systems of the form

$$E_1\dot{x}_1(t) = A_1x_1(t) + A_2x_2(t) + B_1u(t) \tag{7.1a}$$
$$0 = A_3x_1(t) + A_4x_2(t) + B_2u(t) \tag{7.1b}$$
$$y(t) = C_1x_1(t) + C_2x_2(t), \tag{7.1c}$$

where $x_1(t) \in \mathbb{R}^{n_1}$ is a vector of differential variables, $x_2(t) \in \mathbb{R}^{n_2}$ is a vector of algebraic variables, $u \in \mathbb{R}^m$ are the inputs, and $y \in \mathbb{R}^p$ are the outputs. In matrix-vector form, the system can be represented as

$$\begin{aligned} \underbrace{\begin{bmatrix} E_1 & 0 \\ 0 & 0 \end{bmatrix}}_{\mathcal{E}} \begin{bmatrix} \dot{x}_1(t) \\ \dot{x}_2(t) \end{bmatrix} &= \underbrace{\begin{bmatrix} A_1 & A_2 \\ A_3 & A_4 \end{bmatrix}}_{\mathcal{A}} \begin{bmatrix} x_1(t) \\ x_2(t) \end{bmatrix} + \underbrace{\begin{bmatrix} B_1 \\ B_2 \end{bmatrix}}_{\mathcal{B}} u(t) \\ y(t) &= \underbrace{\begin{bmatrix} C_1 & C_2 \end{bmatrix}}_{\mathcal{C}} \begin{bmatrix} x_1(t) \\ x_2(t) \end{bmatrix}, \end{aligned} \tag{7.2}$$

in which $\mathcal{E}$, $\mathcal{A}$, $\mathcal{B}$ and $\mathcal{C}$ are all sparse matrices with appropriate dimensions. Since the matrix $\mathcal{E}$ is singular, the system is called descriptor system. Assume that the block matrix A_4 is invertible and therefore the descriptor system has an index 1 form. Index 1 descriptor systems arise in various disciplines such as in electrical networks [3], fluid dynamics [181], constraint mechanics [64], power systems [71], and so on.

In many applications, models of the above structures can become very large and complex. When this happens, it is very expensive to simulate, control, and optimize the system. Therefore, we want to reduce the complexity of the model by applying model order reduction (MOR), i.e., we seek an approximation to the original model that well-approximates the behavior of the original model but that is much faster to evaluate.

The two most frequently applied modern MOR methods are balanced truncation [125, 163] and rational interpolation of the transfer function by the iterative rational Krylov algorithm (IRKA) [85]. Both approaches have also been discussed in Chapter 4 for the systems governed by ordinary differential equations.

In principle, we are converting the differential-algebraic equation (DAE) system in (7.1) into ordinary differential equation (ODE) system (see the next section) so we can apply both methods to the converted system. However, in this case, the system loses the sparsity pattern. For a large-scale system, such a conversion should be avoided to minimize the computational complexity as well as computer memory requirements.

This chapter discusses the model reduction methods for the model reduction of index 1 system (7.1) without explicitly forming the ODE system. We will also discuss methods for efficiently solving the Lyapunov equations which are raised from the underlying system using the low-rank alternating direction implicit (LR-ADI) algorithm. The proposed algorithms will be applied to the Brazilian power system models. Numerical results are discussed to show the efficiency and capability of the proposed method.

7.2 Reformulation of dynamical system

This section reformulates the index 1 system (7.1) to convert it into an equivalent form of ODE system. Equation (7.1b) yields

$$A_4 x_2 = -A_3 x_1 - B_2 u.$$

Since the block matrix A_4 is nonsingular this implies

$$x_2 = -A_4^{-1} A_3 x_1 - A_4^{-1} B_2 u.$$

Inserting $x_2(t)$ into (7.1a) and (7.1c) descriptor system (7.2) can be rewritten as

$$\begin{aligned} E_1 \dot{x}_1 &= (A_1 - A_2 A_4^{-1} A_3) x_1(t) + (B_1 - A_2 A_4^{-1} B_2) u(t), \\ y(t) &= C_1 - C_2 A_4^{-1} A_3) x_1(t) - C_2 A_4^{-1} B_2 u(t), \end{aligned}$$

which can be compared with the generalized state space system (as in (4.1)) of the form

$$\begin{aligned} E\dot{x}(t) &= Ax(t) + Bu(t), \\ y(t) &= Cx(t) + D_a u(t), \end{aligned} \tag{7.3}$$

where

$$\begin{aligned} &x(t) := x_1(t), \quad E = E_1, \quad A := A_1 - A_2 A_4^{-1} A_3, \\ &B := B_1 - A_2 A_4^{-1} B_2, \quad C := (C_1 - C_2 A_4^{-1} A_3, \quad \text{and} \quad D_a := -C_2 A_4^{-1} B_2. \end{aligned}$$

The following observation shows dynamical system (7.1) or (7.2) and (7.3) are equivalent.

Theorem 15. *The transfer functions of the system (7.2) and (7.3) are equal.*

Proof. The transfer function of the system (7.2) is defined by

$$\begin{aligned} G(s) &= \mathcal{C}(s\mathcal{E} - \mathcal{A})^{-1}\mathcal{B} \\ &= \begin{bmatrix} C_1 & C_2 \end{bmatrix} \left(s \begin{bmatrix} E_1 & 0 \\ 0 & 0 \end{bmatrix} - \begin{bmatrix} A_1 & A_2 \\ A_3 & A_4 \end{bmatrix} \right)^{-1} \begin{bmatrix} B_1 \\ B_2 \end{bmatrix} \\ &= \begin{bmatrix} C_1 & C_2 \end{bmatrix} \begin{bmatrix} sE_1 - A_1 & -A_2 \\ -A_3 & -A_4 \end{bmatrix}^{-1} \begin{bmatrix} B_1 \\ B_2 \end{bmatrix} \\ &= \begin{bmatrix} C_1 & C_2 \end{bmatrix} \begin{bmatrix} X_1 \\ X_2 \end{bmatrix}, \end{aligned} \tag{7.4}$$

where

$$\begin{bmatrix} X_1 \\ X_2 \end{bmatrix} = \begin{bmatrix} sE_1 - A_1 & -A_2 \\ -A_3 & -A_4 \end{bmatrix}^{-1} \begin{bmatrix} B_1 \\ B_2 \end{bmatrix},$$

solves the linear system

$$\begin{bmatrix} sE_1 - A_1 & -A_2 \\ -A_3 & -A_4 \end{bmatrix} \begin{bmatrix} X_1 \\ X_2 \end{bmatrix} = \begin{bmatrix} B_1 \\ B_2 \end{bmatrix}. \tag{7.5}$$

From (7.5), we have

$$(sE_1 - A_1)X_1 - A_2 X_2 = B_1, \tag{7.6}$$

$$-A_3 X_1 - A_4 X_2 = B_2. \tag{7.7}$$

Equation (7.7) gives $X_2 = -A_4^{-1}B_2 - A_4^{-1}A_3X_1$. Inserting X_2 into (7.6) we have

$$(sE_1 - A_1)X_1 - A_2(-A_4^{-1}B_2 - A_4^{-1}A_3X_1) = B_1,$$

which implies

$$X_1 = (sE_1 - A_1 + A_2A_4^{-1}A_3)^{-1}(B_1 - A_2A_4^{-1}B_2).$$

Therefore,

$$X_2 = -A_4^{-1}B_2 - A_4^{-1}A_3(sE_1 - A_1 + A_2A_4^{-1}A_3)^{-1}(B_1 - A_2A_4^{-1}B_2).$$

Inserting X_1 and X_2 into (7.4)

$$\begin{aligned} G(s) &= \begin{bmatrix} C_1 & C_2 \end{bmatrix} \begin{bmatrix} (sE_1 - A_1 + A_2A_4^{-1}A_3)^{-1}(B_1 - A_2A_4^{-1}B_2) \\ -A_4^{-1}B_2 - A_4^{-1}A_3(sE_1 - A_1 + A_2A_4^{-1}A_3)^{-1}(B_1 - A_2A_4^{-1}B_2) \end{bmatrix} \\ &= C_1(sE_1 - A_1 + A_2A_4^{-1}A_3)^{-1}(B_1 + A_2A_4^{-1}B_2) - \\ &\quad C_2A_4^{-1}B_2 - C_2A_4^{-1}A_3(sE_1 - A_1 - A_2A_4^{-1}A_3)^{-1}(B_1 + A_2A_4^{-1}B_2 \\ &= (C1 - C_2A_4^{-1}A_3)(sE_1 - A_1 + A_2A_4^{-1}A_3)^{-1}(B_1 - A_2A_4^{-1}B_2) - C_2A_4^{-1}B_2 \\ &= C(sE - A)^{-1}B - C_2A_4^{-1}B_2, \end{aligned}$$

where E, A, B and C are defined in (7.3). This completes the proof. □

It can also be proved (see the exercise problems) that ODE system (7.3) preserves all the finite eigenvalues of DAE system (7.1).

> The DAE and ODE systems (7.1) and (7.3) are equivalent since the transfer functions and finite spectrums of both systems are the same.

Although index 0 system (7.3) has lower dimension than the index 1 system (7.1), it is sparse and hence all computation can be implemented efficiently. In the following sections we emphasize the fact that we never form the ODE system explicitly to implement the model reduction

---

7.3 Balancing-based MOR

System (7.3) can now be considered to be in its generalized state space form as in (4.1). Therefore, we can apply the LR-SRM algorithm (i.e., Algorithm 19) for balancing-based model reduction of the system. For this purpose, we need to solve two continuous-time algebraic Lyapunov equations

$$APE^T + EPA^T = -B\,B^T, \tag{7.8a}$$

$$A^TQE + E^TQA = -C\,C, \tag{7.8b}$$

where E, A, B and C are defined in (7.3). We will later discuss how to solve these Lyapunov equations efficiently using the LR-ADI iteration as presented in Chapter 3. Suppose that by solving the Lyapunov equations (7.8a) and (7.8b), we compute the low-rank controllability and observability Gramian

Algorithm 27: LR-SRM for index 1 DAEs.

Input : $E_1,\ A_1,\ A_2, A_3, A_4,\ B_1, B_2\ C_1, C_2$ from (7.1).
Output: $\hat{E},\ \hat{A},\ \hat{B},\ \hat{C},\ \hat{D}_a := D_a$ in (7.9).

1 Compute R and L by solving the Lyapunov equations (7.8a) and (7.8b).

2 Compute the singular value decomposition (SVD)
$$L^T E_1 R = U\Sigma V^T = \begin{bmatrix} U_1 & U_2 \end{bmatrix} \begin{bmatrix} \Sigma_1 & \\ & \Sigma_2 \end{bmatrix} \begin{bmatrix} V_1^T \\ V_2^T \end{bmatrix}.$$

3 Construct $V := LV_1\Sigma_1^{-\frac{1}{2}}, \quad W := RU_1\Sigma_1^{-\frac{1}{2}}$.

4 Form the reduced matrices as
$$\begin{aligned} &\hat{E} = W^T E_1 V, \quad \hat{A} = W^T A_1 V - \hat{A}_2 {A_4}^{-1} \hat{A}_3, \\ &\hat{B} = W^T B_1 - \hat{A}_2 A_4^{-1} B_2,\ \hat{C} = C_1 V - C_2 {A_4}^{-1} \hat{A}_3,\ \hat{D}_a = D_a, \end{aligned}$$
where $\hat{A}_2 = W^T A_2$ and $\hat{A}_3 = A_3 V$.

factors R and L, respectively. Now we define the balancing and truncating transformations
$$V := LV_1\Sigma_1^{-\frac{1}{2}} \quad \text{and } W := RU_1\Sigma_1^{-\frac{1}{2}}$$
by computing the SVD:
$$L^T ER = U\Sigma V^T = \begin{bmatrix} U_1 & U_2 \end{bmatrix} \begin{bmatrix} \Sigma_1 & \\ & \Sigma_2 \end{bmatrix} \begin{bmatrix} V_1^T \\ V_2^T \end{bmatrix}.$$
Next we apply V and W to the system (7.3) for the reduced model
$$\begin{aligned} \hat{E}\dot{\hat{x}}(t) &= \hat{A}\hat{x}(t) + \hat{B}u(t), \\ \hat{y}(t) &= \hat{C}\hat{x}(t) + \hat{D}_a u(t), \end{aligned} \tag{7.9}$$
where $\hat{E},\ \hat{A} \in \mathbb{R}^{r\times r},\ \hat{B} \in \mathbb{R}^{r\times m},\ \hat{C} \in \mathbb{R}^{m\times p},\ \hat{D}_a := D_a$. The reduced matrices are constructed as follows
$$\begin{aligned} &\hat{E} = W^T E_1 V, \quad \hat{A} = W^T A_1 V - \hat{A}_2 {A_4}^{-1} \hat{A}_3, \\ &\hat{B} = W^T B_1 - \hat{A}_2 A_4^{-1} B_2,\ \hat{C} = C_1 V - C_2 {A_4}^{-1} \hat{A}_3,\ \hat{D}_a = D_a, \end{aligned} \tag{7.10}$$
where $\hat{A}_2 = W^T A_2$ and $\hat{A}_3 = A_3 V$. A complete algorithm for the LR-SRM for model reduction of index 1 structural DAEs is presented in Algorithm 27. Note that the algorithm outputs a reduced generalized state space model from the index 1 system input. However, a simple algebraic manipulation represents the ROM (7.9) in the reduced index 1 DAE system setting:
$$\begin{aligned} \begin{bmatrix} \hat{E} & 0 \\ 0 & 0 \end{bmatrix} \begin{bmatrix} \dot{\hat{x}}_1(t) \\ \dot{\hat{x}}_2(t) \end{bmatrix} &= \begin{bmatrix} \hat{A} & \hat{A}_2 \\ \hat{A}_3 & A_4 \end{bmatrix} \begin{bmatrix} \hat{x}_1(t) \\ \hat{x}_2(t) \end{bmatrix} \begin{bmatrix} \hat{B}_1 \\ B_2 \end{bmatrix} u(t), \\ y(t) &= \begin{bmatrix} \hat{C}_1 & C_2 \end{bmatrix} \begin{bmatrix} \hat{x}_1(t) \\ \hat{x}_2(t) \end{bmatrix}. \end{aligned} \tag{7.11}$$

7.4 Solution of the Lyapunov equations by LR-ADI iteration

To implement the balancing based Algorithm 27 for the model reduction of index 1 DAEs (7.1), we need to solve two continuous-time algebraic Lyapunov equations (7.8). This section shows how to solve the Lyapunov equations efficiently recalling the LR-ADI iteration discussed in Chapter 3. To solve the controllability Lyapunov equation (7.8a), we can directly apply Algorithm 15. However, since the matrix A is dense it will lead to a high computational workload. In the following we discuss how to avoid explicit formulation of A.

Recall Algorithm 15. Consider the initial Lyapunov residual factor

$$W_0 = B = B_1 - A_2 A_4^{-1} B_2.$$

In the i-th iteration step, to compute V_i, we need to solve the linear system

$$(A + \mu_i E)V_i = W_{i-1}, \tag{7.12}$$

which is equivalent to

$$(A_1 + \mu_i E_1 - A_2 A_4^{-1} A_3)V_i = W_{i-1}. \tag{7.13}$$

It should be noted that V_i can be computed by solving

$$\begin{bmatrix} A_1 + \mu_i E_1 & A_2 \\ A_3 & A_4 \end{bmatrix} \begin{bmatrix} V_i \\ \star \end{bmatrix} = \begin{bmatrix} W_{i-1} \\ 0 \end{bmatrix}. \tag{7.14}$$

Although the linear system in (7.14) has a larger dimension than that of (7.12), it is sparse and can efficiently be solved by suitable direct [58] or iterative [143] solvers.

The whole procedure to compute the low-rank Gramian factor of the controllability Gramian by solving the Lyapunov equation (7.8a) is presented in Algorithm 28. This same algorithm can be used for solving the Lyapunov equation (7.8b) to compute the low-rank observability Gramian factor. In that case, the inputs $(E_1, A_1, A_2, A_3, A_4, B_1, B_2)$ are replaced by $(E_1^T, A_1^T, A_3^T, A_2^T, A_4^T, C_1^T, C_2^T)$, which will lead to the solution of the linear system (in step 3)

$$\begin{bmatrix} A_1^T + \mu_i E_1^T & A_3^T \\ A_2^T & A_4^T \end{bmatrix} \begin{bmatrix} V_i \\ \star \end{bmatrix} = \begin{bmatrix} W_{i-1} \\ 0 \end{bmatrix}.$$

Algorithm 28: LR-ADI for index 1 DAE system.

Input : E_1, A_1, A_2, A_3, A_4, B_1, B_2, $\{\mu_i\}_{i=1}^{J}$.
Output: $R = Z_i$, such that $P \approx RR^T$.
Set $Z_0 = [\,]$ and $W_0 = B_1 - A_2A_4^{-1}B_2$.
while $\|W_{i-1}^T W_{i-1}\| \geq tol \; or \; i \leq i_{max}$ **do**
  Solve the linear system

$$\begin{bmatrix} A_1 + \mu_i E_1 & A_2 \\ A_3 & A_4 \end{bmatrix} \begin{bmatrix} V_i \\ \star \end{bmatrix} = \begin{bmatrix} W_{i-1} \\ 0 \end{bmatrix},$$

  for V_i.
  if $\mathrm{Im}\,(\mu_i) = 0$ **then**
    $Z_i = \begin{bmatrix} Z_{i-1} & \sqrt{-2\mu_i}V_i \end{bmatrix}$,
    $W_i = W_{i-1} - 2\mu_i E_1 V_i$
  else
    $\gamma = -2\,\mathrm{Re}\,(\mu_i)$, $\delta = \frac{\mathrm{Re}\,(\mu_i)}{\mathrm{Im}\,(\mu_i)}$,
    $Z_{i+1} = \begin{bmatrix} Z_{i-1} & \sqrt{2\gamma}(\mathrm{Re}\,(V_i) + \delta\,\mathrm{Im}\,(V_i)) & \sqrt{2\gamma}\sqrt{(\delta^2+1)}\,\mathrm{Im}\,(V_i) \end{bmatrix}$,
    $W_i = W_{i-1} + 2\gamma E_1(\overline{V_i} + 2\delta\,\mathrm{Im}\,(V_i))$.
    $i = i + 1$
  $i = i + 1$

For a better convergence of the algorithm, a set of shift parameters are required; these can be computed either by a heuristic or an adaptive approach as discussed in Section 3.6.

7.5 Tangential interpolation via IRKA

To apply the tangential interpolation method (introduced in Section 4.5) to the index 1 system (7.1), we first convert the system into its equivalent ODE system (7.3). Then Algorithm 21 can directly be used to the ODE system to form a reduced order model as shown in (7.9). In the following, we show how to compute the projector matrices W and V efficiently using IRKA.

Following Algorithm 21, the right and left transformation matrices are defined as

$$V = \left[(\alpha_1 E - A)^{-1}Bb_1, \cdots, (\alpha_r E - A)^{-1}Bb_r\right], \tag{7.15a}$$

$$W = \left[(\alpha_1 E^T - A^T)^{-1}C^Tc_1, \cdots, (\alpha_r E^T - A^T)^{-1}C^Tc_r\right]. \tag{7.15b}$$

where the matrices E, A, B and C are defined in (7.3), α_i are the interpolation points and b_i, c_i are tangential directions. A close observation reveals that to generate each term of V, we have to solve a shifted linear system like

$$(\alpha E - A)v = Bb, \tag{7.16}$$

which is equivalent to

$$(\alpha E_1 - (A_1 - A_2 A_4^{-1} A_3))v = (B_1 - A_2 A_4^{-1} B_2)b.$$

Recalling the strategy in [71], instead of solving the above linear system we can solve the linear system

$$\begin{bmatrix} \alpha E_1 - A_1 & A_2 \\ A_3 & A_4 \end{bmatrix} \begin{bmatrix} v \\ \Lambda \end{bmatrix} = \begin{bmatrix} B_1 \\ B_2 \end{bmatrix} b, \tag{7.17}$$

for v. Note that although the linear system in (7.17) has a larger dimension than that of (7.16), it is sparse and can be efficiently solved by suitable direct [58] or iterative [143] solvers.

Analogously, when we construct the left transformation matrix W to obtain each term of (7.15b), instead of solving a linear system like $(\alpha E^T - A^T)w = C^T c$, we can solve the linear system

$$\begin{bmatrix} \alpha E_1^T - A_1^T & A_3^T \\ A_2^T & A_4^T \end{bmatrix} \begin{bmatrix} w \\ \Lambda \end{bmatrix} = \begin{bmatrix} C_1^T \\ C_2^T \end{bmatrix} c, \tag{7.18}$$

for w. Once we have W and V, then the reduced model (7.9) can be obtained by forming the reduced matrices as in (7.10). The whole procedure of IRKA to construct the reduced model from an index 1 system (7.1) is summarized in Algorithm 29.

> The reduced model that is generated by Algorithm 29 is in the generalized state space form. However, it can be converted into index 1 form as in (7.11).

Algorithm 29: IRKA for index 1 DAE systems.

Input : E_1, A_1, A_2, A_3, A_4, B_1, B_2, C_1, C_2.
Output: $\hat{E}$, $\hat{A}$, $\hat{B}$, $\hat{C}$, $\hat{D}_a = D_a$.

1 Select initial interpolation points $\{\sigma_i\}_{i=1}^r$ and tangential directions $\{b_i\}_{i=1}^r$ and $\{c_i\}_{i=1}^r$.
2 **for** $i = 1, 2, \cdots, r$ **do**
3   Solve the linear systems $\begin{bmatrix} \alpha E_1 - A_1 & A_2 \\ A_3 & A_4 \end{bmatrix} \begin{bmatrix} v_i \\ \Lambda \end{bmatrix} = \begin{bmatrix} B_1 \\ B_2 \end{bmatrix} b_i$ and $\begin{bmatrix} \alpha E_1^T - A_1^T & A_3^T \\ A_2^T & A_4^T \end{bmatrix} \begin{bmatrix} w_i \\ \Lambda \end{bmatrix} = \begin{bmatrix} C_1^T \\ C_2^T \end{bmatrix} c_i$
4   Construct

$$V = \begin{bmatrix} v_1, v_1, \cdots, v_r \end{bmatrix}, \quad W = \begin{bmatrix} w_1, w_1, \cdots, w_r \end{bmatrix},$$

5 **while** *(not converged)* **do**
6   Form $\hat{E}$, $\hat{A}$, $\hat{B}$ and $\hat{C}$ as in (7.10)
7   Compute $\hat{A}z_i = \lambda_i \hat{E} z_i$ and $y^* \hat{A} = \lambda_i y^* \hat{E} z_i$.
8     $\sigma_i \leftarrow -\lambda_i$, $b_i \leftarrow y^* \hat{B}$ and $c_i \leftarrow \hat{C} z_i$.
9   Repeat Step 2 - 4.
10     $i = i + 1$
11 Form the reduced matrices $\hat{E}$, $\hat{A}$, $\hat{B}$, $\hat{C}$ as in (7.10).

7.6 Some useful MATLAB functions

Function name	**Description**
`dss`	create generalized or descriptor state-space models
`hsvd`	computes the Hankel singular values of linear systems
`hsvplot`	plot the Hankel singular values of a linear time invariant (LTI) model
`step`	step response of dynamic systems
`stepplot`	plot step response of linear systems
`tf`	construct transfer function or convert to transfer function of LTI dynamic system
`freqresp`	frequency response of dynamic systems
`sigma`	singular value plot of the frequency response of the dynamic system
`bode`	bode frequency response of dynamic system
`loglog`	log-log scale plot using logarithmic scaled for both the X-axis and Y-axis
`semilogx` (`semilogy`)	log-log scale plot using logarithmic scaled for X- (Y-) axis

7.7 Numerical results

To assess the accuracy and efficiency of the proposed (model reduction) methods of first-order index 1 system, in this section we present some numerical results. For the numerical experiments we consider the Brazilian Interconnected Power System (BIPS) models available in [71]. A brief introduction of the models is given in Appendix A.4. There are several data sets, the one that we apply is the BIPSM-606 with number of differential variables: $n_1 = 606$, algebraic variables: $n_2 = 1142$ and inputs/outputs: $m/p = 4/4$.

All the results were obtained by using MATLAB R2015a (8.5.0.197613) on a board with processor 4×Intel® Core™i5-4460 CPUs running at a clock speed of 2.90 GHz and 16 GB RAM.

To implement the BT method, first we solve the controllability and observability Lyapunov equations by applying the LR-ADI iteration for index 1 DAE system, i.e., Algorithm 28. Figure 7.1 shows the rate of convergence of both the low-rank controllability Gramian factor R and observability Gramian factor L. We have used adaptive shifts for implementing this algorithm. Using these Gramian factors, we have computed a 50 dimensional reduced model via balanced truncation (i.e., Algorithm 27). Using Algorithm 29 which is based on IRKA, the same dimensional reduced model was computed. Figure 7.2 shows the comparisons of the original and the 50 dimensional reduced models obtained by BT and IRKA. From absolute error (in Figure 7.2(b)) and relative error (in Figure 7.2(c)), we can observe that although the overall performance of the BT method is slightly better than the IRKA, in the higher frequency label IRKA shows better accuracy. The comparisons of the BT and IRKA methods also depicted in time domain as shown in Figure 7.3. This figure shows the step responses (Fig. 7.3(a)) of the original and reduced models together with their absolute (Fig. 7.3(b)) and relative (Fig. 7.3(c)) deviations. Note that to compute the step response, the implicit Euler method with a fixed time step size of 5 was applied. Both methods yield satisfactory results, but the IRKA is showing a marginally better accuracy than the BT in the higher time domain.

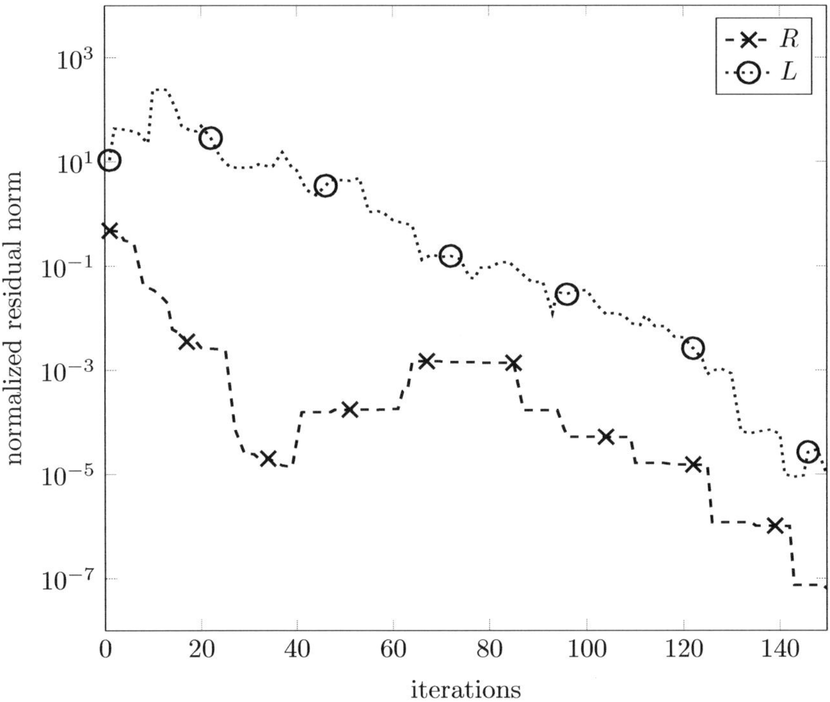

Figure 7.1: Convergence rate of LR-ADI methods for index 1 DAE system.

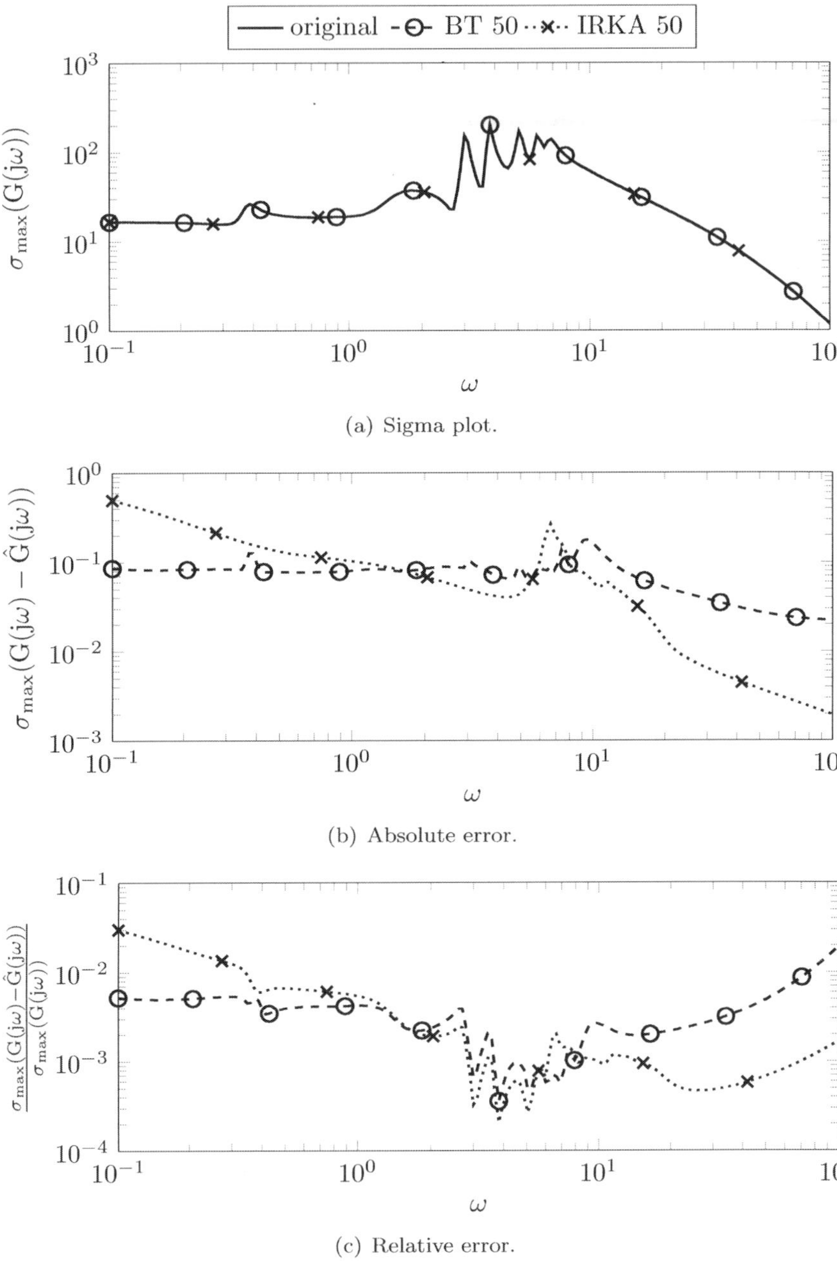

Figure 7.2: Frequency-domain comparison between the original and reduced models obtained by balanced truncation and IRKA.

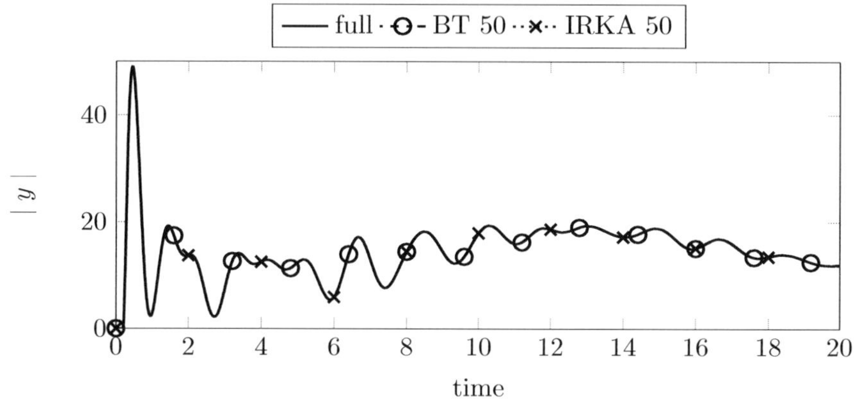

(a) Step response.

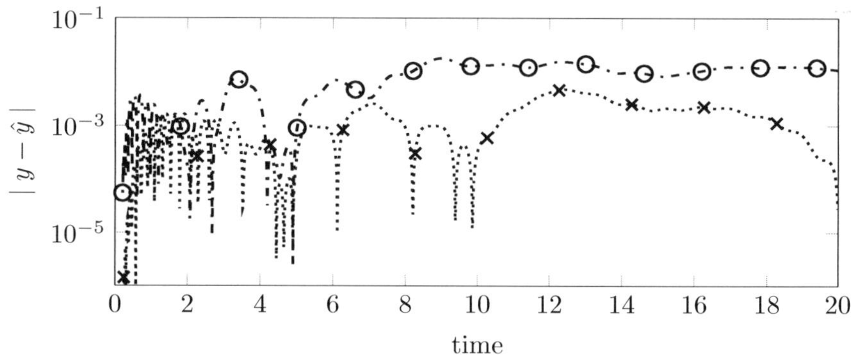

(b) Absolute deviation.

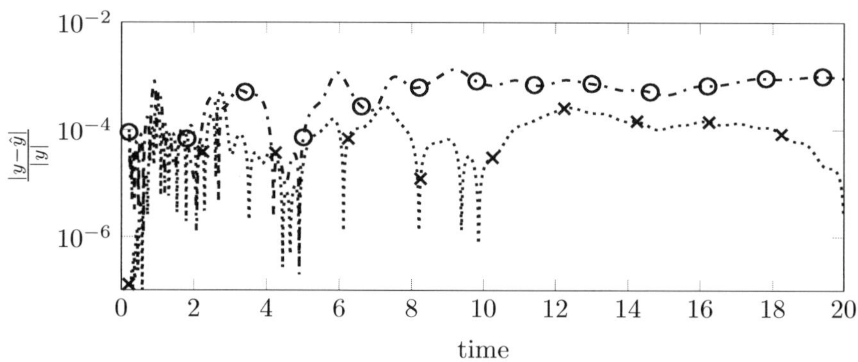

(c) Relative deviation.

Figure 7.3: Time-domain comparison between the original and reduced models obtained by balanced truncation and IRKA.

■ Exercises:

7.1 Number of finite eigenvalues of the index 1 system (7.1) is equal to the number of eigenvalues of the generalized state space system (7.3). Verify this using the data of BIPSM-606 (as in Appendix A.4.1). For this model, compute all finite eigenvalues using the MATLAB command `eig` and plot them in a figure. Using the same command, find the eigenvalues of the converted ODE system and plot them onto the same figure. Do the eigenvalues converge in both cases?

7.2 In this chapter it is claimed that the index 1 system (7.1) and the converted ODE system (7.3) are equivalent. Verify this for BIPSM-606 (see Appendix A.4.1). By using the results, confirm that the following identities match in both representations.

   i) System Hankel singular values (HSVs).
   ii) Frequency responses.
   iii) Step on impulse responses.
   iv) Finite eigenvalues.

7.3 Although the converted ODE system (7.3) has a lower dimension than the index 1 system (7.1), the ODEs formulation should be avoided due to memory restrictions and CPU usage. Using the data of BIPSM-606 show the sparsity of the system matrices for both formulations and discuss their memory requirements.

7.4 Following the discussion in this chapter, find the reduced models of the power system model BIPSM-1693 by applying balanced truncation and interpolatory projection methods via IRKA. Discuss the numerical results to show the efficiency of the methods.

7.5 By applying Algorithm 28, find suitable reduced models for BIPSM-1142, BIPSM-1450, BIPSM-1693, BIPSM-1998, BIPSM-2476 and BIPSM-3078. For all the reduced models, compute $\|G - \hat{G}\|_{\mathcal{H}_2}$ where G and $\hat{G}$ represent the transfer functions of the original and reduced model, respectively. Show the results in a table.

7.6 Consider the following first-order index 1 DAE system

$$\begin{aligned} E_1\dot{x}_1(t) + E_2\dot{x}_2(t) &= A_1x_1(t) + A_2x_2(t) + B_1u(t) \\ 0 &= A_3x_1(t) + A_4x_2(t) + B_2u(t) \\ y(t) &= C_1x_1(t) + C_2x_2(t) + D_au(t). \end{aligned}$$

   i) Eliminating the algebraic part, convert the system into ODE system.

ii) Applying the data of supersonic engine inlet (SEI) (as in Appendix A.4.2), show that ODE system and DAE system have the same responses. Plot the frequency responses on a suitable frequency range for both systems on a single figure.

7.7 Following the discussion in Section 7.5, develop an IRKA for the model reduction of the index 1 DAEs given in Exercise 7.6. Then assess the efficiency of the algorithm by applying it to the data of SEI (in Appendix A.4.2).

7.8 Develop a balancing-based algorithm for the model from Exercise 7.6. Then discuss the efficiency of the algorithm by applying it to the data of SEI (as in Appendix A.4.2).

Chapter 8

Model Reduction of First-Order Index 2 Descriptor Systems

8.1 Introduction

In this chapter we study model reduction of a class of structured index 2 descriptor systems of the form

$$\underbrace{\begin{bmatrix} E_1 & 0 \\ 0 & 0 \end{bmatrix}}_{E} \begin{bmatrix} \dot{x}_1(t) \\ \dot{x}_2(t) \end{bmatrix} = \underbrace{\begin{bmatrix} A_1 & A_2 \\ A_2^T & 0 \end{bmatrix}}_{A} \begin{bmatrix} x_1(t) \\ x_2(t) \end{bmatrix} + \underbrace{\begin{bmatrix} B_1 \\ 0 \end{bmatrix}}_{B} u(t), \tag{8.1a}$$

$$y(t) = \underbrace{\begin{bmatrix} C_1 & C_2 \end{bmatrix}}_{C} \begin{bmatrix} x_1(t) \\ x_2(t) \end{bmatrix}, \tag{8.1b}$$

where $x_1(t) \in \mathbb{R}^{n_1}$ is a vector of differential variables, $x_2(t) \in \mathbb{R}^{n_2}$ ($n_1 > n_2$) is a vector of algebraic variables, $u \in \mathbb{R}^m$ are the inputs and $y \in \mathbb{R}^p$ are the outputs and in which E, A, B and C are all sparse matrices with appropriate dimensions. Assume that E_1 is a symmetric positive definite matrix. Such models arise in disciplines like fluid dynamics: from a semi-discretization (applying FEM) of the linearized stokes [160], Navier stokes [39] or Oseen equations [93]; electrical networks: modified nodal analysis of RL circuits [3]; and mechanics: the damped mass-spring system with a holonomic constraint at the velocity level [64, Chapter 2].

In many applications, models of the above structures can become very large and complex. Under these circumstances, simulation, control and optimization of such large-scale systems become very expensive. Therefore, we want to reduce the complexity of the model by applying model order reduction (MOR), i.e., we seek an approximation to the original model that well approximates the behavior of the original model but which is much faster to evaluate.

The two most frequently applied modern MOR methods are balanced truncation [125, 163] or rational interpolation of the transfer function by the iterative rational Krylov algorithm (IRKA) [85]. Both approaches have also been discussed in Chapter 4 for the systems governed by ordinary differential equations (ODEs).

In principle, by projecting the system onto the hidden manifold (i.e., onto the null-space of A_2^T) we can convert the system (8.1) into an ODE system. However, this will cause the system to lose its sparsity and hence such an ODE formulation must be avoided for a large-scale system due to the memory restriction and computational complexity.

Like the index 1 case discussed in the previous chapter, in order to implement the model reduction of an index 2 descriptor system, we first convert the system into its equivalent form of ODE system. However, unlike the index 1 system, here we project the system onto the hidden manifold to remove the algebraic part of the system. Model reduction methods can then be applied to the projected system. At the end we can show that to perform the MOR, projected systems should not be formed explicitly.

In this chapter we will discuss how to implement the model reduction of projected systems without computing the projectors explicitly. Both the balanced truncation and the interpolatory projection methods will be discussed for the index 2 system (8.1). We will also discuss efficient solution of projected Lyapunov equations arising from the index 2 DAE system to compute the low-rank Gramian factors applying the LR-ADI iteration. Note that the Gramian factors are the important ingredients of the balancing-based model reductions. The efficiency of the proposed methods will be confirmed by numerical experiments.

8.2 Reformulation of dynamical system

In this section we discuss how to convert the index 2 system into an equivalent ODE system. First rewrite the system as

$$E_1\dot{x}_1(t) = A_1x_1(t) + A_2x_2(t) + B_1u(t), \tag{8.2a}$$

$$0 = A_2^T x_1(t), \tag{8.2b}$$

$$y(t) = C_1x_1(t) + C_2x_2(t). \tag{8.2c}$$

From (8.2b) we obtain $A_2^T\dot{x}_1(t) = 0$. Hence, multiplying both sides of (8.2a) by $A_2^T E_1^{-1}$ we obtain

$$0 = A_2^T E_1^{-1} A_1 x_1(t) + A_2^T E_1^{-1} A_2 x_2(t) + A_2^T E_1^{-1} B_1 u(t)$$

which yields

$$x_2(t) = -(A_2^T E_1^{-1} A_2)^{-1} A_2^T E_1^{-1} A_1 x_1(t) - (A_2^T E_1^{-1} A_2)^{-1} A_2^T E_{11}^{-1} B_1 u(t). \tag{8.3}$$

Substituting (8.3) into (8.2a) and (8.2c) and manipulating the system, we obtain

$$E_1\dot{x}(t) = \Pi A_1 x(t) + \Pi B_1 u(t), \tag{8.4a}$$
$$y(t) = Cx(t) + D_a u(t), \tag{8.4b}$$

where

$$\begin{aligned}
x(t) &= x_1(t),\\
C &= C_1 - C_2(A_2^T E_1^{-1} A_2)^{-1} A_2^T E_1^{-1} A_1,\\
D_a &= -C_2(A_2^T E_1^{-1} A_2)^{-1} A_2^T E_1^{-1} B_1,\\
\Pi &= I - A_2(A_2^T E_1^{-1} A_2)^{-1} A_2^T E_1^{-1}.
\end{aligned}$$

It can be shown that

$$\begin{aligned}
\Pi^2 &= (I - A_2(A_2^T E_1^{-1} A_2)^{-1} A_2^T E_1^{-1})(I - A_2(A_2^T E_1^{-1} A_2)^{-1} A_2^T E_1^{-1})\\
&= I - 2A_2(A_2^T E_1^{-1} A_2)^{-1} A_2^T E_1^{-1} + (A_2(A_2^T E_1^{-1} A_2)^{-1}\\
&\quad A_2^T E_1^{-1}) A_2 (A_2^T E_1^{-1} A_2)^{-1} A_2^T E_1^{-1}\\
&= I - 2A_2(A_2^T E_1^{-1} A_2)^{-1} A_2^T E_1^{-1} + (A_2(A_2^T E_1^{-1} A_2)^{-1}\\
&= I - A_2(A_2^T E_1^{-1} A_2)^{-1} A_2^T E_1^{-1}\\
&= \Pi.
\end{aligned}$$

Therefore, Π is a projector. The projector Π satisfies the following properties.

Proposition 2. *Let Π be the projector defined above. The following conditions must hold.*

1. $\Pi E_1 = E_1 \Pi^T$.
2. $\text{Null}(\Pi) = \text{Range}\left(A_2^T\right)$.
3. $\text{Range}(\Pi) = \text{Null}\left(A_2^T E_1^{-1}\right)$.

Proof. 1. We have

$$\begin{aligned}
\Pi E_1 &= (I - A_2(A_2^T E_1^{-1} A_2)^{-1} A_2^T E_1^{-1}) E_1\\
&= E_1 - A_2(A_2^T E_1^{-1} A_2)^{-1} A_2^T)\\
&= E_1(I - E^{-1} A_2 (A_2^T E_1^{-1} A_2)^{-1} A_2^T)\\
&= E_1 \Pi^T.
\end{aligned}$$

2. Suppose that vector a belongs to the nullspace of Π, i.e., $\Pi a = 0$. By the definition of Π, $(I - A_2(A_2^T E_1^{-1} A_2)^{-1} A_2^T E_1^{-1})a = 0$, which implies $a = A_2(A_2^T E_1^{-1} A_2)^{-1} A_2^T E_1^{-1} a$. So $a = A_2 b$, where $(A_2^T E_1^{-1} A_2)^{-1} A_2^T E_1^{-1} a = b$, which implies that a is in the range of A_2. Therefore,

$$\text{Null}\,(\Pi) \subseteq \text{Range}\,(A_2)\,. \tag{8.5}$$

Conversely, suppose that a is in the range of A_2. Therefore, there exists a non-zero vector b, such that $A_2 b = a$. Multiplying both sides by $A_2^T E_1^{-1}$, $A_2^T E_1^{-1} A_2 b = A_2^T E_1^{-1} a$, which implies $b = (A_2^T E_1^{-1} A_2)^{-1} A_2^T E_1^{-1} a$. Again multiplying both sides by A_2, $A_2 b = A_2(A_2^T E_1^{-1} A_2)^{-1} A_2^T E_1^{-1} a$. So $a = A_2(A_2^T E_1^{-1} A_2)^{-1} A_2^T E_1^{-1} a$, which implies $\Pi a = 0$. Therefore a is also in the nullspace of Π, and hence

$$\text{Range}\,(A_2) \subseteq \text{Null}\,(\Pi)\,. \tag{8.6}$$

Equations (8.5) and (8.6) prove $\text{Null}\,(\Pi) = \text{Range}\,(A_2)$.
3. Again, we assume a is in the range of Π, i.e., $\Pi a = a$, which implies $(I - A_2(A_2^T E_1^{-1} A_2)^{-1} A_2^T E_1^{-1})a = a$, or $A_2(A_2^T E_1^{-1} A_2)^{-1} A_2 E_1^{-1} a = 0$. Let $\Phi b = 0$, where $\Phi = A_2(A_2^T M_1^{-1} A_2)^{-1}$ and $b = A_2 E_1^{-1} a$. Multiplying both sides by Φ^T, we obtain $\Phi^T \Phi b = 0$. Since $\Phi^T \Phi$ is invertible, we see $b = 0$ or $A_2^T E_1^{-1} a = 0$. This proves that a is in the nullspace of $A_2^T E_1^{-1}$. Therefore,

$$\text{Range}\,(\Pi) \subseteq \text{Null}\,\left(A_2^T E_1^{-1}\right). \tag{8.7}$$

Conversely, again suppose that $a \in \text{Null}\,\left(A_2^T E_1^{-1}\right)$, i.e., $A_2^T E_1^{-1} a = 0$. Multiplying both sides by $\Phi^T \Phi$, we get $\Phi^T \Phi A_2^T E_1^{-1} a = 0$. Again multiplying both sides by b^T, $b^T \Phi^T \Phi b = 0$, which implies $(\Phi b)^T (\Phi b) = 0$, hence $\Phi b = 0$. Therefore, $A_2(A_2^T E_1^{-1} A_2)^{-1} A_2^T E_1^{-1} a = 0$, i.e., $a = 0$, and therefore, $\Pi a = a$, such that

$$\text{Null}\,\left(A_2^T E_1^{-1}\right) \subseteq \text{Range}\,(\Pi)\,. \tag{8.8}$$

Therefore, Equations (8.7) and (8.8) yield $\text{Range}\,(\Pi) = \text{Null}\,\left(G_1 M_1^{-1}\right)$. □

Theorem 16. *The vector a is in the null space of A_2^T, i.e., $A_2^T a = 0$ iff $\Pi^T a = a$, where Π is defined in (8.4).*

Proof. Suppose the vector a is in the null space of A_2^T, i.e., $A_2^T a = 0$. Multiplying both sides with $-E_1^{-1} A_2(A_2 E_1^{-1} A_2)^{-1}$, we obtain

$$-E_1^{-1} A_2(A_2^T E_1^{-1} A_2)^{-1} A_2^T a = 0,$$

which is equivalent to $(I - E_1^{-1} A_2(A_2^T E_1^{-1} A_2)^{-1} A_2^T)a = a$, i.e., $\Pi^T a = a$. Conversely, suppose that $\Pi^T a = a$ which implies

$$(I - E_1^{-1} A_2(A_2^T E_1^{-1} A_2)^{-1} A_2^T)a = a.$$

We see $E_1^{-1} A_2(A_2^T E_1^{-1} A_2)^{-1} A_2^T a = 0$. Multiplying both sides by A_2^T we obtain $A_2^T a = 0$. □

Due to the properties of Π we can insert $x(t) = \Pi^T x(t)$ into the system (8.4). Since the projector Π satisfies

$$\Pi A_2 = 0,$$

from (8.4) we obtain the following projected system

$$\Pi E_1 \Pi^T \dot{x}(t) = \Pi A_1 \Pi^T x(t) + \Pi B_1 u(t), \tag{8.9a}$$

$$y(t) = C\Pi^T x(t) + D_a u(t). \tag{8.9b}$$

The system dynamics of (8.9) are projected onto the $m_1 := n_1 - n_2$ dimensional subspace Range $\left(\Pi^T\right)$ [93]. This subspace is however still represented in the coordinates of the n_1 dimensional space. The m_1 dimensional representation can be made explicit by employing the *thin* singular value decomposition (SVD)

$$\Pi = \begin{bmatrix} U_1 & U_2 \end{bmatrix} \begin{bmatrix} S_1 & 0 \\ 0 & 0 \end{bmatrix} \begin{bmatrix} V_1^T \\ V_2^T \end{bmatrix} = U_1 S_1 V_1^T = \Theta_l \Theta_r^T, \tag{8.10}$$

where $\Theta_l = U_1$ and $\Theta_r = V_1$ and in which $U_1,\ V_1 \in \mathbb{R}^{n_1 \times m_1}$ consist of the corresponding leading m_1 columns of $U, V \in \mathbb{R}^{n_1 \times n_1}$. Moreover, Θ_l, Θ_r satisfy

$$\Theta_l^T \Theta_r = I_{m_1}. \tag{8.11}$$

This representation is always possible since Π is a projector and therefore, $S_1 = I_{m_1}$. Inserting the decomposition of Π as in (8.10) into (8.9) and considering $\tilde{x}(t) = \Theta_l^T x(t)$, we get

$$\begin{aligned} \Theta_r^T E_1 \Theta_r \dot{\tilde{x}}(t) &= \Theta_r^T A_1 \Theta_r \tilde{x}(t) + \Theta_r^T B_1 u(t), \\ y(t) &= C\Theta_r \tilde{x}(t) + D_a u(t). \end{aligned} \tag{8.12}$$

System (8.12) is practically system (8.9) with the redundant equations removed by the Θ_r projection. We observe that the dynamical systems (8.2), (8.9) and (8.12) are equivalent in the sense that their finite spectrum is the same [64, Theorem 2.7.3] and the input-output relations are the same, i.e., they realize the same transfer function.

8.3 Balancing-based MOR and low-rank ADI iteration

In this section we will discuss how to avoid forming (8.12) explicitly to perform the model reduction. Suppose that we want to apply balanced truncation to system (8.12). To this end, we need to solve the Lyapunov equations

$$\begin{aligned} \Theta_r^T A_1 \Theta_r \tilde{P} \Theta_r^T E_1^T \Theta_r + \Theta_r^T E_1 \Theta_r \tilde{P} \Theta_r^T A_1^T \Theta_r &= -\Theta_r^T B_1 B_1^T \Theta_r, \\ \Theta_r^T A_1^T \Theta_r \tilde{Q} \Theta_r^T E_1 \Theta_r + \Theta_r^T E_1^T \Theta_r \tilde{Q} \Theta_r^T A_1 \Theta_r &= -\Theta_r^T C^T C \Theta_r, \end{aligned} \tag{8.13}$$

$\tilde{P} \in \mathbb{R}^{m_1 \times m_1}$, $\tilde{Q} \in \mathbb{R}^{m_1 \times m_1}$ are the corresponding projected controllability and observability Gramians. Multiplying (8.13) by Θ_l from the left and Θ_l^T from the right and exploiting $\Theta_r = \Pi^T \Theta_r$ (e.g., due to (8.10) and (8.11)) we obtain

$$\begin{aligned} \Pi A_1 \Pi^T P \Pi E_1^T \Pi^T + \Pi E_1 \Pi^T P \Pi A_1^T \Pi^T &= -\Pi B_1 B_1^T \Pi^T, \\ \Pi A_1^T \Pi^T Q \Pi E_1 \Pi^T + \Pi E_1^T \Pi^T Q \Pi A_1 \Pi^T &= -\Pi C^T C \Pi^T, \end{aligned} \tag{8.14}$$

where $P = \Theta_r \tilde{P} \Theta_r^T$ and $Q = \Theta_r \tilde{Q} \Theta_r^T$. These are the respective controllability and observability Lyapunov equations for realization (8.9) and the solutions P, $Q \in \mathbb{R}^{n_1 \times n_1}$ are the corresponding controllability and observability Gramians. The system (8.14) is singular due to the fact that Π is a projection. Uniqueness of solutions is re-established by the condition that the solutions satisfy $P = \Pi^T P \Pi$ and $Q = \Pi^T Q \Pi$.

It is also an easy exercise to go back to (8.13) from (8.14). Let us consider

$$P \approx RR^T, \; Q \approx LL^T \text{ and } \tilde{P} \approx \tilde{R}\tilde{R}^T, \tilde{Q} \approx \tilde{L}\tilde{L}^T.$$

Then R, L, $\tilde{R}$ and $\tilde{L}$ are called approximate low-rank factors. They fulfill the relation

$$R = \Theta_r \tilde{R} \qquad \text{and} \qquad L = \Theta_r \tilde{L}.$$

For large-scale problems, however, computing Θ_r is usually impossible due to memory limitations. Therefore, R and L are computed by solving (8.14). The balancing and truncating transformations for (8.12) are

$$\tilde{W} = \tilde{R} U_k \Sigma_k^{-\frac{1}{2}}, \quad \tilde{V} = \tilde{L} V_k \Sigma_k^{-\frac{1}{2}},$$

where U_k, $V_k, \in \mathbb{R}^{n_m \times k}$ consist of the corresponding leading k columns of U, $V \in \mathbb{R}^{n_m \times n_m}$, and $\Sigma_k \in \mathbb{R}^{k \times k}$ is the upper left $k \times k$ block of Σ in the SVD

$$\tilde{R}^T \Theta_r^T E_1 \Theta_r \tilde{L} = U \Sigma V^T.$$

Observing further that $R^T \Pi E_1 \Pi^T L = \tilde{R}^T \Theta_r^T E_1 \Theta_r \tilde{L} = U \Sigma V^T$, the projection matrices for the system (8.9) can be formed as

$$W = R U_k \Sigma_k^{-\frac{1}{2}} \quad \text{and} \quad V = L V_k \Sigma_k^{-\frac{1}{2}}. \tag{8.15}$$

As in [93], we find that

$$\begin{aligned} W &= R U_k \Sigma_k^{-\frac{1}{2}} = \Theta_r \tilde{R} U_k \Sigma_k^{-\frac{1}{2}} = \Theta_r \tilde{W} = \Theta_r \Theta_l^T \Theta_r \tilde{W} = \Pi^T W, \\ V &= L V_k \Sigma_k^{-\frac{1}{2}} = \Theta_r \tilde{L} V_k \Sigma_k^{-\frac{1}{2}} = \Theta_r \tilde{V} = \Theta_r \Theta_l^T \Theta_r \tilde{V} = \Pi^T V. \end{aligned} \tag{8.16}$$

Now we apply the transformations W and V in (8.9) to find the reduced order model as

$$\begin{aligned} \hat{E}\dot{\hat{x}}(t) &= \hat{A}\hat{x}(t) + \hat{B}u(t) \\ \hat{y}(t) &= \hat{C}\hat{x}(t) + D_a u(t), \end{aligned} \tag{8.17}$$

Algorithm 30: LR-SRM for unstable index 2 DAE system.

Input : E_1, A_1, B_1, C.
Output: $\hat{E}$, $\hat{A}$, $\hat{B}$, $\hat{C}$, $\hat{D}_a = D_a$.
1 Compute R and L by solving the projected Lyapunov equations (8.14).
2 Construct W and V as in (8.16).
3 Form $\hat{E} = W^T E_1 V$, $\hat{A} = W^T A_1 V$, $\hat{B} = W^T B_1$ and $\hat{C} = CV$.

---

where

$$\hat{E} = W^T \Pi E_1 \Pi^T V, \;\; \hat{A} = W^T \Pi A_1 \Pi^T V, \;\; \hat{B} = W^T \Pi B_1 \text{ and } \hat{C} = C\Pi^T V.$$

Due to (8.16), we can avoid the explicit usage of Π and find

$$\hat{E} = W^T E_1 V, \;\; \hat{A} = W^T A_1 V, \;\; \hat{B} = W^T B_1 \text{ and } \hat{C} = CV.$$

> Eventually, we see that the reduced order model (8.17) is obtained without forming the projected system (8.9).

In the next section, we will show how to compute R and L using the LR-ADI iteration without using Π explicitly. The above procedure to compute the ROM for the unstable index 2 DAE system is summarized in Algorithm 30.

8.4 Solution of the projected Lyapunov equations by LR-ADI iteration and related issues

In order to apply the aforementioned balancing-based MOR, we need to solve the projected Lyapunov equations (8.14). We have previously seen that the Π^T invariant solution factors enable us to compute the corresponding truncating transformations. The approach here is different from the spectral projection-based work by Stykel [160] in that we are applying the E_1-orthogonal projection to the hidden manifold, where Stykel used the orthogonal projection (in the Euclidean sense) onto the eigenspaces corresponding to the finite poles of the system. In fact both methods project to the same subspace considering orthogonality in different inner products. Here we are concerned with two main issues. First, we discuss the reformulation of the basic low-rank ADI iteration for the projected Lyapunov equation that ensures the invariance of the solution factor and the computation of the correct corresponding residual factors. We are lifting the ideas of [93] to reformulate the LR-ADI in Algorithm 15. For the spectral projection methods, the analogue procedure has been discussed in [43]. In the second part, we address the important issue of ADI shift parameter computation. There, the main issue in the DAE setting is to avoid the subspaces corresponding to infinite eigenvalues in order to correctly compute the large magnitude Ritz values involved in many parameter choices. The crucial point in both parts is to provide methods that use the projection Π^T at most implicitly and never form the projected system (8.9).

8.4.1 LR-ADI for index 2 systems

Here, we are concerned with the efficient solution of the Lyapunov equations in (8.14) to compute the low-rank Gramian factors using LR-ADI as discussed in Chapter 3. We consider the projected controllability equation elaborately. The observability equation can be handled analogously. For convenience we rewrite the controllability Lyapunov equation in (8.14) as

$$\tilde{A}P\tilde{E}^T + \tilde{E}P\tilde{A}^T = -\tilde{B}\tilde{B}^T \tag{8.18}$$

with

$$\tilde{E} = \Pi E_1 \Pi^T,\ \tilde{A} = \Pi A_1 \Pi^T \quad \text{and } \tilde{B} = \Pi B_1. \tag{8.19}$$

To compute the low-rank controllability Gramian factor R by solving (8.18) such that $P \approx RR^T$ we follow Algorithm 15. In the i-th iteration step of the ADI, the residual of the controllability Lyapunov equation (8.18) can be written as

$$\mathcal{R}(P_i) = \tilde{A}P_i\tilde{E}^T + \tilde{E}P_i\tilde{A}^T + \tilde{B}\tilde{B}^T = \tilde{W}_i\tilde{W}_i^T,$$

where

$$\tilde{W}_i = \prod_{k=1}^{i}(\tilde{A} - \overline{\mu}_i\tilde{E})(\tilde{A} + \mu_i\tilde{E})^{-1}\tilde{B}.$$

In the i-th iteration step, V_i is computed by solving the linear system

$$(\tilde{A} + \mu_i\tilde{E})V_i = \tilde{W}_{i-1}, \tag{8.20}$$

which enables us to update the residual factor according to

$$\tilde{W}_i = (\tilde{A} - \mu^*\tilde{E})V_i = \tilde{W}_{i-1} - 2\,\mathrm{Re}\,(\mu_i)\tilde{E}V_i. \tag{8.21}$$

The following observation enables us to solve this linear system efficiently.

Theorem 17. *The matrix X satisfies $X = \Pi^T X$ and $\Pi(E_1 + \mu A_1)\Pi^T X = \Pi F$ iff*

$$\begin{bmatrix} E_1 + \mu A_1 & A_2 \\ A_2^T & 0 \end{bmatrix} \begin{bmatrix} X \\ \Lambda \end{bmatrix} \begin{bmatrix} F \\ 0 \end{bmatrix} \tag{8.22}$$

Proof. Suppose X satisfies

$$X = \Pi^T X \tag{8.23}$$

and

$$\Pi(E_1 + \mu A_1)\Pi^T X = \Pi F. \tag{8.24}$$

Plugging (8.23) into (8.24) which implies $\Pi\{(E_1 + \mu A_1)X - F\} = 0$. That is, the columns of $(E_1 + \mu A_1)X - F$ are in the null space of Π. From Proposition 4(2) $\mathrm{Null}(\Pi) = \mathrm{Range}(A_2)$. Hence there exists Λ such that $(E_1 + \mu A_1)X - F = -A_2\Lambda$ which gives the first line of (8.22). Due to Proposition 4 (1), from (8.23) we obtain $A_2^T X = 0$ which is the second line of (8.22).

On the other hand, suppose X and Λ satisfy (8.22), then the second line of (8.22), i.e., $A_2^T X = 0$ implies

$$X = \Pi^T X.$$

First line gives $(E_1 + \mu A_1)X + A_2\Lambda = F$. Multiplying both sides of this equation by Π we have $\Pi(E_1 + \mu A_1)X = \Pi F$ (since $\Pi A_2 = 0$). Again since $X = \Pi^T X$ we have

$$\Pi(E_1 + \mu A_1)\Pi^T X = \Pi F.$$

□

Following Theorem 17 instead of solving (8.20), one can compute V_i by solving

$$\begin{bmatrix} A_1 + \mu_i E_1 & A_2 \\ A_2^T & 0 \end{bmatrix} \begin{bmatrix} V_i \\ \star \end{bmatrix} = \begin{bmatrix} \tilde{W}_{i-1} \\ 0 \end{bmatrix}. \tag{8.25}$$

The computation of the initial residual factor $\tilde{W}_0$ is discussed in detail below. Again the computed V_i in (8.25) satisfies $V_i = \Pi^T V_i$. Therefore, the correct projected residual factor in (8.21) can be obtained by

$$\tilde{W}_i = \tilde{W}_{i-1} - 2\,\mathrm{Re}\,(\mu_i) E_1 V_i, \tag{8.26}$$

since we have $\Pi E_1 = E_1 \Pi^T$.

In order to really compute the correct residual, the initial residual must be computed as $\tilde{W}_0 = \Pi B_1$ to ensure $\tilde{W}_0 = \Pi\tilde{W}_0$. This can be performed cheaply using the following lemma.

Lemma 5. *The matrix Ξ satisfies $\Xi = \Pi^T \Xi$ and $E_1 \Xi = \Pi B_1 \Leftrightarrow$*

$$\begin{bmatrix} E_1 & A_2 \\ A_2^T & 0 \end{bmatrix} \begin{bmatrix} \Xi \\ \Lambda \end{bmatrix} = \begin{bmatrix} B_1 \\ 0 \end{bmatrix}. \tag{8.27}$$

Proof. If $\Xi = \Pi^T \Xi$, then $E_1 \Xi = \Pi B_1$ implies $\Pi(E_1 \Xi - B_1) = 0$. Since $\text{Null}(\Pi) = \text{Range}(A_2)$, there exists Λ such that $E_1 \Xi - B_1 = -A_2 \Lambda$, or $E_1 \Xi + A_2 \Lambda = B_1$. Again applying the properties of Π (in Proposition 4), we have $A_2^T \Xi = 0$. These two relations give (8.27).

Conversely, we assume (8.27) holds. From the first block row of (8.27) we get

$$\Xi = E_1^{-1} B_1 - E_1^{-1} A_2 \Lambda,$$

and thus from the second row it follows that

$$0 = A_2^T \Xi = A_2^T E_1^{-1} B_1 - A_2^T E_1^{-1} A_2 \Lambda,$$

such that

$$\Lambda = (A_2^T E_1^{-1} A_2)^{-1} A_2^T E_1^{-1} B_1.$$

Inserting this in the first block row, we get as desired

$$E_1 \Xi = B_1 A_2 (A_2^T E_1^{-1} A_2)^{-1} A_2^T E_1^{-1} B_1 = \Pi B_1.$$

□

This especially ensures $\Xi = \Pi^T \Xi$, since

$$E_1 \Xi = \Pi B_1 = \Pi E_1 \Xi = E_1 \Pi^T \Xi,$$

and thus using $\tilde{W}_0 = E_1 \Xi$, we get the desired invariance $\tilde{W}_0 = \Pi \tilde{W}_0$.

The whole procedure of computing the low-rank factor R of the controllability Gramian is summarized in Algorithm 31. The above findings on the residual factor can be summarized as the following lemma.

Lemma 6. *The residual factor in every step of Algorithm 31 fulfills the relation*

$$\tilde{W}_i = \Pi \tilde{W}_i.$$

The shifts that guarantee fast convergence of the algorithm are closely related to the spectrum of the original pencil. We discuss how these can be computed in the text that follows.

8.4.2 ADI shift parameters selection

The appropriate shift parameter selection is one of the crucial tasks for the fast convergence of the LR-ADI iteration. In recent time, most of the papers are following the *heuristic* procedure introduced by Penzl [133] to compute the suboptimal ADI shift parameters μ_i, $i = 1, 2, \ldots, J$, for a large-scale problem. Very recently, new shift computation ideas considering adaptive and automatic computation of shifts during the iteration [26, 184] have come up. We present the basic ideas to adapt both the classic and the new methods to our framework in the following two paragraphs.

Algorithm 31: LR-ADI for index 2 DAE system.

Input : E_1, A_1, A_2, B_1, $\{\mu_i\}_{i=1}^{J}$.
Output: $R = Z_i$, such that $P \approx RR^T$.

Set $Z_0 = [\,]$.
Solve the linear system (8.27) for Ξ and compute $\tilde{W}_0 = E_1\Xi$
$i = 1$
while $\|\tilde{W}_{i-1}^T \tilde{W}_{i-1}\| \geq tol$ *or* $i \leq i_{max}$ **do**
  Solve the linear system (8.25) for V_i.
  if $\mathrm{Im}\,(\mu_i) = 0$ **then**
    $Z_i = \begin{bmatrix} Z_{i-1} & \sqrt{-2\mu_i} V_i \end{bmatrix}$,
    $\tilde{W}_i = \tilde{W}_{i-1} - 2\mu_i E_1 V_i$
  else
    $\gamma = -2\,\mathrm{Re}\,(\mu_i)$, $\delta = \frac{\mathrm{Re}\,(\mu_i)}{\mathrm{Im}\,(\mu_i)}$,
    $Z_{i+1} = \begin{bmatrix} Z_{i-1} & \sqrt{2\gamma}(\mathrm{Re}\,(V_i) + \delta\,\mathrm{Im}\,(V_i)) & \sqrt{2\gamma}\sqrt{(\delta^2+1)}\,\mathrm{Im}\,(V_i) \end{bmatrix}$,
    $\tilde{W}_i = \tilde{W}_{i-1} + 2\gamma E_1(\overline{V_i} + 2\delta\,\mathrm{Im}\,(V_i))$.
    $i = i + 1$
  $i = i + 1$

Heuristic shift selection. The main ingredient of the heuristic method is the computation of a number of large and small magnitude Ritz values. In the case of DAE systems, the computation of Ritz values of large magnitude cause difficulties due to the existence of infinite eigenvalues. We need to make sure that the infinite eigenvalues are avoided. This can be achieved by the following corollary that is a direct consequence of [52, Theorem 3.1].

Corollary 1. *The matrix pencil*

$$\mathcal{P}_\psi(\lambda) = \lambda \begin{bmatrix} E_1 & \psi A_2 \\ \psi A_2^T & 0 \end{bmatrix} - \begin{bmatrix} A_1 & A_2 \\ A_2^T & 0 \end{bmatrix} \tag{8.28}$$

transforms all infinite eigenvalues of the pencil $\lambda E - A$ to $\frac{1}{\psi}$ while at the same time preserving its finite eigenvalues.

Thus from the pencil $\mathcal{P}_\psi$ we can compute the desired Ritz values of large magnitude via an Arnoldi iteration [142]. The parameter δ can be easily determined after the small Ritz values β_i have been computed with respect to the original pencil. In order to ensure that $\frac{1}{\psi}$ is avoided by the Arnoldi process for the large magnitude Ritz values and thus only finite eigenvalues of the original pencil are approximated, one could for example set $\psi = \frac{1}{\min\limits_i \mathrm{Re}\,(\beta_i)}$.

Adaptive shift selection. A second shift computation strategy we use in the numerical experiments follows the lines of the adaptive shift strategy proposed in Benner et al. [26]. There, the shifts are initialized by the eigenvalues of the pencil projected to the span of W_0. Then, whenever all shifts in the current set have been used, the pencil is projected for example to the span of the current V_i and the eigenvalues are used as the next set of shifts. Here, we use the same initialization. For the update step, however, we extend the subspace to all the V_i generated with the current set of shifts and then choose the next shifts following Penzl's heuristic with the ritzvalues replaced by the projected eigenvalues computed with respect to this larger subspace. Note that in lack of the conditions for Bendixon's theorem, we cannot guarantee that the projected eigenvalues will be contained in $\mathbb{C}^-$ (left complex half-plane). Should any of them end up in the wrong half-plane, we neglect them. Note further that the problem with the infinite eigenvalues does not exist in this case. Since we have $\Pi^T Z = Z$, for any orthogonal basis U of a set of columns of Z, we also have $\Pi^T U = U$. As an immediate result of this fact, the projected pencil $(U^T A_1 U - \lambda U^T E_1 U)$ automatically resides on the hidden manifold and can thus only have finite eigenvalues.

8.5 Interpolatory projection method via IRKA

Analogous to the balanced truncation, we can apply the iterative rational Krylov algorithm (IRKA) as introduced in Chapter 4 on system (8.2). Interpolatory projection method for structured index 2 DAE system is also discussed in [89]. Once the index 2 system (8.2) is converted into an ODE system (8.13), Algorithm 21 can be applied for the model reduction of the ODE system. But, due to the complexity, we cannot form the ODE system explicitly. In Section 8.2 we have seen that the dynamical systems (8.2), (8.9) and (8.12) are equivalent. Hence it can be shown that instead of applying IRKA onto system (8.12), we can apply it onto the projected system (8.9). This section shows how to reformulate Algorithm 21 for the model reduction of (8.9) without forming the projector Π explicitly. To follow Algorithm 21 for the model reduction of the projected system (8.9), at each iteration we need to construct the right and left transformation matrices:

$$V = \left[(\alpha_1 \tilde{E} - \tilde{A})^{-1}\tilde{B}b_1, \cdots, (\alpha_r \tilde{E} - \tilde{A})^{-1}\tilde{B}b_r\right], \tag{8.29a}$$

$$W = \left[(\alpha_1 \tilde{E} - \tilde{A})^{-T}\tilde{C}^T c_1, \cdots, (\alpha_r \tilde{E} - \tilde{A})^{-T}\tilde{C}^T c_r\right], \tag{8.29b}$$

where

$$\tilde{E} = \Pi E_1 \Pi^T,\ \tilde{A} = \Pi A_1 \Pi^T, \tilde{B} = \Pi B_1 \quad \text{and } \tilde{C} = C\Pi^T.$$

A close observation reveals that at each column of V, we have to solve a shifted linear system like

$$(\alpha \tilde{E} - \tilde{A})v = \tilde{B}b, \tag{8.30}$$

which is equivalent to

$$\Pi(\alpha E_1 - A_1)\Pi^T v = \Pi B_1 b.$$

Recalling Theorem 17, instead of solving the above linear system we can solve the linear system

$$\begin{bmatrix} \alpha E_1 - A_1 & A_2 \\ A_2^T & 0 \end{bmatrix} \begin{bmatrix} v \\ \Lambda \end{bmatrix} = \begin{bmatrix} B_1 b \\ 0 \end{bmatrix}, \tag{8.31}$$

for v.

> Note that although the linear system in (8.31) has a larger dimension than that of (8.30), it is sparse and can be solved efficiently by suitable direct [58] or iterative solvers [143].

Analogously, when we construct the left transformation matrix W in (8.29), at each column we need to solve a linear system like $(\alpha\tilde{E}^T - \tilde{A}^T)w = \tilde{C}^T c$. Instead, we solve the linear system

$$\begin{bmatrix} \alpha E_1^T - A_1^T & A_2 \\ A_2^T & 0 \end{bmatrix} \begin{bmatrix} w \\ \Lambda \end{bmatrix} = \begin{bmatrix} C^T c \\ 0 \end{bmatrix}, \tag{8.32}$$

for w. In this way, the transformations V and W are constructed without using Π. Applying the transformations into the projected system (8.9), the reduced model (8.17) is formed; the reduced matrices are constructed as

$$\hat{E} = W^T \Pi E_1 \Pi^T V, \ \ \hat{A} = W^T \Pi A_1 \Pi^T V, \ \ \hat{B} = W^T \Pi B_1 \ \text{and} \ \hat{C} = C\Pi^T V.$$

Due to (8.16) we can avoid the explicit usage of Π and find

$$\hat{E} = W^T E_1 V, \ \ \hat{A} = W^T A_1 V, \ \ \hat{B} = W^T B_1 \ \text{and} \ \hat{C} = CV. \tag{8.33}$$

The whole procedure to obtain the reduced ODE system (8.17), for a given index 2 descriptor system (8.1), is shown in Algorithm 32.

8.6 Numerical results

For assessing the accuracy and efficiency of the proposed (model reduction) methods discussed in this chapter, this section presents some numerical results. For the numerical experiments, we consider the semi-discretized linearized stokes model from [160]. The model is briefly introduced in Appendix A.4.4.

Algorithm 32: IRKA for index 2 DAE system.

Input : E_1, A_1, A_2, B_1, C .

Output: $\hat{E}$, $\hat{A}$, $\hat{B}$, $\hat{C}$.

1 Select initial interpolation points $\{\sigma_i\}_{i=1}^r$ and tangent directions $\{b_i\}_{i=1}^r$ and $\{c_i\}_{i=1}^r$.

2 **for** $i = 1, 2, \cdots, r$ **do**

3 Solve the linear systems $\begin{bmatrix} \alpha E_1 - A_1 & A_2 \\ A_2^T & 0 \end{bmatrix} \begin{bmatrix} v_i \\ \Lambda \end{bmatrix} = \begin{bmatrix} B_1 b \\ 0 \end{bmatrix}$, for v and $\begin{bmatrix} \alpha E_1^T - A_1^T & A_2 \\ A_2^T & 0 \end{bmatrix} \begin{bmatrix} w_i \\ \Lambda \end{bmatrix} = \begin{bmatrix} C^T c \\ 0 \end{bmatrix}$, for w.

4 Construct

$$V = \begin{bmatrix} v_1, v_1, \cdots, v_r \end{bmatrix}, \quad W = \begin{bmatrix} w_1, w_1, \cdots, w_r \end{bmatrix},$$

5 **while** *(not converged)* **do**

6 Form $\hat{E}$, $\hat{A}$, $\hat{B}$ and $\hat{C}$ as in (8.33)

7 Compute $\hat{A}z_i = \lambda_i \hat{E} z_i$ and $y^* \hat{A} = \lambda_i y^* \hat{E}$.

8 $\sigma_i \leftarrow -\lambda_i$, $b_i \leftarrow y^* \hat{B}$ and $c_i \leftarrow \hat{C} z_i$.

9 Repeat Step 2 - 4.

10 $i = i + 1$

11 Form the reduced matrices as in (8.33).

All the results are obtained by using MATLAB R2015a (8.5.0.197613) on a board with processor 4×Intel® Core™i5-4460 CPUs with a 2.90 GHz clock speed and 16 GB RAM.

First we compute the low-rank controllability and observability Gramian factors (i.e., $\tilde{R}$ and $\tilde{L}$) by solving the Lyapunov equations as in (8.14). We apply Algorithm 31 using the adaptive shift parameters for computing the Gramian factors. The algorithm takes 24 and 25 iteration steps for computing low-rank controllability and observability Gramian factors, respectively. The rates of convergence for computing both Gramian factors are shown in Figure 8.1.

When the Gramian factors are used in Algorithm 30, we obtain a 32 dimensional reduced model with a 10^{-5} truncation tolerance. We have also computed a 32 dimensional reduced model via IRKA for the same model. The performance of the BT and the IRKA is shown in Figure 8.2. In both cases the transfer functions (as shown in Figure 8.2) of the reduced models are matching accurately with the original model. The absolute and relative errors are shown in Figures 8.2 and 8.2, respectively. From the (absolute and relative) errors we can observe that the BT method yields a better approximate for the reduced model than IRKA.

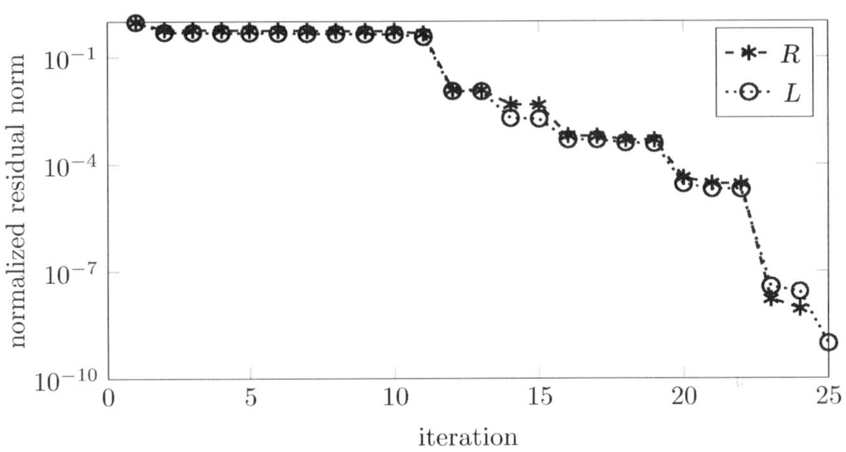

Figure 8.1: Convergence rate of low-rank controllability and observability Gramian factors.

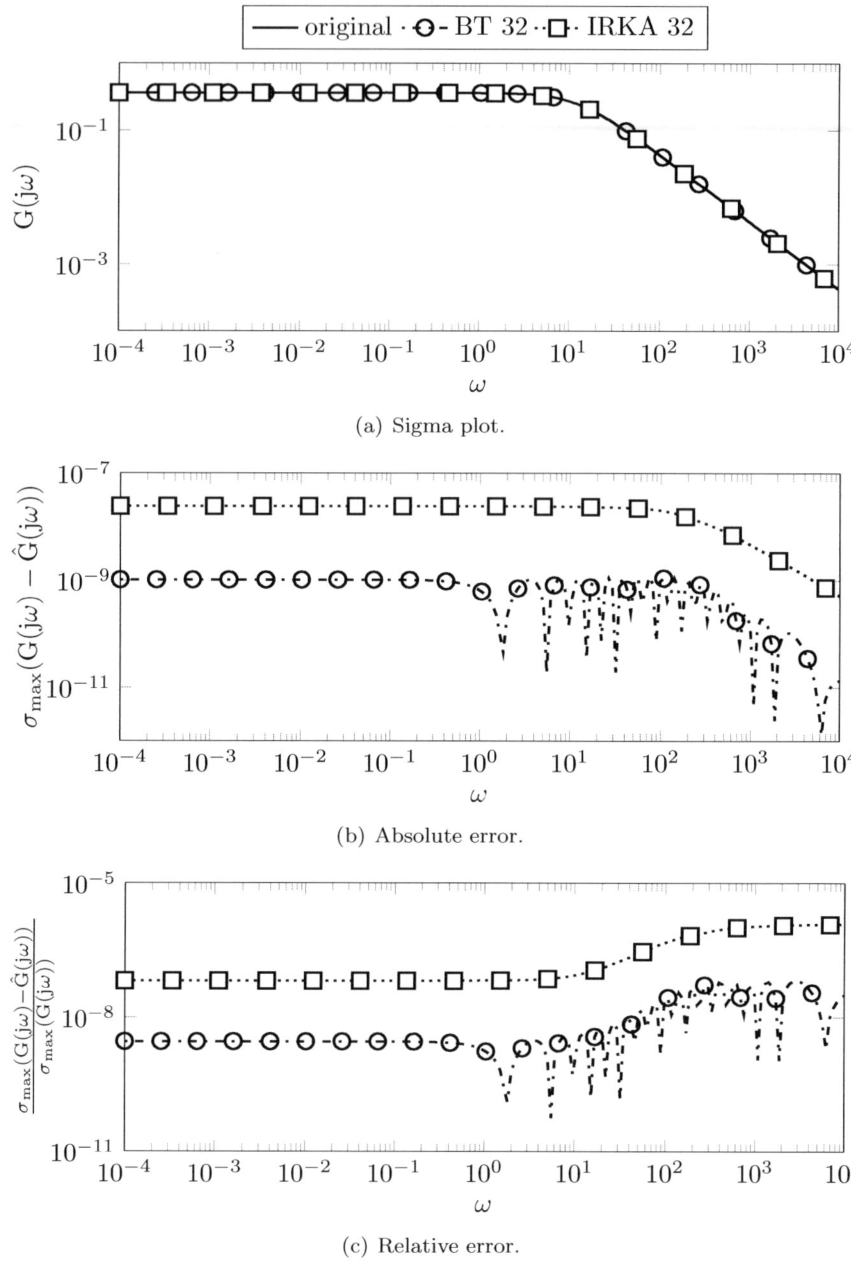

(a) Sigma plot.

(b) Absolute error.

(c) Relative error.

Figure 8.2: Comparison between BT and IRKA based model reduction methods using stokes model.

Exercises:

8.1 For the first-order index 2 system (8.1) the $\Pi-$projector is defined by $\Pi = I - A_2(A_2^T E_1^{-1} A_2)^{-1} A_2^T E_1^{-1} A_1$. For the Semi-discretized linearized Naviar stokes model (SLNSM) as in Appendix A.4.3, construct Π and then verify that it satisfies (i) $\Pi^2 = \Pi$ and (ii) $\Pi E_1 = E_1 \Pi^T$.

8.2 Show that the matrix pencil $\mathcal{P}(\lambda) = \lambda E - A$ system (8.1) is regular iff A_2 has full (column) rank.

8.3 The number of finite eigenvalues of system (8.1) is equal to $n - 2n_1$ where n is the dimension of the system and n_1 is the number of algebraic equations that describe the system. Using the MATLAB command `eig` find the eigenvalues of the model SLNSM as in Appendix A.4.3 and verify this identity.

8.4 For the Semi-discretized linearized Naviar stokes model (SLNSM) as in Appendix A.4.3, compute Π-projected system $(\Pi E_1 \Pi^T, \Pi A_1 \Pi^T, \Pi B_1, C_1 \Pi^T)$ and Θ-projected system $(\Theta_r^T E_1 \Theta_r, \Theta_r^T A_1 \Theta_r, \Theta_r^T B_1, C_1 \Theta_r)$. Then show that the frequency response and step response of both projected systems are the same as the original system.

8.5 Apply the balanced truncation to the Π-projected system (avoiding the projector as discussed in this chapter) and the $\Theta-$projected system computed in Exercise 8.4 to the system SLNSM and compare the computational time between the two systems.

8.6 Apply Algorithm 31 to the system SLNSM for computing low-rank (controllability and observability) Gramian factors using both the heuristic and the adaptive shift parameters; show the performance of both the shift computational strategies in terms of convergence rate and computational time.

8.7 Following the discussion in Section 8.2, construct a Π and Θ projected system for the given system

$$\left(\begin{bmatrix} E_1 & 0 \\ 0 & 0 \end{bmatrix}, \begin{bmatrix} A_1 & A_2 \\ A_3 & 0 \end{bmatrix}, \begin{bmatrix} B_1 \\ B_2 \end{bmatrix}, \begin{bmatrix} C_1 & C_2 \end{bmatrix} \right).$$

8.8 Write the Lyapunov equations for the Π projected systems constructed in Exercise 8.7. Then generalize the LR-ADI iteration for solving the controllability (or observability) Lyapunov equation of Π projected system.

8.9 Develop balanced truncation and IRKA-based algorithms for the model mentioned in Exercise 8.7. If possible, construct an artificial data to verify the efficiency of the algorithms.

Chapter 9

Model Reduction of First-Order Index 2 Unstable Descriptor Systems

9.1 Introduction

In this chapter we study model reduction of a class of structured index 2 descriptor systems [41, 93] of the form

$$\underbrace{\begin{bmatrix} E_1 & 0 \\ 0 & 0 \end{bmatrix}}_{\breve{E}} \begin{bmatrix} \dot{v}(t) \\ \dot{p}(t) \end{bmatrix} = \underbrace{\begin{bmatrix} A_1 & A_2 \\ A_2^T & 0 \end{bmatrix}}_{\breve{A}} \begin{bmatrix} v(t) \\ p(t) \end{bmatrix} + \underbrace{\begin{bmatrix} B_1 \\ 0 \end{bmatrix}}_{\breve{B}} u(t), \tag{9.1a}$$

$$y(t) = \underbrace{\begin{bmatrix} C_1 & 0 \end{bmatrix}}_{\check{C}} \begin{bmatrix} v(t) \\ p(t) \end{bmatrix}, \tag{9.1b}$$

where $v(t) \in \mathbb{R}^{n_1}$, $p(t) \in \mathbb{R}^{n_2}$ ($n_1 > n_2$) are the states, $u \in \mathbb{R}^m$ are the inputs and $y \in \mathbb{R}^p$ are the outputs and in which $\bar{E}$, $\bar{A}$, $\bar{B}$ and $\bar{C}$ are all sparse matrices with appropriate dimensions. We assume that some of the eigenvalues of the matrix pencil, $\lambda\bar{E} - \bar{A}$ lie in $\mathbb{C}^+$ which makes the system (9.1) unstable. Such models arise, for instance, from a spatial discretization of the Navier stokes equations with a moderate Reynolds number ($Re \geq 300$) using the finite element method (see Appendix A.4.3 for details).

If the model becomes very large simulation, control and optimization of such large-scale systems are very expensive. Therefore, we want to reduce the complexity of the model by applying model order reduction (MOR), i.e., we seek an approximation to the original model that well approximates the behavior of the original model but which is much faster to evaluate.

In this chapter we mainly focus on balancing-based model reduction techniques for the system in (9.1). To obtain the IRKA-based reduced models, we can directly follow the approaches as discussed in Chapter 8 because IRKA does not rely on the stability of the system. In principle, one can apply the balanced truncation technique to this model by stabilizing the system first using a proper stabilizing feedback matrix (SFM) and then following the approach in Chapter 8. But here we apply the balancing and truncating transformations computed with respect to the stabilized system to the original unstable system. In order to perform the balanced truncation, we need to compute the low-rank controllability and observability Gramian factors by solving two projected algebraic Lyapunov equations of the stabilized system. This chapter also updates the LR-ADI algorithm to solve the projected Lyapunov equations for the Bernoulli stabilized system. Moreover, here we show that a Riccati-based boundary feedback stabilization matrix [15] for the original model can be computed using a reduced order model.

The proposed method is applied to data for a spatially FEM semi-discretized linearized Navier stokes model. Numerical results are discussed for both the BT model reduction as well as the reduced-order model-based stabilization.

9.2 BT for unstable systems

In Chapter 4 we have not introduced the idea of balanced truncation for an unstable generalized state space system. Therefore, in this section we concentrate on BT for unstable systems

$$\begin{aligned} E\dot{x}(t) &= Ax(t) + Bu(t), \\ y(t) &= Cx(t), \end{aligned} \tag{9.2}$$

via *Bernoulli stabilization.* All the matrices and vectors are defined in (4.1). The helpful feature of our investigated example is that the number of anti-stable eigenvalues is still very small. This is exactly the property we exploit for fast computation of the ROMs and ROM-based feedback matrices. In Chapter 4 we have recalled classical (Lyapunov-based) balancing for stable systems. The main ingredients there are the two Gramians (e.g., [7])

$$\begin{aligned} P &= \int_0^\infty e^{E^{-1}At} E^{-1}BB^TE^{-T} e^{A^TE^{-T}t}\,dt, \\ Q &= \int_0^\infty e^{(E^{-1}A)^Tt} E^{-T}C^TCE^{-1} e^{E^{-1}At}\,dt, \end{aligned}$$

which obviously do not exist if the system's unstable poles are controllable; this is in fact the desired case in our motivating example. In [194], the authors use the frequency domain representations of these integrals

$$P = \frac{1}{2\pi}\int_{-\infty}^{\infty}(\imath\omega E - A)^{-1}BB^T(-\imath\omega E^T - A^T)^{-1}\,d\omega$$
$$Q = \frac{1}{2\pi}E^T\{\int_{-\infty}^{\infty}(\imath\omega E^T - A^T)^{-1}C^TC(-\imath\omega E - A)^{-1}\,d\omega\}E$$

to extend the definition of the Gramians to systems with no poles on the imaginary axis.

Following the theory in [194], the generalized controllability and observability Gramians P_s, Q_s for such systems can be computed by solving the algebraic Lyapunov equations

$$\begin{aligned}(A - B\mathcal{K}_c^{fm})P_sE^T + EP_s(A - B\mathcal{K}_c^{fm})^T &= -BB^T,\\ (A - \mathcal{K}_o^{fm}C)^TQ_sE + E^TQ_s(A - \mathcal{K}_o^{fm}C) &= -C^TC,\end{aligned} \tag{9.3}$$

where $\mathcal{K}_c^{fm} = B^TX_cE$ and $\mathcal{K}_o^{fm} = EX_oC^T$ are called Bernoulli stabilizing feedback matrices; this is due to the fact that the matrices X_c and X_o are the respective stabilizing solutions of the generalized algebraic Bernoulli equations

$$\begin{aligned}E^TX_cA + A^TX_cE &= E^TX_cBB^TX_cE,\\ AX_oE^T + EX_oA^T &= EX_oC^TCX_oE^T.\end{aligned} \tag{9.4}$$

Since the Bernoulli stabilization only mirrors the anti-stable eigenvalues across the imaginary axis, it is sufficient to solve these Bernoulli equations only on the invariant subspaces corresponding to those eigenvalues. That is, for orthogonal matrices T_c, $T_o \in \mathbb{R}^{n\times l}$ respectively spanning the left and right eigenspaces corresponding to the anti-stable eigenvalues, we define the Petrov-Galerkin projected system $(\breve{E}, \breve{A}, \breve{B}, \breve{C})$ by

$$\breve{E} := T_o^TET_c, \quad \breve{A} := T_o^TAT_c, \quad \breve{B} := T_o^TB, \quad \breve{C} := CT_c,$$

where $\breve{E}$, $\breve{A} \in \mathbb{R}^{l\times l}$, $\breve{B} \in \mathbb{R}^{l\times m}$, and $\breve{C} \in \mathbb{R}^{p\times l}$. The size of these projected matrices is very small, since we have considered a few anti-stable eigenvalues. Therefore, we solve very small-sized projected Bernoulli equations

$$\begin{aligned}\breve{E}^T\breve{X}_c\breve{A} + \breve{A}^T\breve{X}_c\breve{E} &= \breve{E}^T\breve{X}_c\breve{B}\breve{B}^T\breve{X}_c\breve{E},\\ \breve{A}\breve{X}_o\breve{E}^T + \breve{E}\breve{X}_o\breve{A}^T &= \breve{E}\breve{X}_o\breve{C}^T\breve{C}\breve{X}_oE^T,\end{aligned} \tag{9.5}$$

and construct $\mathcal{K}_c^{fm} = B^TT_c\breve{X}_cT_c^TE$ and $\mathcal{K}_o^{fm} = ET_o^T\breve{X}_oT_oC^T$. The projected Bernoulli equations in (9.5) can be solved by the *Matrix Sign Function* method presented in Barrachina et al. [17].

The low-rank factors of P_s and Q_s can also be computed by solving (9.3) using the LR-ADI iteration introduced in Chapter 3, but to avoid forming the closed-loop matrices stays crucial. We discuss this issue in more detail in Section 9.4. Using these Gramian factors from the balancing and truncating transformations and applying them to the original unstable system, we can compute an unstable ROM that satisfies the error bound as in (4.27) but with the $\mathcal{H}_\infty$-norm replaced by the $\mathcal{L}_\infty$-norm. Therefore, this error bound cannot be translated to a global time domain error bound, as in the classic setting, due to the lack of a Parseval-identity-like result. In fact, in the numerical experiments we observed that the error may very well grow over time.

9.3 BT for index 2 unstable descriptor systems

To apply the balancing-based model reduction to system (9.1), first we convert the system into an equivalent ODE system in order to make it fit into the framework for the BT-based model order reduction as discussed in the previous section. Recalling the strategy from Chapter 8 (Section 8.2), let us consider a projector of the form

$$\Pi_2 = I_{n_1} - A_2(A_2^T E_1^{-1} A_2)^{-1} A_2^T E_1^{-1}, \tag{9.6}$$

which satisfies $\mathrm{Null}\,(\Pi_2) = \mathrm{Range}\,(A_2)$, $\mathrm{Range}\,(\Pi_2) = \mathrm{Null}\left(A_2^T E_1^{-1}\right)$ and $\Pi_2 E_1 = E_1 \Pi_2^T$. These properties imply

$$A_2^T Y = 0 \quad \text{if and only if} \quad \Pi_2^T Y = Y, \tag{9.7}$$

i.e., the image of Π_2^T is exactly the subspace where the algebraic condition of the DAE system is satisfied. Now we apply the projector to (9.1) and by exploiting property (9.7) we obtain the following projected system

$$\Pi_2 E_1 \Pi_2^T \dot{v}(t) = \Pi_2 A_1 \Pi_2^T v(t) + \Pi_2 B_1 u(t), \tag{9.8a}$$

$$y(t) = C_1 \Pi_2^T v(t). \tag{9.8b}$$

The system dynamics of (9.8) are projected onto the $m_1 := n_1 - n_2$ dimensional subspace $\mathrm{Range}\left(\Pi_2^T\right)$ [93]. This subspace is still, however, represented in the coordinates of the n_1 dimensional space. The m_1 dimensional representation can be made explicit by employing the *thin* singular value decomposition (SVD)

$$\Pi_2 = \begin{bmatrix} U_1 & U_2 \end{bmatrix} \begin{bmatrix} S_1 & 0 \\ 0 & 0 \end{bmatrix} \begin{bmatrix} V_1^T \\ V_2^T \end{bmatrix} = U_1 S_1 V_1^T = \Theta_{2,l} \Theta_{2,r}^T, \tag{9.9}$$

where $\Theta_{2,l} = U_1$ and $\Theta_{2,r} = V_1$ and in which $U_1,\ V_1 \in \mathbb{R}^{n_1 \times m_1}$ consist of the corresponding leading m_1 columns of $U,\ V \in \mathbb{R}^{n_1 \times n_1}$. Moreover, $\Theta_{2,l}, \Theta_{2,r}$ satisfy

$$\Theta_{2,l}^T \Theta_{2,r} = I_{m_1}. \tag{9.10}$$

This representation is always possible since Π_2 is a projector and therefore, $S_1 = I_{m_1}$. Inserting the decomposition of Π_2 as in (9.9) into (9.8) and considering $\tilde{v}(t) = \Theta_{2,l}^T v(t)$, we get

$$\begin{aligned} \Theta_{2,r}^T E_1 \Theta_{2,r} \dot{\tilde{v}}(t) &= \Theta_{2,r}^T A_1 \Theta_{2,r} \tilde{v}(t) + \Theta_{2,r}^T B_1 u(t), \\ y(t) &= C_1 \Theta_{2,r} \tilde{v}(t). \end{aligned} \tag{9.11}$$

System (9.11) is practically system (9.2) with the redundant equations removed by the $\Theta_{2,r}$ projection. We observe that the dynamical systems (9.1), (9.8) and (9.11) are equivalent in the sense that their finite spectrum is the same [64, Theorem 2.7.3] and the input-output relations are the same, i.e., they realize the same transfer function. In the text that follows, we will discuss how to avoid forming (9.11) explicitly to perform the model reduction.

Suppose that we want to apply balanced truncation to system (9.11). To this end, we need to solve the Lyapunov equations

$$\begin{aligned} \Theta_{2,r}^T A_c \Theta_{2,r} \tilde{P} \Theta_{2,r}^T E_1^T \Theta_{2,r} + \Theta_{2,r}^T E_1 \Theta_{2,r} \tilde{P} \Theta_{2,r}^T A_c^T \Theta_{2,r} &= -\Theta_{2,r}^T B_1 B_1^T \Theta_{2,r}, \\ \Theta_{2,r}^T A_o^T \Theta_{2,r} \tilde{Q} \Theta_{2,r}^T E_1 \Theta_{2,r} + \Theta_{2,r}^T E_1^T \Theta_{2,r} \tilde{Q} \Theta_{2,r}^T A_o \Theta_{2,r} &= -\Theta_{2,r}^T C_1^T C_1 \Theta_{2,r}, \end{aligned} \tag{9.12}$$

where $A_c = A_1 - B_1 K_c^{fm}$, $A_o = A_1 - K_o^{fm} C_1$ and $\tilde{P} \in \mathbb{R}^{m_1 \times m_1}$, $\tilde{Q} \in \mathbb{R}^{m_1 \times m_1}$ are the corresponding projected controllability and observability Gramians. Again, K_c^{fm} and K_o^{fm} are the Bernoulli stabilizing feedback matrices and can be computed as described in Section 9.2. The solutions $\tilde{P}$, $\tilde{Q}$ of (9.12) are unique since we are assured that the respective dynamical system is asymptotically stable and symmetric positive (semi-)definite since the right-hand side is semi-definite.

By multiplying (9.12) with $\Theta_{2,l}$ from the left and $\Theta_{2,l}^T$ from the right and exploiting the fact that $\Theta_{2,r} = \Pi_2^T \Theta_{2,r}$ (e.g., due to (9.9) and (9.10)) we obtain

$$\begin{aligned} \Pi_2 A_c \Pi_2^T P \Pi_2 E_1^T \Pi_2^T + \Pi_2 E_1 \Pi_2^T P \Pi_2 A_c^T \Pi_2^T &= -\Pi_2 B_1 B_1^T \Pi_2^T, \\ \Pi_2 A_o^T \Pi_2^T Q \Pi_2 E_1 \Pi_2^T + \Pi_2 E_1^T \Pi_2^T Q \Pi_2 A_o \Pi_2^T &= -\Pi_2 C_1^T C_1 \Pi_2^T, \end{aligned} \tag{9.13}$$

where $P = \Theta_{2,r} \tilde{P} \Theta_{2,r}^T$ and $Q = \Theta_{2,r} \tilde{Q} \Theta_{2,r}^T$. These are the respective controllability and observability Lyapunov equations for realization (9.8) and the solutions P, $Q \in \mathbb{R}^{n_1 \times n_1}$ are the corresponding controllability and observability Gramians. System (9.13) is singular due to the fact that Π_2 is a projector. Uniqueness of solutions is reestablished by the condition that the solutions satisfy $P = \Pi_2^T P \Pi_2$ and $Q = \Pi_2^T Q \Pi_2$.

It is also an easy exercise to go back to (9.12) from (9.13). Let us consider $P \approx RR^T$, $Q \approx LL^T$ and $\tilde{P} \approx \tilde{R}\tilde{R}^T$, $\tilde{Q} \approx \tilde{L}\tilde{L}^T$. Then R, L, $\tilde{R}$ and $\tilde{L}$ are called approximate low-rank factors. They fulfill the relation

$$R = \Theta_{2,r} \tilde{R} \qquad \text{and} \qquad L = \Theta_{2,r} \tilde{L}.$$

For large-scale problems, however, computing $\Theta_{2,r}$ is usually impossible due to memory limitations. Therefore, R and L are computed by solving (9.13). The balancing and truncating transformations for (9.11) are

$$\tilde{W} = \tilde{R}U_k\Sigma_k^{-\frac{1}{2}}, \quad \tilde{V} = \tilde{L}V_k\Sigma_k^{-\frac{1}{2}},$$

where $U_k, V_k \in \mathbb{R}^{n_m \times k}$ consist of the corresponding leading k columns of $U, V \in \mathbb{R}^{n_m \times n_m}$, and $\Sigma_k \in \mathbb{R}^{k \times k}$ is the upper left $k \times k$ block of Σ in the SVD

$$\tilde{R}^T \Theta_r^T E_1 \Theta_r \tilde{L} = U\Sigma V^T.$$

Observing further that $R^T \Pi_2 E_1 \Pi_2^T L = \tilde{R}^T \Theta_{2,r}^T E_1 \Theta_{2,r} \tilde{L} = U\Sigma V^T$, the projection matrices for the system (9.8) can be formed as

$$W = RU_k\Sigma_k^{-\frac{1}{2}} \quad \text{and} \quad V = LV_k\Sigma_k^{-\frac{1}{2}}. \tag{9.14}$$

As in [93] we find that

$$\begin{aligned} W &= RU_k\Sigma_k^{-\frac{1}{2}} = \Theta_{2,r}\tilde{R}U_k\Sigma_k^{-\frac{1}{2}} = \Theta_{2,r}\tilde{W} = \Theta_{2,r}\Theta_{2,l}^T\Theta_{2,r}\tilde{W} = \Pi_2^T W, \\ V &= LV_k\Sigma_k^{-\frac{1}{2}} = \Theta_{2,r}\tilde{L}V_k\Sigma_k^{-\frac{1}{2}} = \Theta_{2,r}\tilde{V} = \Theta_{2,r}\Theta_{2,l}^T\Theta_{2,r}\tilde{V} = \Pi_2^T V. \end{aligned} \tag{9.15}$$

Now we apply the transformations W and V in (9.8) to find the reduced order model as

$$\begin{aligned} \hat{E}\dot{\hat{v}}(t) &= \hat{A}\hat{v}(t) + \hat{B}u(t) \\ \hat{y}(t) &= \hat{C}\hat{v}(t), \end{aligned} \tag{9.16}$$

where

$$\hat{E} = W^T \Pi_2 E_1 \Pi_2^T V, \;\; \hat{A} = W^T \Pi_2 A_1 \Pi_2^T V, \;\; \hat{B} = W^T \Pi_2 B_1 \;\text{ and }\; \hat{C} = C_1 \Pi_2^T V.$$

Due to (9.15) we can avoid the explicit usage of Π_2 and find

$$I = \hat{E} = W^T E_1 V, \;\; \hat{A} = W^T A_1 V, \;\; \hat{B} = W^T B_1 \;\text{ and }\; \hat{C} = C_1 V.$$

Eventually, we see that the reduced order model (9.16) is obtained without forming the projected system (9.8). In the next section, we will show how to compute R and L using a tailored version of the LR-ADI iteration without using Π_2 explicitly. The above procedure to compute the ROM for the unstable index 2 DAE system is summarized in Algorithm 33.

9.4 Solution of the projected Lyapunov equations

In order to apply the aforementioned balancing-based MOR, we need to solve the projected Lyapunov equations (9.13). We have seen that the Π_2^T

Algorithm 33: LR-SRM for unstable index 2 DAEs.

Input : E_1, A_1, B_1, C_1 from (9.1).
Output: $\hat{E}$, $\hat{A}$, $\hat{B}$, $\hat{C}$ in (9.16).
1 Compute R and L by solving the projected Lyapunov equations (9.13).
2 Construct W and V as in (9.15).
3 Form $I = \hat{E} = W^T E_1 V$, $\hat{A} = W^T A_1 V$, $\hat{B} = W^T B_1$ and $\hat{C} = C_1 V$.

---

invariant solution factors enable us to compute the corresponding truncating transformations. The approach here is different from the spectral projection based work by Stykel in that we are applying the E_1-orthogonal projection to the hidden manifold where Stykel used the orthogonal projection (in the Euclidean sense) onto the eigenspaces corresponding to the finite poles of the system. In fact both methods project to the same subspace considering orthogonality in different inner products. Here we are concerned with two main issues. First, we discuss the reformulation of the basic low-rank ADI iteration for the projected Lyapunov equation that ensures the invariance of the solution factor and the computation of the correct corresponding residual factors. We are using the ideas of Chapter 8 to reformulate the LR-ADI iteration. Here, we are concerned with the efficient solution of the Lyapunov equations in (9.13) to compute the low-rank Gramian factors using LR-ADI as discussed in Chapter 3. Here, we consider the projected controllability equation elaborately. The observability equation can be handled analogously. For convenience, we rewrite the Lyapunov equations (9.13) as

$$\begin{aligned} \tilde{A}P\tilde{E}^T + \tilde{E}P\tilde{A}^T &= -\tilde{B}\tilde{B}^T, \\ \tilde{A}^T Q\tilde{E} + \tilde{E}^T Q\tilde{A} &= -\tilde{C}^T\tilde{C}, \end{aligned} \tag{9.17}$$

with $\tilde{E} = \Pi_2 E_1 \Pi_2^T$, $\tilde{A} = \Pi_2 A_c \Pi_2^T$, $\tilde{B} = \Pi_2 B_1$ and $\tilde{C} = C_1 \Pi_2^T$.

In the i-th iteration step of the ADI, the residual of the controllability Lyapunov equation (9.17) can be written as

$$\mathcal{R}(P_i) = \tilde{A}P_i\tilde{E}^T + \tilde{E}P_i\tilde{A}^T + \tilde{B}\tilde{B}^T = \tilde{W}_i\tilde{W}_i^T,$$

where

$$\tilde{W}_i = \prod_{k=1}^{i}(\tilde{A} - \overline{\mu}_i\tilde{E})(\tilde{A} + \mu_i\tilde{E})^{-1}\tilde{B}.$$

To compute the low-rank controllability Gramian factor R, we follow Algorithm 31. In the i-th iteration step, V_i is computed from

$$(\tilde{A} + \mu_i\tilde{E})V_i = \tilde{W}_{i-1}, \tag{9.18}$$

which enables us to update the residual factor according to

$$\tilde{W}_i = (\tilde{A} - \mu^*\tilde{E})V_i = \tilde{W}_{i-1} - 2\,\mathrm{Re}\,(\mu_i)\tilde{E}V_i. \tag{9.19}$$

In complete analogy to Theorem 17 in Chapter 8, we observe that instead of solving (9.18), one can compute V_i by solving

$$\begin{bmatrix} A_c + \mu_i E_1 & A_2 \\ A_2^T & 0 \end{bmatrix} \begin{bmatrix} V_i \\ \star \end{bmatrix} = \begin{bmatrix} \tilde{W}_{i-1} \\ 0 \end{bmatrix}. \tag{9.20}$$

The computation of the initial residual factor $\tilde{W}_0$ (for the special case i = 1) is discussed in detail below.

Inserting $A_c = A_1 - B_1 K_c^{fm}$ in (9.20),

$$\begin{bmatrix} A_1 + \mu_i E_1 - B_1 K_c^{fm} & A_2 \\ A_2^T & 0 \end{bmatrix} \begin{bmatrix} V_i \\ \star \end{bmatrix} = \begin{bmatrix} \tilde{W}_{i-1} \\ 0 \end{bmatrix},$$

implies

$$\left(\underbrace{\begin{bmatrix} A_1 + \mu_i E_1 & A_2 \\ A_2^T & 0 \end{bmatrix}}_{\underline{A}} - \underbrace{\begin{bmatrix} B_1 \\ 0 \end{bmatrix}}_{\underline{B}} \underbrace{\begin{bmatrix} K_c^{fm} & 0 \end{bmatrix}}_{\underline{K}} \right) \begin{bmatrix} V_i \\ \star \end{bmatrix} = \begin{bmatrix} \tilde{W}_{i-1} \\ 0 \end{bmatrix}. \tag{9.21}$$

In this equation, the inversion of $(\underline{A} - \underline{B}\,\underline{K})$ should be computed using the *Sherman-Morrison-Woodbury formula* (SMWF) (see, e.g., [78]):

$$(\underline{A} - \underline{B}\,\underline{K})^{-1} = \underline{A}^{-1} + \underline{A}^{-1}\underline{B}(I - \underline{K}\,\underline{A}^{-1}\underline{B})^{-1}\underline{K}\,\underline{A}^{-1}, \tag{9.22}$$

to avoid the explicit formulation of the (usually dense) matrix $\underline{A} - \underline{B}\,\underline{K}$. In accordance with Theorem 17, the computed V_i in (9.20) satisfies $V_i = \Pi_2^T V_i$. Therefore, the correct projected residual factor in (9.19) can be obtained by

$$\tilde{W}_i = \tilde{W}_{i-1} - 2\,\mathrm{Re}\,(\mu_i) E_1 V_i, \tag{9.23}$$

since we have $\Pi_2 E_1 = E_1 \Pi_2^T$.

In order to really compute the correct residual, the initial residual must be computed as $\tilde{W}_0 = \Pi_2 B_1$ to ensure $\tilde{W}_0 = \Pi_2 \tilde{W}_0$. This can be performed cheaply using Lemma 5 (in Chapter 8) as follows. Solve the linear system

$$\begin{bmatrix} E_1 & A_2 \\ A_2^T & 0 \end{bmatrix} \begin{bmatrix} \Xi \\ \Lambda \end{bmatrix} = \begin{bmatrix} B_1 \\ 0 \end{bmatrix}. \tag{9.24}$$

for Ξ. This $\Xi = E_1^{-1}\Pi_2 B_1$ and thus using $\tilde{W}_0 = E_1 \Xi$, we get the desired initial Lyapunov residual.

The above findings on the residual factor can be summarized as the following lemma.

Lemma 7. *The residual factor in every step of Algorithm 34 fulfills the relation*

$$\tilde{W}_i = \Pi_2 \tilde{W}_i.$$

Algorithm 34: LR-ADI for unstable index 2 DAE system.

Input : E_1, A_1, A_2, B_1, K_c^{fm}, $\{\mu_i\}_{i=1}^J$.
Output: $R = Z_i$, such that $P \approx RR^T$.

Set $Z_0 = [\,]$.
Solve the linear system (9.24) for Ξ and compute $\tilde{W}_0 = E_1\Xi$
$i = 1$
while $\|\tilde{W}_{i-1}^T \tilde{W}_{i-1}\| \geq tol$ *or* $i \leq i_{max}$ **do**
  Solve the linear system (9.21) for V_i.
  if $\mathrm{Im}\,(\mu_i) = 0$ **then**
    $Z_i = \begin{bmatrix} Z_{i-1} & \sqrt{-2\mu_i}V_i \end{bmatrix}$,
    $\tilde{W}_i = \tilde{W}_{i-1} - 2\mu_i E_1 V_i$
  else
    $\gamma = -2\,\mathrm{Re}\,(\mu_i)$, $\delta = \frac{\mathrm{Re}\,(\mu_i)}{\mathrm{Im}\,(\mu_i)}$,
    $Z_{i+1} = \begin{bmatrix} Z_{i-1} & \sqrt{2\gamma}(\mathrm{Re}\,(V_i) + \delta\,\mathrm{Im}\,(V_i)) & \sqrt{2\gamma}\sqrt{(\delta^2+1)}\,\mathrm{Im}\,(V_i) \end{bmatrix}$,
    $\tilde{W}_i = \tilde{W}_{i-1} + 2\gamma E_1(\overline{V_i} + 2\delta\,\mathrm{Im}\,(V_i))$.
    $i = i + 1$
  $i = i + 1$

The whole procedure of computing the low-rank factor of the controllability Gramian $\tilde{R}$ is summarized in Algorithm 34. Selection of appropriate shift parameters is one of the crucial tasks for fast convergence of Algorithm 34. Following the discussion in Chapter 8 (Subsection 8.4.2), we investigate both the *heuristic* and the *adaptive* approaches to compute suboptimal ADI shift parameters μ_i, $i = 1, 2, \ldots, J$ for the underlying index 2 DAE system.

> One can follow the same procedure to develop the LR-ADI algorithm for solving the projected observability Lyapunov equation.

9.5 Riccati-based feedback stabilization from ROM

Stabilization of the nonstationary incompressible Navier stokes equations around a stationary solution using a Riccati-based feedback has received considerable attention regarding theory as well as numerical methods during recent years. In the Riccati-based boundary feedback stabilization technique, the most challenging task is to solve the corresponding algebraic Riccati equation (ARE) for the full dimensional model. The key problem in the linear-quadratic regulator (LQR) approach for the model under investigation is to compute the boundary feedback stabilization matrix K_f (see, e.g., [14]) such that the stabilized system has the following form:

$$\begin{aligned} E_1\dot{v}(t) &= (A_1 - B_1K_f)v(t) + A_2p(t) + B_1u(t), \\ A_2^T v(t) &= 0. \end{aligned} \tag{9.25}$$

The authors in [15] presented an algorithm (see [15, Algorithm 2]) to compute K_f which is based on the standard LQR approach [56, 155] for a projected semi-discretized state space system. The most challenging part of this algorithm is to solve the (usually) very large, generalized projected algebraic Riccati equation (GPARE) based on the full order semi-discretized model. We employ the reduced-order model (9.16) to compute an approximation to the optimal LQR feedback matrix of the full system. The main advantage of this approach is that we only need to solve two projected algebraic Lyapunov equations in order to derive the reduced-order model instead of one Lyapunov equation per Newton step in the solver for the GPARE, which are usually significantly more [43].

Based on the reduced model (9.16), the GPARE

$$\hat{A}^T\hat{X} + \hat{X}\hat{A} - \hat{X}\hat{B}\hat{B}^T\hat{X} = -\hat{C}^T\hat{C} \tag{9.26}$$

is now much smaller in dimension. It can thus easily be solved for $\hat{X}$ using classical solvers like the MATLAB `care` command. The stabilizing feedback matrix for the reduced model (9.16) is then

$$\hat{K}_f = \hat{B}^T\hat{X}.$$

The ROM-based approximation to the SFM for the full order model can now be retrieved as

$$K_f = \hat{B}^T\hat{X}W^TE_1 = \hat{K}_fW^TE_1 \tag{9.27}$$

where W is the left balancing and truncating transformation (see Section 9.3) used to compute the reduced-order model.

9.6 Numerical results

To assess the performance of the techniques, this section presents some numerical tests. The proposed techniques are applied to a data of linearized semi-discretized Navier stokes equations as described in Appendix A.4.3.

All the computations were carried out using MATLAB 7.11.0 (R2010b) on a board with 2 Intel® Xeon® X5650 CPUs with a 2.67-GHz clock speed.

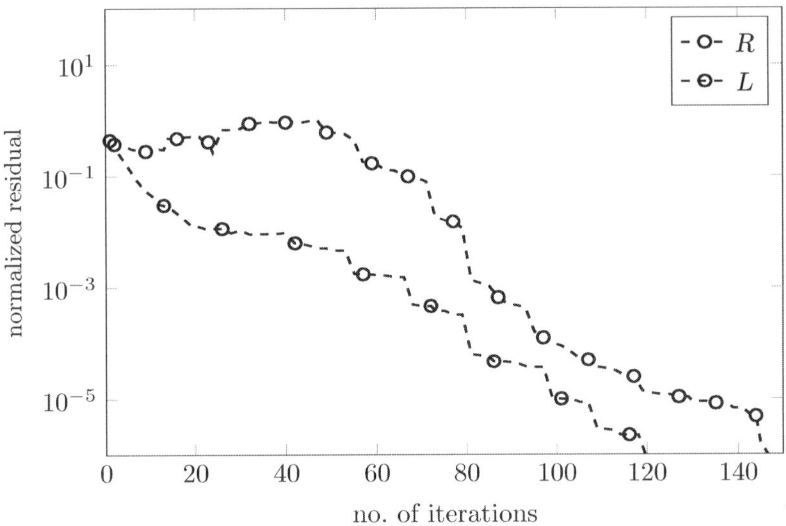

Figure 9.1: Convergence rate of the LR-ADI iteration for computing the low-rank Gramian factors by Algorithm 34.

Table A.2 (in Appendix A) shows different-sized models of linearized semi-discretized Navier stokes equations. The models are generated by the authors of [15] using the Reynolds number $R_e = 500$. For the numerical experiments, we apply the model reduction techniques to the largest model, i.e., SLNSM-5.

Applying the LR-ADI iteration (i.e., Algorithm 34), we compute the low-rank factors $\tilde{R}$ and $\tilde{L}$ considering a tolerance of 10^{-6}. Note that to implement this algorithm, the Bernoulli stabilizing feedback matrices K_c^{fm} and K_o^{fm} are computed by applying the procedure from [4] and [15, Section 2]. Algorithm 34 uses adaptive shifts where each cycle uses 10 proper shift parameters. The shifts are selected by following the procedure discussed above. For computing the initial shifts, first we project the pencil $(A_1 - \lambda E_1)$ onto the column space of a $n_1 \times 100$ random matrix. The convergence rate of the LR-ADI iteration is shown in Figure 9.1 for computing both the Gramian factors. Using the Gramian factors, we apply Algorithm 33 which yields 183 dimensional reduced models with the truncation tolerance 10^{-5}. The frequency domain error analysis is shown in Figure 9.2. In Figure 9.2(a) we see the frequency responses of the full and the 183 dimensional reduced-order models with a good match. The absolute and relative deviations between full and reduced-order models are shown in Figures 9.2(b) and 9.2(c), respectively. Here, we can see that the absolute error is bounded by the prescribed truncation tolerance of 10^{-5}. For higher frequencies, the relative error is slightly increasing since the frequency response is decreasing more rapidly than the absolute error can. Figure 9.3 depicts time domain simulation of the full and reduced-order models. This figure shows the step responses from Input 1 to Output 1 together with

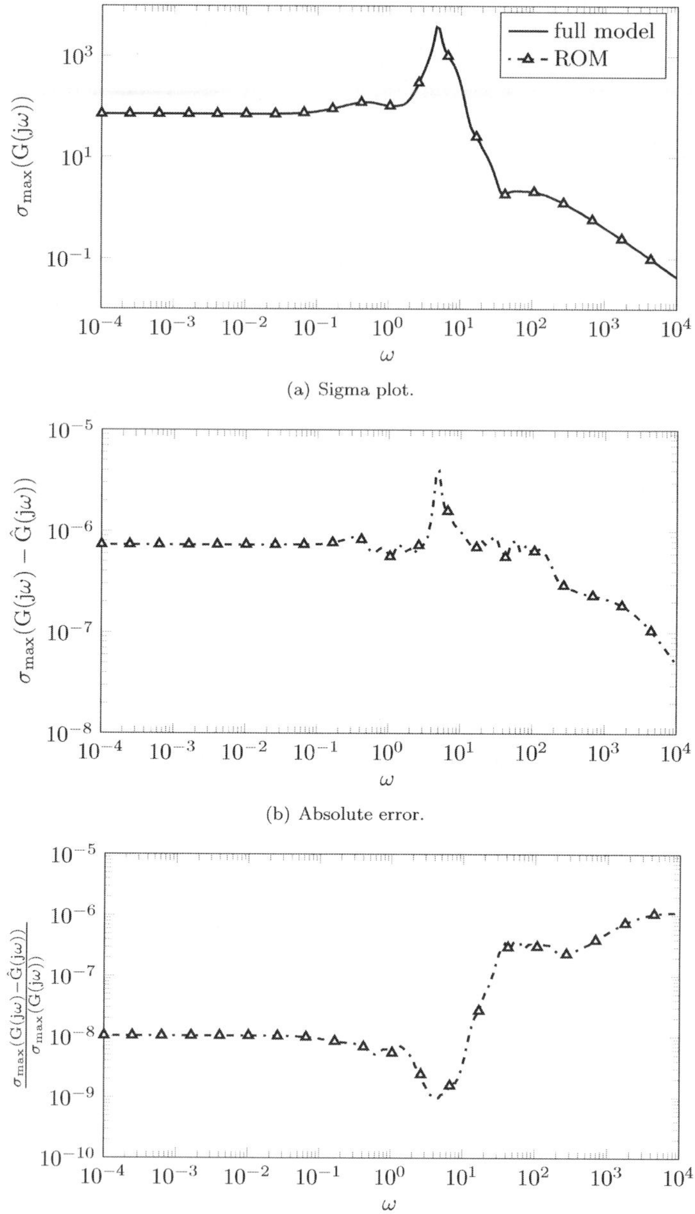

(a) Sigma plot.

(b) Absolute error.

(c) Relative error.

Figure 9.2: Comparison of the full and reduced models in frequency domain.

their absolute deviations. To compute the step response, we use an implicit Euler method with a fixed time step size 10^{-2}. Initially, the imposed control is kept inactive; therefore, the responses for both (full and reduced) models are constant within the range 0 to 15. Switching the control to constant unit actuation on Input 1, the responses are oscillating with increasing amplitude in the higher time domain caused by the instability of the model. Here we also see the issue with the balanced truncation error bound for unstable systems since the absolute error is increasing gradually with increasing time.

Numerical experiments for the stabilized system: In Section 9.5, we mentioned that the stabilizing feedback matrix for the full model can be computed from the reduced order model. To this end, we solve the corresponding algebraic Riccati equation for the reduced order model (9.16) arising from the linear quadratic regulator approach using the MATLAB `care` command and compute the optimal stabilizing feedback matrix $\hat{K}_f$. The ROM based approximation to the stabilizing feedback matrix for the full order model (9.1) is then computed by (9.27). Figure 9.4 shows the step response (from 1st input to 1st output) of closed loop full and reduced order models and their absolute error. For the generation of the step response, the same procedure has been followed as for the unstable case above. Note that for a stabilizing feedback, the step response system has to be viewed as that of an asymptotically stable system with a constant source term. Thus the outputs stabilize at constant non-zero values.

Comparison of BT with interpolatory technique: We compute the 50 dimensional reduced models applying both balanced truncation and the interpolatory method via IRKA. For the interpolatory based approach, we exactly follow Algorithm 32. The quality of the IRKA-based reduced model also depends on the number of cycles (i.e., how many times the interpolation points are updated). Although taking more cycles ensures a better reduced model, it gets more expensive. Here to construct the ROM, we restricted the updates to 10 cycles. Figure 9.5 shows the error comparisons of the sigma plots (as in Figure 9.2(a)) between full system and 50 dimensional reduced systems. From this figure one can notice that in the higher frequency range, the interpolatory method performs better than balanced truncation.

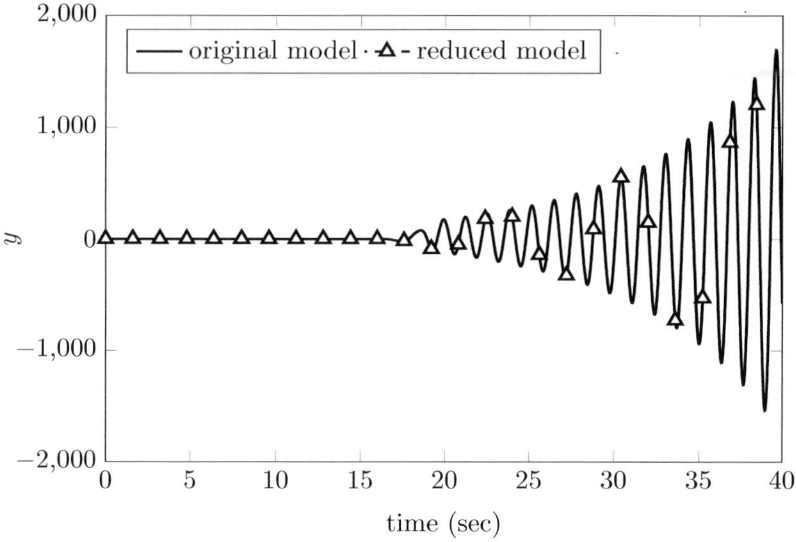

(a) Step response.

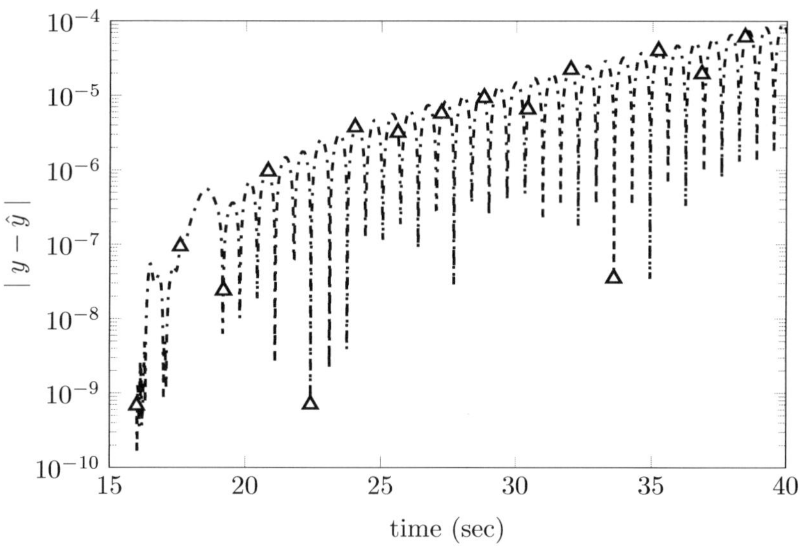

(b) Absolute error.

Figure 9.3: Step responses of 1st input to 1st output of the unstable full and reduced-order models and respective absolute deviations.

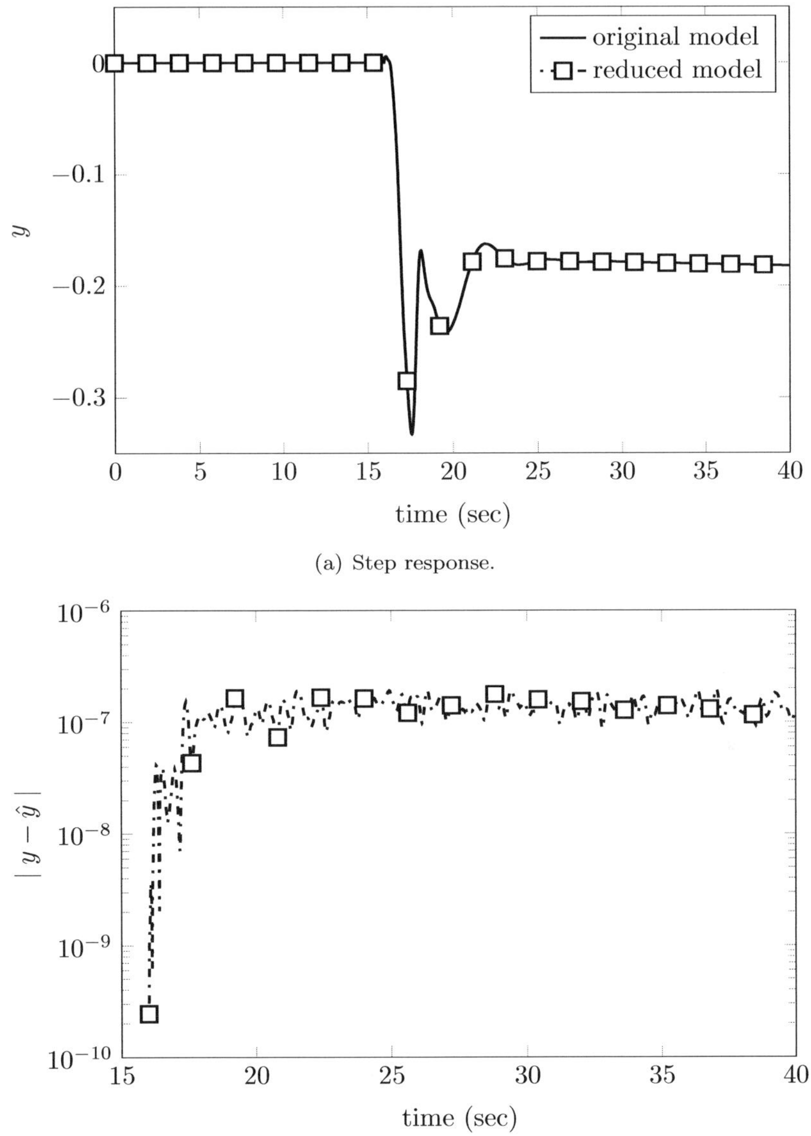

(a) Step response.

(b) Absolute error.

Figure 9.4: Step responses of 1st input to 1st output of stabilized full and reduced-order models and respective absolute deviations.

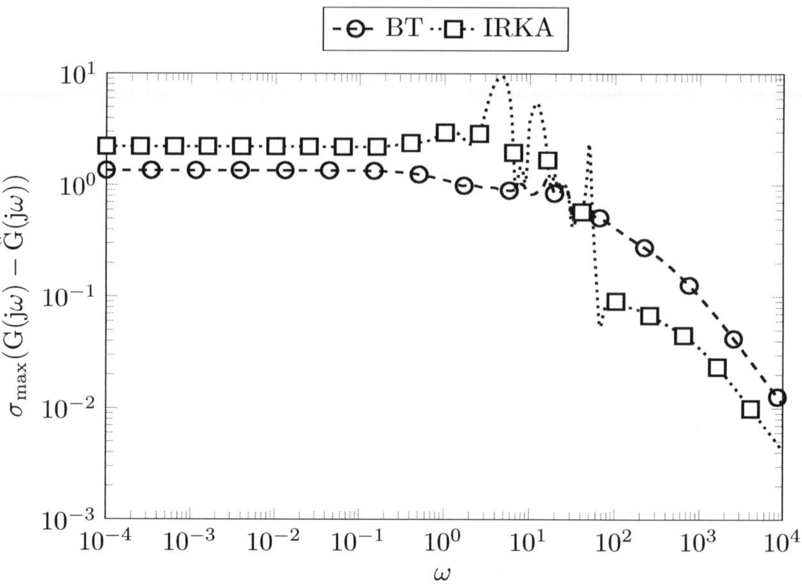

(a) Absolute error.

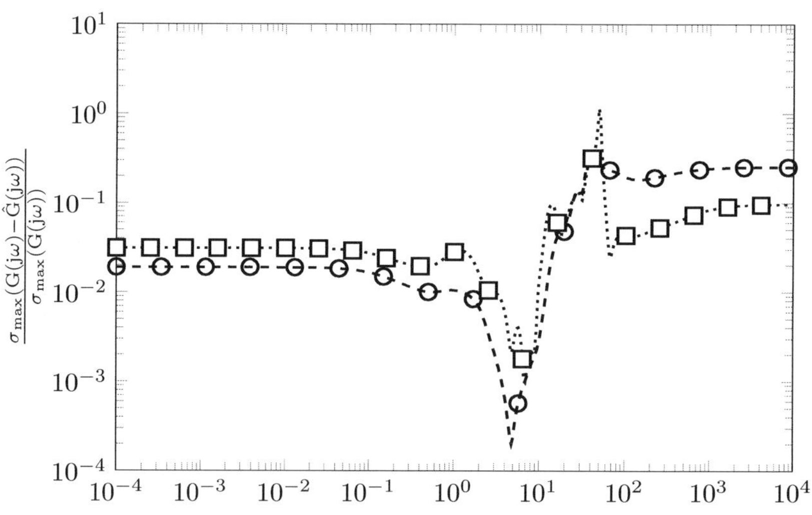

(b) Relative error.

Figure 9.5: Errors between the full and 50 dimensional reduced systems computed by BT and IRKA.

■ Exercises:

9.1 Compute all finite eigenvalues of the unstable semi-discretized linearized Naviar stokes model (USLNSM) as in Appendix A.4.3 and plot them using the MATLAB function `plot`. By applying the feedback matrix K_1, form the stabilized system and compute all finite eigenvalues for the stable system and then plot them on the same figure. How do you interpret the changes of the eigenvalues between the unstable and the stabilized systems?

9.2 For the unstable semi-discretized linearized Naviar stokes model (USLNSM), as in Appendix A.4.3, compute the Π-projected and Θ-projected systems as discussed in Section 9.3 . Then show that the frequency and step responses of both the projected systems coincide with the original system.

9.3 Using the initial feedback matrix K_1, the system (9.1) can be stabilized as

$$\begin{bmatrix} E_1 & 0 \\ 0 & 0 \end{bmatrix} \begin{bmatrix} \dot{x}_1(t) \\ \dot{x}_2(t) \end{bmatrix} = \begin{bmatrix} A_1 - B_1K_1 & A_2 \\ A_3 & 0 \end{bmatrix} \begin{bmatrix} x_1(t) \\ x_2(t) \end{bmatrix} + \begin{bmatrix} B_1 \\ 0 \end{bmatrix} u(t),$$
$$y(t) = \begin{bmatrix} C_1 & 0 \end{bmatrix} \begin{bmatrix} x_1(t) \\ x_2(t) \end{bmatrix}.$$

Following the discussion in this chapter, develop a balancing-based algorithm including the LR-ADI iteration for the stabilized system by preserving the sparsity of the original system.

9.4 Verify the efficiency of the algorithm that was developed in Exercises 9.3 by applying it to the data of SLNSM-1.

9.5 Using the data of SLNSM-4, construct the Riccati-based optimal feedback from the ROM obtained by the algorithm in Exercise 9.3.

9.6 Compare the numerical results of the balancing-based model reduction as discussed in this chapter and Exercise 9.3. You may use any data of SLNSM to compare the results.

9.7 Following the discussion in this chapter, construct balancing-based algorithm for the following unstable DAE system

$$\left(\begin{bmatrix} E_1 & 0 \\ 0 & 0 \end{bmatrix}, \begin{bmatrix} A_1 & A_2 \\ A_2^T & 0 \end{bmatrix}, \begin{bmatrix} B_1 \\ B_2 \end{bmatrix}, \begin{bmatrix} C_1 & C_2 \end{bmatrix} \right).$$

Chapter 10

Model Reduction of First-Order Index 3 Descriptor Systems

10.1 Introduction

In this chapter we consider a class of first-order differential-algebraic equation (DAE) systems [9, 122]

$$\dot{x}_1(t) = x_2(t) \tag{10.1a}$$

$$M\dot{x}_2(t) = Kx_1(t) + Dx_2(t) - G_1^T x_3(t) + Hu(t), \tag{10.1b}$$

$$G_1 x_1(t) = 0, \tag{10.1c}$$

$$y(t) = Lx_1(t), \tag{10.1d}$$

where $x_1(t) \in \mathbb{R}^{n_1}, x_2(t) \in \mathbb{R}^{n_1}$ and $x_3(t) \in \mathbb{R}^{n_2}$ are states, $u(t) \in \mathbb{R}^m$ are the inputs and $y(t) \in \mathbb{R}^p$ are the outputs. The matrices $M, D, K \in \mathbb{R}^{n_1 \times n_1}$ are sparse, $G_1 \in \mathbb{R}^{n_1 \times n_2}$ ($n_2 < n_1$) is known as the constraint matrix, while $H \in \mathbb{R}^{n_1 \times m}$ and $L \in \mathbb{R}^{p \times n_1}$ are known as input and output matrices, respectively. We assume that M is symmetric and positive definite and

$$\operatorname{rank}(G_1) = n_2,$$

i.e., G_1 has a full rank. Considering $x = \begin{bmatrix} x_1(t)^T, & x_2(t)^T, & x_3(t)^T \end{bmatrix}^T \in \mathbb{R}^n$ where $n = 2n_1 + n_2$, dynamical system (10.1) can be converted into its first-order form

$$\begin{aligned} E\dot{x}(t) &= Ax(t) + Bu(t), \\ y(t) &= Cx(t), \end{aligned} \tag{10.2}$$

where

$$E = \begin{bmatrix} I & 0 & 0 \\ 0 & M & 0 \\ 0 & 0 & 0 \end{bmatrix}, A = \begin{bmatrix} 0 & I & 0 \\ K & D & -G_1^T \\ G_1 & 0 & 0 \end{bmatrix}, B = \begin{bmatrix} 0 \\ H \\ 0 \end{bmatrix} \quad \text{and}\, C = \begin{bmatrix} L & 0 & 0 \end{bmatrix}.$$

The system in (10.1) or (10.2) was described as an index 3 descriptor system in other literature [45, 64, 122]. Such structure models arise in a large variety of applications like in the modeling of constrained vibrational systems, multibody systems or electrical networks (see, e.g., [64, 122, 136, 138]).

In many applications, models of the above structures can become very large and complex. In vibrational analysis simulations, for instance, the models are often generated by the finite element method (FEM) [53]. If the grid resolution gets very fine, because many geometrical details have to be resolved, then the number n can get very large. When this happens, it is very expensive to simulate, control, and optimize the system. Therefore, we want to reduce the complexity of the model by applying model order reduction (MOR), i.e., we seek for an approximation to the original model that well approximates the behavior of the original model but which is much faster to evaluate.

This chapter will discuss both balanced truncation and interpolatory projection methods for the model reduction of index 3 descriptor systems. In recent time, both methods have become prominent and are frequently used for the model reduction of large-scale dynamical systems. Like the index 2 descriptor system we discussed in Chapter 8, here we will convert the index 3 system into its equivalent ODE system by projecting it onto the hidden manifold where solution of the system lies. We will discuss how the MOR methods can be applied to the projected system without forming the projected system explicitly. To perform the balancing-based method, we require Gramian factors which are obtained by solving projected Lyapunov equations. This chapter also discusses efficient solutions of projected Lyapunov equations arising from the underlying index 3 DAE system by using the LR-ADI iteration. Numerical results will be discussed to show the efficiency of the proposed methods.

10.2 Equivalent reformulation of the dynamical system

This section will show how to convert an index 3 descriptor system of the form (10.1) into its equivalent ODE system via projection of the system onto the *hidden manifold* on which the solution of the DAE system evolves. First, we focus on the construction of the projector by exploiting the structure of the system. Secondly, we prove that the finite spectra of the original and projected systems are the same.

10.2.1 Projector for index 3 system

Substituting (10.1a) into (10.1c) we obtain

$$G_1\dot{x}_2(t) = 0.$$

Inserting this identity after multiplying both sides of (10.1b) by G_1M^{-1}, we find

$$0 = G_1M^{-1}Kx_1(t) + G_1M^{-1}Dx_2(t) - G_1M^{-1}G_1^T x_3(t) + G_1M^{-1}Hu(t). \tag{10.3}$$

Since $G_1M^{-1}G_1^T$ is invertible, it implies

$$\begin{aligned} x_3(t) =& (G_1M^{-1}G_1^T)^{-1}G_1M^{-1}Kx_1(t) + (G_1M^{-1}G_1^T)^{-1}G_1M^{-1}Dx_2(t) + \\ & (G_1M^{-1}G_1^T)^{-1}G_1M_1^{-1}Hu(t). \end{aligned} \tag{10.4}$$

By inserting $x_3(t)$ into (10.1b) and simplifying it, we obtain

$$M\dot{x}_2(t) = \Pi Kx_1(t) + \Pi Dx_2(t) + \Pi Hu(t), \tag{10.5}$$

where

$$\Pi := I_{n_1} - G_1^T(G_1M^{-1}G_1^T)^{-1}G_1M^{-1}, \tag{10.6}$$

in which I_{n_1} is an identity matrix of size n_1. In fact, Π is a projector since

$$\begin{aligned} \Pi^2 &= (I_{n_1} - G_1^T(G_1M^{-1}G_1^T)^{-1}G_1M^{-1})(I_{n_1} - G_1^T(G_1M^{-1}G_1^T)^{-1}G_1M^{-1}) \\ &= I_{n_1} - 2G_1^T(G_1M^{-1}G_1^T)^{-1}G_1M^{-1} + G_1^T(G_1M^{-1}G_1^T)^{-1}G_1M^{-1} \\ &= I_{n_1} - G_1^T(G_1M^{-1}G_1^T)^{-1}G_1M^{-1} = \Pi. \end{aligned}$$

The projector Π satisfies the following properties.

Proposition 3. *Let Π be the projector defined above. The following conditions must hold.*

1. $\Pi M = M\Pi^T$.
2. $\mathrm{Null}\,(\Pi) = \mathrm{Range}\,(G_1)$.
3. $\mathrm{Range}\,(\Pi) = \mathrm{Null}\left(G_1M^{-1}\right)$.

Proof. For the proof, see Proposition 4 in Chapter 8. □

Theorem 18. *The vector z is in the null space of G_1, i.e., $G_1 z = 0$ iff $\Pi^T z = z$ where Π is defined in (10.6).*

Proof. Suppose the vector z is in the null space of G_1, i.e., $G_1 z = 0$. Multiplying both sides by $-M^{-1}G_1^T(G_1 M^{-1} G_1^T)^{-1}$, we obtain

$$-M^{-1}G_1^T(G_1 M^{-1}G_1^T)^{-1}G_1 z = 0,$$

which is equivalent to

$$(I_{n_1} - M^{-1}G_1^T(G_1 M^{-1}G_1^T)^{-1}G_1)z = z,$$

i.e., $\Pi^T z = z$. Conversely, suppose that $\Pi^T z = z$ which implies

$$(I_{n_1} - M^{-1}G_1^T(G_1 M^{-1}G_1^T)^{-1}G_1)z = z.$$

We see $M^{-1}G_1^T(G_1 M^{-1}G_1^T)^{-1}G_1 z = 0$. Multiplying both sides by G_1, we obtain $G_1 z = 0$. □

10.2.2 Formulation of projected system

Following Theorem 18, we can write

$$\Pi^T x_1(t) = x_1(t) \quad \text{and} \quad \Pi^T x_2(t) = x_2(t), \tag{10.7}$$

$$\Pi^T \dot{x}_1(t) = \dot{x}_1(t) \quad \text{and} \quad \Pi^T \dot{x}_2(t) = \dot{x}_2(t). \tag{10.8}$$

Inserting these identities into (10.1a), (10.1b) and also multiplying the resulting equations by Π from left, we obtain

$$\Pi M \Pi^T \dot{x}_1(t) = \Pi M \Pi^T x_2(t), \tag{10.9}$$

$$\Pi M \Pi^T \dot{x}_2(t) = \Pi K \Pi^T x_1(t) + \Pi D \Pi^T x_2(t) + \Pi H u(t). \tag{10.10}$$

Analogously, the output equation (10.3) yields

$$y(t) = L\Pi^T x_1(t).$$

In this way we convert the system in (10.1) into an equivalent projected system which can be written in matrix vector form as

$$\begin{aligned} \tilde{\Pi} E_1 \tilde{\Pi}^T \dot{x}(t) &= \tilde{\Pi} A_1 \tilde{\Pi}^T x(t) + \tilde{\Pi} B_1 u(t), \\ y(t) &= C_1 \tilde{\Pi}^T x(t), \end{aligned} \tag{10.11}$$

where

$$\begin{aligned} &\tilde{\Pi} = \begin{bmatrix} \Pi & \\ & \Pi \end{bmatrix}, \quad E_1 = \begin{bmatrix} M & 0 \\ 0 & M \end{bmatrix}, \quad A_1 = \begin{bmatrix} 0 & M \\ K & D \end{bmatrix}, \\ &B_1 = \begin{bmatrix} 0 \\ H \end{bmatrix}, \quad C_1 = \begin{bmatrix} L & 0 \end{bmatrix} \quad \text{and} \quad x(t) = \begin{bmatrix} x_1(t) \\ x_2(t) \end{bmatrix}. \end{aligned} \tag{10.12}$$

The system dynamics of (10.1) are now coincided into the projected system (10.11). In system (10.11), all the coefficient matrices are singular since $\tilde{\Pi}$ has rank deficiency (due to the singularity of Π). This means the system contains redundant elements. To remove the redundant elements, let us decompose $\tilde{\Pi}$ as

$$\tilde{\Pi} = \bar{\Theta}_l \bar{\Theta}_r^T, \tag{10.13}$$

with $\bar{\Theta}_l^T \bar{\Theta}_r = I_k$, where $\bar{\Theta}_l, \bar{\Theta}_r \in \mathbb{R}^{2n_\xi \times k}$ and $k = \text{rank}\left(\tilde{\Pi}\right)$. Now applying the decomposition of $\tilde{\Pi}$ from (10.13) and defining $\tilde{x}_1(t) := \bar{\Theta}_l^T x_1(t)$, we obtain

$$\begin{aligned} \bar{\Theta}_r^T E_1 \bar{\Theta}_r \dot{\tilde{x}}_1(t) &= \bar{\Theta}_r^T A_1 \bar{\Theta}_r \tilde{x}_1(t) + \bar{\Theta}_r^T B_1 u(t), \\ y(t) &= C_1 \bar{\Theta}_r \tilde{x}_1(t). \end{aligned} \tag{10.14}$$

The systems in (10.1), (10.11) and (10.14) are in fact equivalent since they have the same transfer functions and their finite spectrum are the same. Moreover, system (10.14) can be compared with the generalized state space system (4) and hence one can directly apply a naive approach of MOR techniques (balanced truncation and IRKA) as discussed in Chapter 4. However, for large-scale index 3 DAE system, construction of (10.14) should be avoided. This formulation loses the sparsity of the original system and leads to additional computational complexity in implementation. In the following section, we will show how to find the reduced model without computing the projected system explicitly.

10.3 Model reduction with the balanced truncation avoiding the formulation of projected system

On the theoretical level, we know exactly how to perform the MOR now. First we project the system to the inherent ordinary differential system and then apply classical MOR methods. The main focus, however, is to show how the MOR can be carried out without explicitly forming the projected systems or the projection operators.

For this purpose, the Lyapunov equations

$$\bar{\Theta}_r^T A_1 \bar{\Theta}_r \bar{P} \bar{\Theta}_r^T E_1^T \bar{\Theta}_r + \bar{\Theta}_r^T E_1 \bar{\Theta}_r \bar{P} \bar{\Theta}_r^T A_1^T \bar{\Theta}_r = -\bar{\Theta}_r^T B_1 B_1^T \bar{\Theta}_r, \tag{10.15a}$$

$$\bar{\Theta}_r^T A_1^T \bar{\Theta}_r \bar{Q} \bar{\Theta}_r^T E_1 \bar{\Theta}_r + \bar{\Theta}_r^T E_1^T \bar{\Theta}_r \bar{Q} \bar{\Theta}_r^T A_1 \bar{\Theta}_r = -\bar{\Theta}_r^T C_1^T C_1 \bar{\Theta}_r, \tag{10.15b}$$

are to be solved where $\bar{P}$ and $\bar{Q}$ are known as the controllability Gramian and observability Gramian, respectively. The solutions $\bar{P}$, $\bar{Q}$ of the Lyapunov equations are unique because the corresponding system is asymptotically stable and symmetric positive (semi-)definite since the right-hand side is semidefinite. Multiplying both equations in (10.15) by $\bar{\Theta}_l$ from the left and $\bar{\Theta}_l^T$ from the right and exploiting the property of Π, the resulting equations become

$$\tilde{\Pi}A_1\tilde{\Pi}^T\tilde{P}\tilde{\Pi}E_1^T\tilde{\Pi}^T + \tilde{\Pi}E_1\tilde{\Pi}^T\tilde{P}\tilde{\Pi}A_1^T\tilde{\Pi}^T = -\tilde{\Pi}B_1B_1^T\tilde{\Pi}^T, \tag{10.16a}$$

$$\tilde{\Pi}A_1^T\tilde{\Pi}^T\tilde{Q}\tilde{\Pi}E_1\tilde{\Pi}^T + \tilde{\Pi}E_1^T\tilde{\Pi}^T\tilde{Q}\tilde{\Pi}A_1\tilde{\Pi}^T = -\tilde{\Pi}C_1^TC_1\tilde{\Pi}^T, \tag{10.16b}$$

where $\tilde{P} = \bar{\Theta}_r\bar{P}\bar{\Theta}_r^T, \quad \tilde{Q} = \bar{\Theta}_r\bar{Q}\bar{\Theta}_r^T$. The Lyapunov equations in (10.16) are nothing but the Lyapunov equations of the projected system (10.11) where $\tilde{P}$, $\tilde{Q} \in \mathbb{R}^{2n_\xi \times 2n_\xi}$ are the system controllability and observability Gramians, respectively. Under condition (10.13), it can be shown that $\tilde{P}$ and $\tilde{Q}$ satisfy

$$\tilde{P} = \tilde{\Pi}\tilde{P}\tilde{\Pi}^T \quad \text{and} \quad \tilde{Q} = \tilde{\Pi}\tilde{Q}\tilde{\Pi}^T, \tag{10.17}$$

which ensures that the solutions are unique; although the equations in (10.16) are singular due to the singular projectors. These projected Lyapunov equations can be solved by LR-ADI iteration without computing the projected system explicitly. The avoidance of the explicit computation of the projectors in solving the Lyapunov equations will be discussed in the next section. Let $\tilde{R}$ and $\tilde{L}$ be the low-rank factors of the controllability and observability Gramians of system (10.11) such that

$$\tilde{P} \approx \tilde{R}\tilde{R}^T, \quad \tilde{Q} \approx \tilde{L}\tilde{L}^T, \tag{10.18}$$

and let $\bar{R}$ and $\bar{L}$ be the low-rank factors of the controllability and observability Gramians of the system (10.14) such that

$$\bar{P} \approx \bar{R}\bar{R}^T, \quad \bar{Q} \approx \bar{L}\bar{L}^T.$$

Then the controllability Gramian factor and the observability Gramian factor of the systems (10.11) and (10.14) are related by

$$\tilde{R} = \bar{\Theta}_r\bar{R} \quad \text{and} \quad \tilde{L} = \bar{\Theta}_r\bar{L}.$$

This relation can easily be obtained since

$$\begin{aligned} \tilde{R}\tilde{R}^T &\approx \tilde{P} = \bar{\Theta}_r\bar{P}\bar{\Theta}_r^T \approx \bar{\Theta}_r\bar{R}\bar{R}^T\bar{\Theta}_r^T \quad \text{and} \\ \tilde{L}\tilde{L}^T &\approx \tilde{Q} = \bar{\Theta}_r\bar{Q}\bar{\Theta}_r^T \approx \bar{\Theta}_r\bar{L}\bar{L}^T\bar{\Theta}_r^T. \end{aligned}$$

Consider the singular value decomposition of $\bar{L}^T\bar{\Theta}_r^TE_1\bar{\Theta}_r\bar{R}$, as

$$\bar{L}^T\bar{\Theta}_r^TE_1\bar{\Theta}_r\bar{R} = \begin{bmatrix} U_1 & U_2 \end{bmatrix} \begin{bmatrix} \Sigma_1 & \\ & \Sigma_2 \end{bmatrix} \begin{bmatrix} V_1^T \\ V_2^T \end{bmatrix}.$$

Now construct the left and right *balancing and truncating* transformations $\bar{W}$ and $\bar{V}$ as

$$\bar{W} = \bar{L}U_1\Sigma_1^{-\frac{1}{2}}, \quad \bar{V} = \bar{R}V_1\Sigma_1^{-\frac{1}{2}},$$

where U_1, V_1 consist of the corresponding leading l ($l \ll k$) columns of U, V, and Σ_1 is the first leading $l \times l$ block of Σ. Again, considering the singular value decomposition

$$\tilde{L}^T \tilde{\Pi} E_1 \tilde{\Pi}^T \tilde{R} = \bar{L}^T \bar{\Theta}_r^T E_1 \bar{\Theta}_r \bar{R} = \begin{bmatrix} U_1 & U_2 \end{bmatrix} \begin{bmatrix} \Sigma_1 & \\ & \Sigma_2 \end{bmatrix} \begin{bmatrix} V_1^T \\ V_2^T \end{bmatrix}. \tag{10.19}$$

The left- and right-balancing and truncating transformations can be constructed as

$$\tilde{W} = \tilde{L} U_1 \Sigma_1^{-\frac{1}{2}}, \quad \tilde{V} = \tilde{R} V_1 \Sigma_1^{-\frac{1}{2}}. \tag{10.20}$$

It can be observed that

$$\begin{aligned} \tilde{W} &= \tilde{L} U_1 \Sigma_1^{-\frac{1}{2}} = \bar{\Theta}_r \bar{L} U_1 \Sigma_1^{-\frac{1}{2}} = \bar{\Theta}_r \bar{W} = \bar{\Theta}_r \bar{\Theta}_l^T \bar{\Theta}_r \bar{W} = \tilde{\Pi}^T \tilde{W}, \\ \tilde{V} &= \tilde{R} U_1 \Sigma_1^{-\frac{1}{2}} = \bar{\Theta}_r \bar{R} U_1 \Sigma_1^{-\frac{1}{2}} = \bar{\Theta}_r \bar{V} = \bar{\Theta}_r \bar{\Theta}_l^T \bar{\Theta}_r \bar{V} = \tilde{\Pi}^T \tilde{V}. \end{aligned} \tag{10.21}$$

By applying the balancing and truncating transformations $\bar{W}$ and $\bar{V}$ to the system (10.14), construct the reduced-order model

$$\begin{aligned} \hat{E}\dot{\hat{x}}(t) &= \hat{A}\hat{x}(t) + \hat{B}u(t), \\ \hat{y}(t) &= \hat{C}\hat{x}(t), \end{aligned} \tag{10.22}$$

where the coefficient matrices are formed by

$$\begin{aligned} \hat{E} &= \bar{W}^T \bar{\Theta}_r^T E_1 \bar{\Theta}_r \bar{V} = \tilde{W}^T \tilde{\Pi} E_1 \tilde{\Pi}^T \tilde{V} = \tilde{W}^T E_1 \tilde{V}, \\ \hat{A} &= \bar{W}^T \bar{\Theta}_r^T A_1 \bar{\Theta}_r \bar{V} = \tilde{W}^T \tilde{\Pi} A_1 \tilde{\Pi}^T \tilde{V} = \tilde{W}^T A_1 \tilde{V}, \\ \hat{B} &= \bar{W}^T \bar{\Theta}_r^T B_1 = \tilde{W}^T \tilde{\Pi} B_1 = \tilde{W}^T B_1, \\ \hat{C} &= C_1 \bar{\Theta}_r \bar{V} = C_1 \tilde{\Pi}^T \tilde{V} = C_1 \tilde{V}. \end{aligned} \tag{10.23}$$

From the above discussion, it is clear that in order to obtain the reduced model (10.22), explicit formulation of the system (10.11) or (10.14) is not required. Forming only the balancing and truncating transformations $\tilde{W}$ and $\tilde{V}$ as given in (10.20), the reduced matrices can be constructed as

$$\hat{E} = \tilde{W}^T E_1 \tilde{V}, \ \hat{A} = \tilde{W}^T A_1 \tilde{V}, \ \hat{B} = \tilde{W}^T B_1 \quad \text{and} \quad \hat{C} = C_1 \tilde{V}. \tag{10.24}$$

The procedure to compute the reduced generalized state-space system (10.22) from the index 3 DAEs (10.1) is summarized in Algorithm 35.

10.4 Solution of projected Lyapunov equations

To implement Algorithm 35, the projected Lyapunov equations (10.16) need to be solved for computing the low-rank Gramian factors $\tilde{R}$ and $\tilde{L}$. This

Algorithm 35: SRM for first-order index 3 systems.

Input : E_1, A_1, B_1 and C_1 from (10.12).
Output: $\hat{E}$, $\hat{A}$, $\hat{B}$ and $\hat{C}$ as in (10.22).
1 Compute the low-rank Gramian factors $\tilde{R}$ and $\tilde{L}$ by solving the projected Lyapunov equations (10.16).
2 Construct the balancing and truncating transformations $\tilde{W}$ and $\tilde{V}$ by performing (10.19)-(10.20).
3 Form the reduced matrices using (10.24).

section discusses an efficient technique to solve such projected Lyapunov equations using the LR-ADI iteration introduced in Chapter 3. This section shows the computation of the low-rank controllability Gramian factor by solving the Lyapunov equation (10.16a) elaborately. The same procedure is applicable for solving the Lyapunov equation (10.16b) also. Algorithm 15 can be applied for solving the projected Lyapunov equation (10.16a). The text that follows highlights a few issues of computing the projector explicitly.

10.4.1 Initial residual factor

In the formulation of the LR-ADI in Algorithm 15, the initial residual factor is $W_0 = B$. In this case, $W_0 = \tilde{\Pi} B_1$ which can be partitioned as

$$W_0 = \tilde{B} = \tilde{\Pi} B_1 = \begin{bmatrix} \Pi & \\ & \Pi \end{bmatrix} \begin{bmatrix} 0 \\ H \end{bmatrix} = \begin{bmatrix} 0 \\ \Pi H \end{bmatrix} = \begin{bmatrix} W_0^{(1)} \\ W_0^{(2)} \end{bmatrix},$$

i.e., $W_0^{(1)} = 0$ and $W_0^{(2)} = \Pi H$. One can observe that the linear system

$$\begin{bmatrix} M & G_1^T \\ G_1 & 0 \end{bmatrix} \begin{bmatrix} \Theta \\ \Gamma \end{bmatrix} = \begin{bmatrix} H_1 \\ 0 \end{bmatrix}, \tag{10.25}$$

can compute $\Theta = M^{-1}\Pi H$. Hence, $W_0^{(2)}$ can be computed by solving (10.25) for Θ and multiplying it by M. Thus, in the initial residual, $W_0^{(2)}$ can be computed without explicit usage of Π. After this initialization, the updated formula can be used without additional treatment since the updates are already projected by construction and thus all subsequent residual factors will automatically be invariant under projection with $\tilde{\Pi}$.

10.4.2 Solutions of linear systems and update of residual factors

To solve the Lyapunov equation (10.16a) using the LR-ADI iteration (see, e.g., Algorithm 15), at the i-th iteration step, one has to solve the linear system

$$(\tilde{A} + \mu_i \tilde{E})V_i = \tilde{W}_{i-1}, \tag{10.26}$$

where $\tilde{W}_{i-1}$ is the ADI residual factor computed from the $(i-1)$-st iteration. Partitioning $\tilde{W}_{i-1}$ as $\tilde{W}_{i-1} = \begin{bmatrix} \tilde{W}_{i-1}^{(1)} \\ \tilde{W}_{i-1}^{(2)} \end{bmatrix}$, equation (12.57) can be written as

$$\begin{bmatrix} \mu_i \Pi M \Pi^T & \Pi M \Pi^T \\ \Pi K \Pi^T & \Pi D \Pi^T + \mu_i \Pi M \Pi^T \end{bmatrix} \begin{bmatrix} V_i^{(1)} \\ V_i^{(2)} \end{bmatrix} = \begin{bmatrix} \tilde{W}_{i-1}^{(1)} \\ \tilde{W}_{i-1}^{(2)} \end{bmatrix}. \tag{10.27}$$

The following observation will enable us to solve the linear system without using Π explicitly.

Theorem 19. *The matrices X_1, X_2 satisfy*

$$X_1 = \Pi^T X_1, \quad X_2 = \Pi^T X_2$$

and the linear system

$$\begin{bmatrix} \mu \Pi M \Pi^T & \Pi M \Pi^T \\ \Pi K \Pi^T & \Pi D \Pi^T + \mu \Pi K \Pi^T \end{bmatrix} \begin{bmatrix} X_1 \\ X_2 \end{bmatrix} = \begin{bmatrix} \Pi F_1 \\ \Pi F_2 \end{bmatrix}, \tag{10.28}$$

if and only if they satisfy the linear system

$$\begin{bmatrix} \mu M & M & -G_1^T & 0 \\ K & D + \mu M & 0 & -G_1^T \\ G_1 & 0 & 0 & 0 \\ 0 & G_1 & 0 & 0 \end{bmatrix} \begin{bmatrix} X_1 \\ X_2 \\ \Gamma_1 \\ \Gamma_2 \end{bmatrix} = \begin{bmatrix} F_1 \\ F_2 \\ 0 \\ 0 \end{bmatrix}. \tag{10.29}$$

Proof. Suppose that the matrices X_1 and X_2 satisfy $X_1 = \Pi^T X_1$ and $X_2 = \Pi^T X_2$. Recalling Theorem 18 these properties imply that

$$G_1 X_1 = 0, \quad \text{and} \tag{10.30}$$

$$G_1 X_2 = 0. \tag{10.31}$$

Since $\Pi^T X_1 = X_1$, $\Pi^T X_2 = X_2$, the first line of (10.28) can be written as

$$\Pi(\mu M X_1 + M X_2 - F_1) = 0,$$

i.e., $\mu M X_1 + M X_2 - F_1$ is in the null space of Π. Since $\text{Null}(\Pi) = \text{Range}(G_1^T)$ (by Proposition 3(2.)), there exists a Γ_1 such that

$$\mu M X_1 + M X_2 - F_1 = G_1^T \Gamma_1,$$

which implies

$$\mu M X_1 + M X_2 - G_1^T \Gamma_1 = F_1. \tag{10.32}$$

Similarly, the second line of (10.28) gives

$$\Pi(KX_1 + DX_2 + \mu MX_2 - F_2) = 0,$$

i.e., $KX_1 + DX_2 + \mu MX_2 - F_2$ is in the null space of Π. Again, using Proposition 3 (2.) there exists a Γ_2 such that

$$KX_1 + DX_2 + \mu MX_2 - F_2 = G_1^T\Gamma_2,$$

which implies

$$KX_1 + DX_2 + \mu MX_2 - G_1^T\Gamma_2 = F_2. \tag{10.33}$$

Equations (10.32), (10.33), (10.30) and (10.31) produce the linear system (10.29).

Conversely, suppose that X_1 and X_2 satisfy the linear system (10.29). Then the third and fourth lines respectively imply,

$$X_1 = \Pi^T X_1 \quad \text{and} \quad X_2 = \Pi^T X_2.$$

Applying these properties, the first and second lines of (10.29) respectively give,

$$\begin{aligned} \mu M\Pi^T X_1 + M\Pi^T X_2 - G_1^T\Gamma_1 &= F_1, \\ K\Pi^T X_1 + (D + \mu M)\Pi^T X_2 - G_2^T\Gamma_2 &= F_2. \end{aligned}$$

Multiplying both equations from left side by Π yields

$$\begin{aligned} \mu\Pi M\Pi^T X_1 + \Pi M\Pi^T X_2 &= \Pi F_1, \\ \Pi K\Pi^T X_1 + \Pi(D + \mu M)\Pi^T X_2 &= \Pi F_2. \end{aligned}$$

which form the linear system in (10.28). □

According to Theorem 19, instead of solving the linear system (10.27) the linear system

$$\begin{bmatrix} \mu_i M & M & -G_1^T & 0 \\ K & D + \mu_i M & 0 & -G_1^T \\ G_1 & 0 & 0 & 0 \\ 0 & G_1 & 0 & 0 \end{bmatrix} \begin{bmatrix} V_i^{(1)} \\ V_i^{(2)} \\ \Gamma_1 \\ \Gamma_2 \end{bmatrix} = \begin{bmatrix} \tilde{W}_{i-1}^{(1)} \\ \tilde{W}_{i-1}^{(2)} \\ 0 \\ 0 \end{bmatrix}, \tag{10.34}$$

can be solved for $\begin{bmatrix} V_i^{(1)^T} & V_i^{(2)^T} \end{bmatrix}^T$.

Note that although the matrix in (10.34) has larger dimensions, it is highly sparse and can be solved efficiently using any sparse solver.

The matrix (vector) $\begin{bmatrix} \tilde{W}_{i-1}^{(1)} \\ \tilde{W}_{i-1}^{(2)} \end{bmatrix}$ is updated in each iteration which is computed from the ADI residual factor of the previous step. At each iteration, the ADI residual factor can be computed by (see Algorithm 15)

$$\tilde{W}_i = \tilde{W}_{i-1} - 2\,\mathrm{Re}\,(\mu_i)\tilde{E}V_i, \tag{10.35}$$

which can be partitioned as

$$\begin{aligned} \begin{bmatrix} \tilde{W}_i^{(1)} \\ \tilde{W}_i^{(2)} \end{bmatrix} &= \begin{bmatrix} \tilde{W}_{i-1}^{(1)} \\ \tilde{W}_{i-1}^{(2)} \end{bmatrix} - 2\,\mathrm{Re}\,(\mu_i) \begin{bmatrix} \Pi M \Pi^T & 0 \\ 0 & \Pi M \Pi^T \end{bmatrix} \begin{bmatrix} V_i^{(1)} \\ V_i^{(2)} \end{bmatrix}. \\ &= \begin{bmatrix} \tilde{W}_{i-1}^{(1)} - 2\,\mathrm{Re}\,(\mu_i)\Pi M \Pi^T V_i^{(1)} \\ \tilde{W}_{i-1}^{(2)} - 2\,\mathrm{Re}\,(\mu_i)\Pi M \Pi^T V_i^{(2)} \end{bmatrix}. \end{aligned}$$

Exploiting the properties of Π, i.e., $\Pi M = M\Pi^T$ and $\Pi^T V_i^{(1)} = V_i^{(1)}$ and $\Pi^T V_i^{(2)} = V_i^{(2)}$, the above equation results in

$$\begin{aligned} \tilde{W}_i^{(1)} &= \tilde{W}_{i-1}^{(1)} - 2\,\mathrm{Re}\,(\mu_i) M V_i^{(1)}, \\ \tilde{W}_i^{(2)} &= \tilde{W}_{i-1}^{(2)} - 2\,\mathrm{Re}\,(\mu_i) M V_i^{(2)}. \end{aligned}$$

Therefore, (10.35) becomes

$$\tilde{W}_i = \tilde{W}_{i-1} - 2\,\mathrm{Re}\,(\mu_i)E_1 V_i, \tag{10.36}$$

If the two consecutive shift parameters are complex conjugates of each other, i.e., $\{\mu_i, \mu_{i+1} := \overline{\mu_i}\}$, then recalling Algorithm 15,

$$\tilde{W}_{i+1} = \tilde{W}_{i-1} - 4\,\mathrm{Re}\,(\mu_i)\tilde{E}\left(\mathrm{Re}\,(V_i) + \delta\,\mathrm{Im}\,(V_i)\right),$$

where $\delta = \frac{\mathrm{Re}\,(\mu_i)}{\mathrm{Im}\,(\mu_i)}$. A little algebraic manipulations will enable us to write this as

$$\tilde{W}_{i+1} = \tilde{W}_{i-1} - 4\,\mathrm{Re}\,(\mu_i)E_1\left(\mathrm{Re}\,(V_i) + \delta\,\mathrm{Im}\,(V_i)\right). \tag{10.37}$$

Based on the above discussion, Algorithm 36 is presented to compute the low-rank controllability Gramian factor by solving the controllability Lyapunov equation (10.16a). The same procedure can be applied to solve the observability Lyapunov equation (10.16b).

10.4.3 Computation of ADI shift parameters

Algorithm 36 relies on certain shift parameters that are crucial to get fast convergence of the method. In recent years, *Penzl's heuristic* approach introduced in Penzl [133] has become one of the most commonly used approaches to compute the ADI shift parameters for a large-scale system. This

Algorithm 36: LR-ADI for projected Lyapunov equation for first-order index 3 system.

Input : $E_1,\ A_1,\ B_1, \{\mu_i\}_{i=1}^{J}$, a tolerance $0 < \tau \ll 1$ for the normalized residual.

Output: $\tilde{R} = Z_i$, such that $\tilde{P} \approx \tilde{R}\tilde{R}^T$.

Set $Z_0 = [\,]$, $i = 1$ and $\tilde{W}_0 = \tilde{B}$ as in (10.25)

while $\frac{\|\tilde{W}_{i-1}^T \tilde{W}_{i-1}\|}{\|\tilde{B}\tilde{B}^T\|} \geq \tau$ **do**

  Solve the linear system (10.34) for $V_i = \begin{bmatrix} V_i^{(1)^T} & V_i^{(2)^T} \end{bmatrix}^T$

  if $\mathrm{Im}\,(\mu_i) = 0$ **then**

    $Z_i = \begin{bmatrix} Z_{i-1} & \sqrt{-2\mu_i}\,\mathrm{Re}\,(V_i) \end{bmatrix}$

    Update, $\tilde{W}_i$ as in (10.36)

  else

    $\gamma = -2\,\mathrm{Re}\,(\mu_i)$, $\delta = \frac{\mathrm{Re}\,(\mu_i)}{\mathrm{Im}\,(\mu_i)}$ and $\beta = \sqrt{(\delta^2 + 1)}$

    $Z_i = \sqrt{2\gamma}\begin{bmatrix} \mathrm{Re}\,(V_i) + \delta\,\mathrm{Im}\,(V_i) & \beta\,\mathrm{Im}\,(V_i) \end{bmatrix}$

    $Z_{i+1} = \begin{bmatrix} Z_{i-1} & Z_i \end{bmatrix}$,

    Compute $V_{i+1} = \mathrm{Re}\,(V_i) + \delta\,\mathrm{Im}\,(V_i)$

    Update $\tilde{W}_{i+1}$ as in (10.37).

    $i = i + 1$

  $i = i + 1$

approach is investigated in many research articles (e.g., [71, 159, 168]) to solve the Lyapunov equations of large-scale systems with the LR-ADI methods. In this approach, however, one often needs a set of approximate finite eigenvalues which consist of some large magnitude and small magnitude Ritz values of the matrix pencil corresponding to the underlying system (see details in [133]). For the index 3 descriptor system (10.1), the corresponding matrix pencil is

$$\lambda E - A; \quad \forall\, \lambda \in \mathbb{C}, \tag{10.38}$$

where A and E are defined in (10.2). Due to the singularity of E, the matrix pencil features some infinite eigenvalues that prevent the direct usage of Arnoldi's method for the approximation of large magnitude eigenvalues. To overcome this problem, the strategy introduced in [52] (see, also [41]) can be applied by looking at the modified eigenvalue problem of the first-order structured index 2 DAEs system as shown in Chapter 8. In order to follow the strategy, modify the matrix pencil (10.38) as

$$\lambda E_2 - A_2; \quad \forall\, \lambda \in \mathbb{C}, \tag{10.39}$$

where A_2 and E_2 are defined as

$$(E_2, A_2) = \left(\begin{bmatrix} I_{n_1} & 0 & 0 \\ 0 & M & 0 \\ 0 & 0 & 0 \end{bmatrix}, \begin{bmatrix} 0 & I_{n_1} & 0 \\ K & D & -G_1^T \\ 0 & G_1 & 0 \end{bmatrix} \right)$$

According to [64, Theorem 2.7.3], the matrix pencils in (10.39) and (10.38) share the same non-zero finite spectrum. Now the modified matrix pencil

$$\lambda \begin{bmatrix} I_{n_1} & 0 & 0 \\ 0 & M & -\nu G_1^T \\ 0 & \nu G_1 & 0 \end{bmatrix} - \underbrace{\begin{bmatrix} 0 & I_{n_1} & 0 \\ K & D & -G_1^T \\ 0 & G_1 & 0 \end{bmatrix}}_{A_2} \tag{10.40}$$

moves all infinite eigenvalues to $\frac{1}{\nu}$ ($\nu \in \mathbb{R}$), without altering the finite eigenvalues. The parameter ν can be chosen such that $\frac{1}{\nu}$ is close to the smallest magnitude eigenvalues after those have been determined with respect to the original matrices. Note that the matrix A_2 in (10.40) will always be singular, such that the small eigenvalue approximations cannot be computed since the inversion of A_2 would be required. Therefore, small magnitude ritzvalues should be always computed from the matrix pencil (10.38).

10.5 Interpolatory method via IRKA

We concentrate on the interpolatory method via IRKA for model reduction of system (10.1). The method was introduced in Chapter 4. Unlike balanced truncation, to apply IRKA onto the system (10.1), we first convert the system into (10.15). Then Algorithm 21 can be applied to this projected system to obtain the ROM. But we are not to form system (10.15) considering the computational complexity. It can be shown that instead of applying IRKA to (10.15), the method can be applied to system (10.11). However, like balanced truncation discussed above, we do not require to form system (10.11) explicitly. We discuss this in the following text.

Following Algorithm 21, to implement IRKA for system (10.11), we need to construct the right and left projectors as follows

$$\tilde{V} = \left[(\alpha_1 \tilde{E} - \tilde{A})^I \tilde{B} b_1, \cdots, (\alpha_r \tilde{E} - \tilde{A})^I \tilde{B} b_r\right], \tag{10.41a}$$

$$\tilde{W} = \left[(\alpha_1 \tilde{E}^T - \tilde{A}^T)^I \tilde{C}^T c_1, \cdots, (\alpha_r \tilde{E}^T - \tilde{A}^T)^I \tilde{C}^T c_r\right], \tag{10.41b}$$

where $\tilde{E} := \tilde{\Pi} E_1 \tilde{\Pi}^T$, $\tilde{A} := \tilde{\Pi} A_1 \tilde{\Pi}^T$, $\tilde{B} := \tilde{\Pi} B_1$, $\tilde{C} := C_1 \tilde{\Pi}^T$, and $(\alpha_i \tilde{E} - \tilde{A})^I = (\alpha_i \tilde{E}^T - \tilde{A}^T)^I := \bar{\Theta}_r (\bar{\Theta}_r^T E_1 \bar{\Theta}_r - \alpha_i \bar{\Theta}_r^T A_1 \bar{\Theta}_r)^{-1} \bar{\Theta}_r^T$, for $i = 1, \cdots, r$. Recalling

[89, Theorem 6.2], if we can construct $\tilde{V}$ and $\tilde{W}$ as in (10.41), then the reduced model in (10.22) can be formed by computing the reduced matrices as in (10.24). Therefore, to construct the reduced model (10.22) we can avoid the projectors. However, in each term of $\tilde{V}$ and $\tilde{W}$, the projectors are implicitly hidden. The solution of this problem has been shown in [89] based on [93, Theorem 5.2]. In our case, each column of $\tilde{V}$ contains a vector such as

$$v = (\alpha\tilde{E} - \tilde{A})^I \tilde{B} b. \tag{10.42}$$

Following [89, Lemma 6.3], it can be shown that v in (10.42) solves the linear system

$$(\alpha\tilde{E} - \tilde{A})v = \tilde{B}b, \tag{10.43}$$

which is equivalent to

$$\begin{bmatrix} \alpha\Pi M\Pi^T & -\Pi M\Pi^T \\ -\Pi K\Pi^T & \alpha\Pi M\Pi^T - \Pi D\Pi^T \end{bmatrix} \begin{bmatrix} v_1 \\ v_2 \end{bmatrix} = \begin{bmatrix} 0 \\ Hb \end{bmatrix}. \tag{10.44}$$

According to Theorem 19, instead of solving the linear system (10.44) the linear system

$$\begin{bmatrix} \alpha M & -M & G_1^T & 0 \\ -K & \alpha M - D & 0 & G_1^T \\ G_1 & 0 & 0 & 0 \\ 0 & G_1 & 0 & 0 \end{bmatrix} \begin{bmatrix} v_1 \\ v_2 \\ \Gamma_1 \\ \Gamma_2 \end{bmatrix} = \begin{bmatrix} 0 \\ Hb \\ 0 \\ 0 \end{bmatrix}, \tag{10.45}$$

can be solved for $\begin{bmatrix} v_1^T & v_2^T \end{bmatrix}^T$. Analogously, a vector

$$w = \begin{bmatrix} w_1^T & w_2^T \end{bmatrix}^T = (\alpha\tilde{E}^T - \tilde{A}^T)^I \tilde{C}^T c,$$

in each term in $\tilde{W}$ can be computed by solving the linear system

$$\begin{bmatrix} \alpha\Pi M^T\Pi^T & -\Pi K^T\Pi^T \\ -\Pi M^T\Pi^T & \alpha\Pi M^T\Pi^T - \Pi D^T\Pi^T \end{bmatrix} \begin{bmatrix} w_1 \\ w_2 \end{bmatrix} = \begin{bmatrix} L^T c \\ 0 \end{bmatrix}, \tag{10.46}$$

which is again equivalent to solving the linear system

$$\begin{bmatrix} \alpha M^T & -K^T & G_1^T & 0 \\ -M^T & \alpha M^T - D^T & 0 & G_1^T \\ G_1 & 0 & 0 & 0 \\ 0 & G_1 & 0 & 0 \end{bmatrix} \begin{bmatrix} w_1 \\ w_2 \\ \Gamma_1 \\ \Gamma_2 \end{bmatrix} = \begin{bmatrix} L^T c \\ 0 \\ 0 \\ 0 \end{bmatrix}, \tag{10.47}$$

for $\begin{bmatrix} w_1^T & w_2^T \end{bmatrix}^T$. A complete IRKA procedure to compute the reduced model from index 3 system (10.1) is presented in Algorithm 37.

Algorithm 37: IRKA for index 3 systems.

Input : M, D, K, G_1, H, L from (10.1).
Output: $\hat{E}$, $\hat{A}$, $\hat{B}$, $\hat{C}$.

Set up the matrices E_1, A_1, B_1, C_1 as in (10.12).
Select initial interpolation points $\{\sigma_i\}_{i=1}^{r}$ and tangent directions $\{b_i\}_{i=1}^{r}$ and $\{c_i\}_{i=1}^{r}$.
while *(not converged)* **do**
  for $i = 1, 2, \cdots, r$ **do**

$$\begin{bmatrix} \alpha M & -M & G_1^T & 0 \\ -K & \alpha M - D & 0 & G_1^T \\ G_1 & 0 & 0 & 0 \\ 0 & G_1 & 0 & 0 \end{bmatrix} \begin{bmatrix} v_i^{(1)} \\ v_i^{(2)} \\ \Gamma_1 \\ \Gamma_2 \end{bmatrix} = \begin{bmatrix} 0 \\ Hb_i \\ 0 \\ 0 \end{bmatrix},$$

$$\begin{bmatrix} \alpha M^T & -K^T & G_1^T & 0 \\ -M^T & \alpha M^T - D^T & 0 & G_1^T \\ G_1 & 0 & 0 & 0 \\ 0 & G_1 & 0 & 0 \end{bmatrix} \begin{bmatrix} w_i^{(1)} \\ w_i^{(2)} \\ \Gamma_1 \\ \Gamma_2 \end{bmatrix} = \begin{bmatrix} L^T c_i \\ 0 \\ 0 \\ 0 \end{bmatrix},$$

  Form $\mathrm{v}_i = \begin{bmatrix} v_i^{(1)} \\ v_i^{(1)} \end{bmatrix}$, $\mathrm{w}_i = \begin{bmatrix} w_i^{(1)} \\ w_i^{(2)} \end{bmatrix}$.
  Construct $\tilde{V} = \begin{bmatrix} \mathrm{v}_1, \mathrm{v}_1, \cdots, \mathrm{v}_r \end{bmatrix}$, $\tilde{W} = \begin{bmatrix} \mathrm{w}_1, \mathrm{w}_1, \cdots, \mathrm{w}_r \end{bmatrix}$.
  $\hat{E} = \tilde{W}^T E_1 \tilde{V}$, $\hat{A} = \tilde{W}^T A_1 \tilde{V}$, $\hat{B} = \tilde{W}^T B_1$ and $\hat{C} = C_1 \tilde{V}$
  Compute $\hat{A} z_i = \lambda_i \hat{E} z_i$ and $y^* \hat{A} = \lambda_i y^* \hat{E}$.
  $\alpha_i \leftarrow -\lambda_i$, $b_i^* \leftarrow y^* \hat{B}$ and $c_i \leftarrow \hat{C} z_i$.
Form $\hat{E}$, $\hat{A}$, $\hat{B}$, and $\hat{C}$ with (10.24).

10.6 Numerical results

To show the efficiency of the proposed model reduction techniques discussed in this chapter, we present some numerical results. For the numerical experiments, we consider a holonomically constrained damped mass-spring system (DSMS) [122] as briefly described in Appendix A.4.5. The dimension of the original system is 20001 including one algebraic constraint. The number of inputs is 1 and the number of outputs is 3.

All the results are obtained by using MATLAB R2015a (8.5.0.197613) on a board with processor 4×Intel®Core™i5-4460 CPUs with a 2.90 GHz clock speed and 16 GB RAM.

We have computed the low-rank controllability and observability Gramian factors, $\tilde{R}$ and $\tilde{L}$, by solving the Lyapunov equations (10.16a) and (10.16b), respectively. The Lyapunov equations are solved by Algorithm 36 using the adaptive shift computation approach. The convergence rate of both Gramian

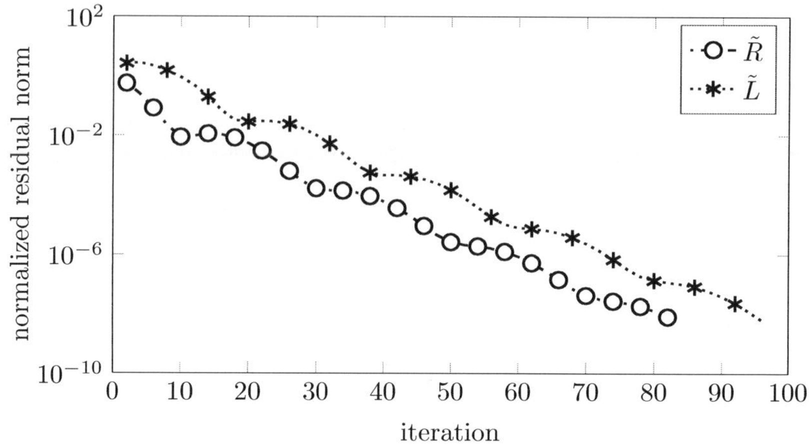

Figure 10.1: Convergence history of the low-rank Gramian factors computed by Algorithm 36.

factors is shown in Figure 10.1. For computing low-rank controllability and observability Gramian factors, the algorithm respectively takes 82 and 96 iteration steps to converge with the tolerance (τ) 10^{-8}. Note that here we have used 20 heuristic shift parameters out of 50 large and 40 small magnitude ritzvalues. These Gramian factors are used in Algorithm 35 which yields an 11 dimensional reduced system with the truncation tolerance 10^{-5}. We have also computed the same dimensional reduced model for the DSMS by applying the interpolatory projection method, i.e., Algorithm 37. Both the model reduction methods (BT and IRKA) generate reduced models with good accuracy which is shown in Figure 10.2. This figure shows the efficiency of both the model reduction methods discussed in this chapter. Figure 10.2(a) shows a good match of the transfer functions of the original and the 11 dimensional reduced models obtained by BT and IRKA. The approximation error as shown in Figure 10.2(b) is below 10^{-5} which was the error bound for the BT method. We have also shown the input 1 to output 1 relation of the original and reduced models in Figure 10.3. Figure 10.3(a) plots the step responses of the original and reduced models, and Figure 10.3(a) shows the absolute deviation of the step responses between the original and reduced models obtained by balanced truncation and IRKA. Although the error for the reduced model with IRKA is increasing in the higher time domain, it is still far below the error bound of 10^{-5}. Note that to compute the step response, the implicit Euler method with a fixed time step size of .5 was applied.

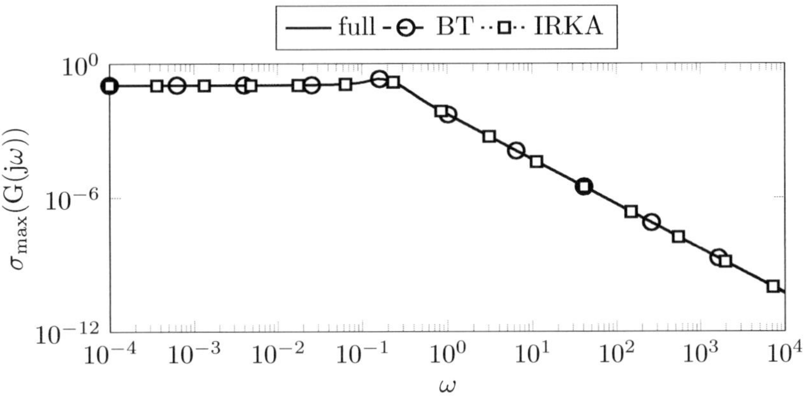

(a) Sigma plot.

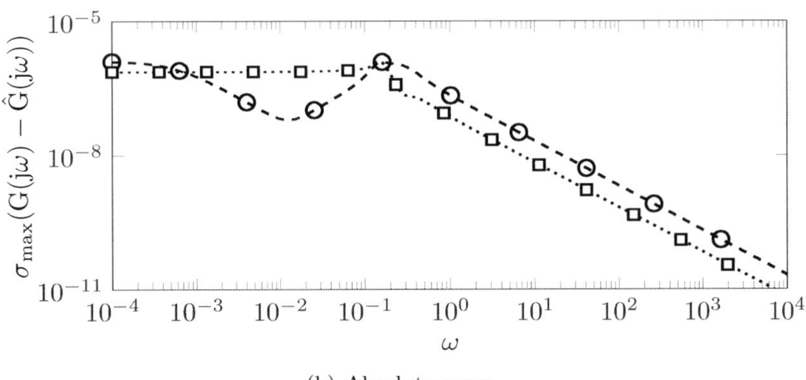

(b) Absolute error.

Figure 10.2: Frequency-domain comparison of the full and the 11 dimensional reduced models obtained by applying balanced truncation and IRKA.

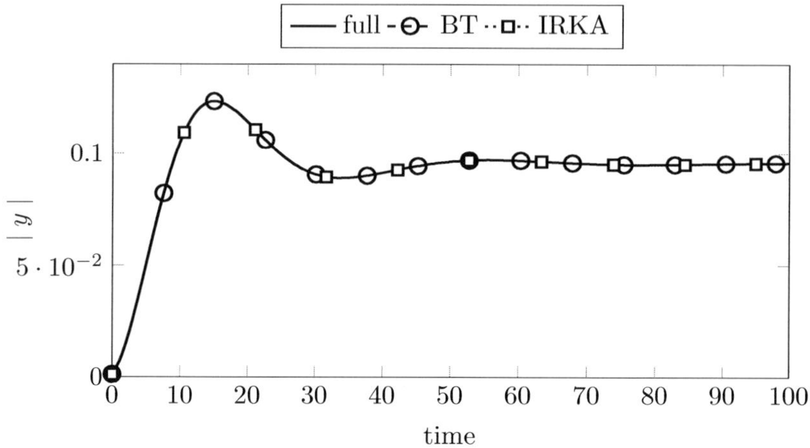

(a) Step response.

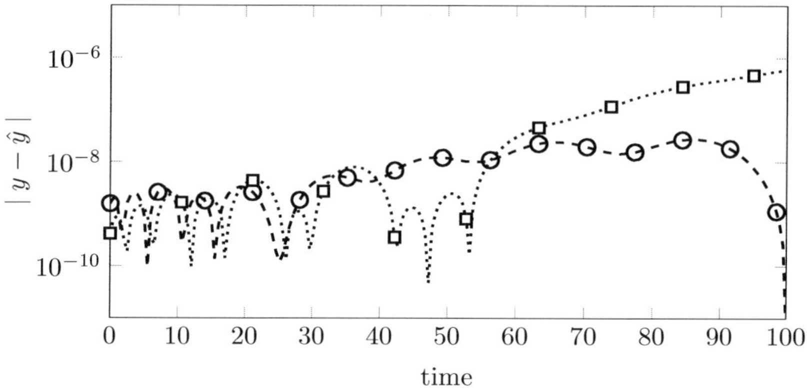

(b) Absolute error.

Figure 10.3: Time-domain comparison of the full and the 11 dimensional reduced models obtained by applying balanced truncation and IRKA.

■ Exercises:

10.1 Consider a first-order index 3 descriptor system (10.1) where

$$M = \begin{bmatrix} 1 & 0 \\ 0 & 2 \end{bmatrix}, \; K = \begin{bmatrix} -5 & 2 \\ 2 & -5 \end{bmatrix}, \; D = 0.001\,M + 5\,K,$$

$$G_1 = \begin{bmatrix} -1 & -1 \end{bmatrix} \text{ and } L^T = H = \begin{bmatrix} 1 \\ 0 \end{bmatrix}.$$

Write the system in matrix-vector form as in (10.2). Using MATLAB function `eig`, find the eigenvalues of the system. Is it stable?

10.2 Show that the matrix pencil $\mathcal{P}(\lambda) = \lambda E - A$ of system (10.2) is regular iff G_1 has full (row) rank.

10.3 The number of finite eigenvalues of system (10.2) is equal to $n - 3n_a$ where n is the dimension and n_a is the number of algebraic variables of the system. Verify this using the model in Exercise 10.1.

10.4 Section 10.2 claims that the transfer function of the first-order index 3 system (10.1) is equal to the transfer functions of projected systems (10.11) and (10.14); verify this using the model in Exercise 10.1. Graph the frequency responses and the absolute deviations of the index 3 system and the projected system on a suitable frequency range.

10.5 The finite spectrum of systems (10.1), (10.11) and (10.14) are equal. Verify this using the data of constrained damped mass-spring (CDMS) system in Appendix A.4.5. (You can create a small system by choosing $n_g = 10$ masses only).

10.6 Write an algorithm for solving the projected observability Lyapunov equation (10.16b) based on the LR-ADI iteration that is discussed in this chapter. Then verify the performance of the algorithm using both heuristic and adaptive shifts for the CDMS model in Appendix A.4.5.

Chapter 11

Model Reduction of Second-Order Index 1 Descriptor Systems

11.1 Introduction

In this chapter we consider second-order index 1 descriptor systems of the form

$$\underbrace{\begin{bmatrix} M_{11} & 0 \\ 0 & 0 \end{bmatrix}}_{\check{M}} \begin{bmatrix} \ddot{\xi}(t) \\ \ddot{\varphi}(t) \end{bmatrix} + \underbrace{\begin{bmatrix} D_{11} & 0 \\ 0 & 0 \end{bmatrix}}_{\check{D}} \begin{bmatrix} \dot{\xi}(t) \\ \dot{\varphi}(t) \end{bmatrix} + \underbrace{\begin{bmatrix} K_{11} & K_{12} \\ K_{21} & K_{22} \end{bmatrix}}_{\check{K}} \begin{bmatrix} \xi(t) \\ \varphi(t) \end{bmatrix} = \underbrace{\begin{bmatrix} H_1 \\ H_2 \end{bmatrix}}_{\check{H}} u(t), \tag{11.1a}$$

$$\underbrace{\begin{bmatrix} L_1 & L_2 \end{bmatrix}}_{\check{L}} \begin{bmatrix} \xi(t) \\ \varphi(t) \end{bmatrix} = y(t), \tag{11.1b}$$

where $\xi(t) \in \mathbb{R}^{n_\xi}$, $\varphi(t) \in \mathbb{R}^{n_\varphi}$ are the states and $n_\xi > n_\varphi$; $u(t) \in \mathbb{R}^m$ are control inputs and the measurement outputs are $y(t) \in \mathbb{R}^p$. Note that the matrices

$\bar{M}$, $\bar{D}$ and $\bar{K}$ are sparse. We assume the block matrix K_{22} to be nonsingular. We call (11.1) an index 1 system due to the analogy of first-order index 1 linear time invariant (LTI) systems [165, 166, 168]. Such dynamical systems usually arise in different branches of engineering such as mechanics [64] where an extra constraint is imposed in order to control the dynamic behavior of the systems and mechatronics [126, 127, 128] where mechanical and electrical components are coupled with each other. In the specific case of the model example we use in the numerical experiments, the index 1 characteristics result from the multiphysics application with very different timescales. See the model example in Appendix A.5.1. This allows us to treat one variable by a stationary analysis, while the other is covered fully dynamically.

If the model is very large, performing the simulation with it has a prohibitively high computational complexity or is simply impossible due to the limited computer memory. Therefore, reducing the size of the system is essential for fast simulation.

In principle, the index 1 system (11.1) can be converted into an ODE system (see, e.g., the next section). Then the naive approaches of model reduction can be applied to the second-order ODE system as discussed in Chapter 5. However, this ODE system formulation loses the sparsity of the system and becomes dense. Therefore, one has to encounter complexity in computation. Therefore, we never compute ODEs explicitly in implementing MOR of the dynamical system (11.1).

This chapter will discuss both second-order-to-first-order and second-order-to-second-order methods for the model reduction of the underlying system. For the second-order-to-first-order reduction, we will apply balanced truncation and interpolatory projection methods. On the other hand, in second-order-to-second-order reduction, we will apply balanced truncation and PDEG (projection onto the dominant eigenspace of the Gramian). Moreover, we will also investigate the LR-ADI iteration and related issues (e.g., ADI shift parameter computation) for the second-order index 1 system.

The proposed techniques will be applied to a piezo-actuated structural FEM model [130, 129] that is used as a certain building block of a parallel kinematic machine tool. Numerical results will illustrate the efficiency of the techniques.

11.2 Second-order-to-first-order reduction techniques

This section discusses second-order-to-first-order reduction techniques of second-order index 1 system (11.1) using both the balanced truncation and interpolatoty projection methods. Although there exists a variety of techniques for transforming of (11.1) into its first order form, we consider the following formulation:

$$\begin{bmatrix} 0 & M_{11} & 0 \\ M_{11} & D_{11} & 0 \\ 0 & 0 & 0 \end{bmatrix} \begin{bmatrix} \ddot{\xi}(t) \\ \dot{\xi}(t) \\ \dot{\varphi}(t) \end{bmatrix} = \begin{bmatrix} M_{11} & 0 & 0 \\ 0 & -K_{11} & -K_{12} \\ 0 & -K_{21} & -K_{22} \end{bmatrix} \begin{bmatrix} \dot{\xi}(t) \\ \xi(t) \\ \varphi(t) \end{bmatrix} + \begin{bmatrix} 0 \\ H_1 \\ H_2 \end{bmatrix} u(t),$$
$$y(t) = \begin{bmatrix} 0 & L_1 & L_2 \end{bmatrix} \begin{bmatrix} \dot{\xi}(t) \\ \xi(t) \\ \varphi(t) \end{bmatrix}. \tag{11.2}$$

Another first-order representation of the underlying second-order index 1 system can be

$$\begin{bmatrix} M_{11} & 0 & 0 \\ 0 & M_{11} & 0 \\ 0 & 0 & 0 \end{bmatrix} \begin{bmatrix} \dot{\xi}(t) \\ \ddot{\xi}(t) \\ \dot{\varphi}(t) \end{bmatrix} = \begin{bmatrix} 0 & M_{11} & 0 \\ -K_{11} & -D_{11} & -K_{12} \\ -K_{21} & 0 & -K_{22} \end{bmatrix} \begin{bmatrix} \xi(t) \\ \dot{\xi}(t) \\ \varphi(t) \end{bmatrix} + \begin{bmatrix} 0 \\ H_1 \\ H_2 \end{bmatrix} u(t),$$
$$y(t) = \begin{bmatrix} L_1 & 0 & L_2 \end{bmatrix} \begin{bmatrix} \xi(t) \\ \dot{\xi}(t) \\ \varphi(t) \end{bmatrix}. \tag{11.3}$$

Systems (11.2) and (11.3) are now in first-order index 1 form (as defined in Chapter 6). This chapter considers the representation in (11.2) which can again be written as

$$\underbrace{\begin{bmatrix} E_1 & 0 \\ 0 & 0 \end{bmatrix}}_{\check{E}} \begin{bmatrix} \dot{z}(t) \\ \dot{\varphi}(t) \end{bmatrix} = \underbrace{\begin{bmatrix} A_1 & A_2 \\ A_3 & A_4 \end{bmatrix}}_{\check{A}} \begin{bmatrix} z(t) \\ \varphi(t) \end{bmatrix} + \underbrace{\begin{bmatrix} B_1 \\ B_2 \end{bmatrix}}_{\check{B}} u(t),$$
$$y(t) = \underbrace{\begin{bmatrix} C_1 & C_2 \end{bmatrix}}_{\check{C}} \begin{bmatrix} z(t) \\ \varphi(t) \end{bmatrix}, \tag{11.4}$$

where

$$E_1 := \begin{bmatrix} 0 & M_{11} \\ M_{11} & D_{11} \end{bmatrix}, \quad A_1 := \begin{bmatrix} M_{11} & 0 \\ 0 & -K_{11} \end{bmatrix}, \quad A_2 := \begin{bmatrix} 0 \\ -K_{12} \end{bmatrix},$$
$$A_3 := \begin{bmatrix} 0 & -K_{21} \end{bmatrix}, \quad A_4 = -K_{22}, \quad B_1 := \begin{bmatrix} 0 \\ H_1 \end{bmatrix}, \quad B_2 := H_2, \tag{11.5}$$
$$C_1 := \begin{bmatrix} 0 & L_1 \end{bmatrix}, \quad C_2 := L_2, \quad z(t) := \begin{bmatrix} \dot{\xi}(t) \\ \xi(t)) \end{bmatrix}.$$

This system can be compared with the first-order index 1 system (7.1) from Chapter 7. Therefore, we can apply the procedure of model reduction as we have introduced in Chapter 7. Since the submatrix A_4 is nonsingular, we can express system (11.4) in a compact form

$$E\dot{z}(t) = Az(t) + Bu(t), \quad y(t) = Cz(t) + D_a u(t), \tag{11.6}$$

with

$$E = E_1, \quad A = A_1 - A_2 A_4^{-1} A_3,$$
$$B = B_1 - A_2 A_4^{-1} B_2, \quad C = C_1 - C_2 A_4^{-1} A_3, \quad D_a = -C_2 A_4^{-1} B_2.$$

The algebraic part of system (11.4) has been removed in (11.6). Hence, one can apply the standard model reduction techniques (e.g., balanced truncation and interpolatory methods as discussed in Chapter 4) to system (11.6) to obtain a substantially reduced system

$$\begin{aligned} \hat{E}\dot{\hat{z}}(t) &= \hat{A}\hat{z}(t) + \hat{B}u(t), \\ \hat{y}(t) &= \hat{C}^T \hat{z}(t) + \hat{D}_a u(t). \end{aligned} \tag{11.7}$$

Again note that matrix A is typically dense, which increases the computational cost and memory requirements in the implementation. Therefore, we should avoid converting system (11.4) into (11.6) explicitly. In the subsections to come, we discuss efficient balanced truncation (BT) and interpolatory methods for the model reduction of system (11.4) while avoiding the explicit formulation of system (11.6).

11.2.1 Balancing-based method

Suppose we want to apply BT onto system (11.4) which is similar to the generalized state space model in (4.1). We therefore recall Algorithm 19 (LR-SRM). For this purpose, we need to solve two continuous-time algebraic Lyapunov equations

$$APE^T + EPA^T = -BB^T, \tag{11.8}$$
$$A^T QE + E^T QA = -C^T C. \tag{11.9}$$

where $P \in \mathbb{R}^{2n_\xi \times 2n_\xi}$ and $Q \in \mathbb{R}^{2n_\xi \times 2n_\xi}$ are the system's controllability and the observability Gramians, respectively; the matrices E, A, B and C are defined in (11.6). We can solve the Lyapunov equations via the low-rank ADI iteration (i.e., Algorithm 15) introduced in Chapter 3 to find the low-rank Gramian factors. Note that later in this section, we will discuss how to solve these Lyapunov equations efficiently using the LR-ADI method.

Suppose R and L are the low-rank factors of P and Q, respectively, which satisfy

$$RR^T \approx P \quad \text{and} \quad LL^T \approx Q. \tag{11.10}$$

Once we have the Gramian factors R and L, the balancing and truncating transformation can be formed by computing the singular value decomposition (SVD)

$$LER^T = LE_1 R^T = \begin{bmatrix} U_1 & U_2 \end{bmatrix} \begin{bmatrix} \Sigma_1 & \\ & \Sigma_2 \end{bmatrix} \begin{bmatrix} V_1^T \\ V_2^T \end{bmatrix}, \tag{11.11}$$

Algorithm 38: LR-SRM for second-order index 1 systems.

Input : M_{11}, D_{11}, K_{11}, K_{12}, K_{21}, K_{22}, H_1, H_2, L_1, L_2.
Output: $\hat{E}$, $\hat{A}$, $\hat{B}$, $\hat{C}$, $\hat{D}_a := D_a$.

1 Form E_1, A_1, A_2, A_3, A_4, B_1, B_2, C_1, C_2 as in (11.4) and $D_a = -C_2^T A_4^{-1} B_2$.

2 Compute R, L by solving the Lyapunov equations:

$$APE^T + EPA^T = -BB^T, \quad A^TQE + E^TQA = -C^TC.$$

where E, A, B and C are defined in (11.6).

3 Compute V and W by performing (11.11) - (11.12).

4 Form the reduced matrices $\hat{E}$, $\hat{A}$, $\hat{B}$ and $\hat{C}$ as in (11.13).

and defining

$$V := LV_1\Sigma_1^{-\frac{1}{2}}, \quad W := RU_1\Sigma_1^{-\frac{1}{2}}. \tag{11.12}$$

The reduced system (11.7) is obtained by constructing the reduced matrices as

$$\begin{aligned} &\hat{E} = W^TEV = W^TE_1V, \ \hat{A} = \hat{A}_1 - \hat{A}_2^T A_4^{-1} \hat{A}_2, \\ &\hat{B} = \hat{B}_1 - \hat{A}_2 A_4^{-1} B_2, \quad \hat{C} = \hat{C}_1 - C_2 A_4^{-1} \hat{A}_3, \quad \hat{D}_a := D_a. \end{aligned} \tag{11.13}$$

where $\hat{A}_1 = W^TA_1V$, $\hat{A}_2 = W^TA_2$, $\hat{A}_3 = A_3V$, $\hat{B}_1 = V^TB_1$, $\hat{C}_1 = C_1V$. The whole procedure to obtain the reduced ODE system (11.7) for a given second-order index 1 system (11.1) is summarized in Algorithm 38. However, we represent (11.7) in the reduced index 1 DAE setting as

$$\begin{aligned} \begin{bmatrix} \hat{E} & 0 \\ 0 & 0 \end{bmatrix} \begin{bmatrix} \dot{\hat{z}}(t) \\ \dot{\varphi}(t) \end{bmatrix} &= \begin{bmatrix} \hat{A}_1 & \hat{A}_2 \\ \hat{A}_3 & A_4 \end{bmatrix} \begin{bmatrix} \hat{z}(t) \\ \varphi(t) \end{bmatrix} + \begin{bmatrix} \hat{B}_1 \\ B_2 \end{bmatrix} u(t), \\ y(t) &= \begin{bmatrix} \hat{C}_1^T & C_2^T \end{bmatrix} \begin{bmatrix} \hat{z}(t) \\ \varphi(t) \end{bmatrix}. \end{aligned} \tag{11.14}$$

Note that the reduced system (11.14) is not very useful if the block matrix A_4 is large, because in that case the reduced model is still large.

11.2.2 Interpolatory projections via IRKA

In this subsection, we discuss the model reduction technique for the second-order index 1 descriptor system (11.1) by applying the interpolatory method via IRKA. We can start with the same procedure as discussed for the balanced truncation. That means, first we convert the second-order DAEs (11.1) into their first-order form (11.4), then to the generalized state space form (11.6). When the ODE in (11.6) has been formed, we can immediately follow Algorithm 21 to construct the ROM for the second-order index 1 system.

Again note that the explicit formulation of (11.6) is prohibitive due to the reasons mentioned earlier. Following the procedure discussed in Chapter 4 (i.e., Algorithm 21), we construct the left transformation (W) and the right transformation (V) as

$$V = \left[(\alpha_1 E - A)^{-1}Bb_1, \cdots, (\alpha_r E - A)^{-1}Bb_r\right], \tag{11.15a}$$

$$W = \left[(\alpha_1 E^T - A^T)^{-1}C^T c_1, \cdots, (\alpha_r E^T - A^T)^{-1}C^T c_r\right], \tag{11.15b}$$

where $\{\alpha_i\}_{i=1}^r$ is the set of interpolation points and $\{b_i\}_{i=1}^r$ and $\{c_i\}_{i=1}^r$ are the set of tangential directions. By using the transformations V and W, we construct the ROM as in (11.7) where the reduced matrices are formed following (11.13) thus completing the method. Now the question of how to efficiently construct V and W in (11.15) is answered in the following text.

In (11.15a) each column of V can be computed by solving a shifted linear system like

$$(\alpha E - A)\chi = Bb, \tag{11.16}$$

which implies

$$(\alpha E_1 - A_1 + A_2 A_4^{-1} A_3)\chi = (B_1 - A_2 A_4^{-1} B_2)b.$$

Undoing the *Schur complement* [192], this linear system leads to

$$\begin{bmatrix} \alpha E_1 - A_1 & -A_2 \\ -A_3 & -A_4 \end{bmatrix} \begin{bmatrix} \chi \\ \Gamma \end{bmatrix} = \begin{bmatrix} B_1 b \\ B_2 b \end{bmatrix}. \tag{11.17}$$

Inserting E_1, A_1, A_2, A_4, B_1 and B_2 from (11.34), the linear system (11.23) becomes

$$\begin{bmatrix} -M_{11} & \alpha M_{11} & 0 \\ \alpha M_{11} & \alpha D_{11} + K_{11} & K_{12} \\ 0 & K_{21} & K_{22} \end{bmatrix} \begin{bmatrix} \chi_1 \\ \chi_2 \\ \Gamma \end{bmatrix} = \begin{bmatrix} 0 \\ H_1 b \\ H_2 b \end{bmatrix}, \tag{11.18}$$

for $\begin{bmatrix}\chi_1^T & \chi_2^T\end{bmatrix}^T$. Although the matrix in (11.18) has larger dimension ($2n_\xi + n_\varphi$), it is sparse and can efficiently be solved by suitable direct (e.g., [58, 63]) or iterative (e.g., [143, 170]) solvers. Further, splitting the linear system (11.18) as

$$-M_{11}\chi_1 + \alpha M_{11}\chi_2 = 0, \tag{11.19a}$$

$$\alpha M_{11}\chi_1 + (\alpha D_{11} + K_{11})\chi_2 + K_{12}\Gamma = H_1 b, \tag{11.19b}$$

$$K_{21}\chi_2 + K_{22}\Gamma = H_2 b, \tag{11.19c}$$

from (11.19a) and (11.19c) we respectively obtain

$$\chi_1 = \alpha\chi_2 \tag{11.20}$$

and

$$\Gamma = K_{22}^{-1} H_2 b - K_{22}^{-1} K_{21} \chi_2.$$

Inserting χ_1 and Γ into (11.19b) yields

$$\alpha^2 M_{11} \chi_2 + (\alpha D_{11} + K_{11}) \chi_2 - K_{12} K_{22}^{-1} K_{21} \chi_2 = H_1 b - K_{12} K_{22}^{-1} H_2 b,$$

which is again equivalent to the solution of the linear system

$$\begin{bmatrix} \alpha^2 M_{11} + \alpha D_{11} + K_{11} & K_{12} \\ K_{21} & K_{22} \end{bmatrix} \begin{bmatrix} \chi_2 \\ \Gamma \end{bmatrix} = \begin{bmatrix} H_1 b \\ H_2 b \end{bmatrix}, \tag{11.21}$$

for χ_2. Therefore, instead of solving (11.18) we can solve (11.21) for χ_2 then compute χ_1 from (11.20).

Again in (11.15b), each column of W can be computed by solving a shifted linear system like

$$(\alpha E^T - A^T)\chi = C^T c, \tag{11.22}$$

which implies

$$(\alpha E_1^T - A_1^T + A_2^T A_4^{-T} A_3^T)\chi = (C_1^T - A_3^T A_4^{-T} C_2^T)c.$$

Undoing the *Schur complement* [192], this linear system leads to

$$\begin{bmatrix} \alpha E_1^T - A_1^T & -A_3^T \\ -A_2^T & -A_4^T \end{bmatrix} \begin{bmatrix} \chi \\ \Gamma \end{bmatrix} = \begin{bmatrix} C_1^T c \\ C_2^T c \end{bmatrix}. \tag{11.23}$$

Inserting E_1, A_1, A_2, A_4, C_1 and C_2 from (11.34), the linear system (11.23) becomes

$$\begin{bmatrix} -M_{11}^T & \alpha M_{11}^T & 0 \\ \alpha M_{11}^T & \alpha D_{11}^T + K_{11}^T & K_{21}^T \\ 0 & K_{12}^T & K_{22}^T \end{bmatrix} \begin{bmatrix} \chi_1 \\ \chi_2 \\ \Gamma \end{bmatrix} = \begin{bmatrix} 0 \\ L_1^T c \\ L_2^T c \end{bmatrix}, \tag{11.24}$$

for $\begin{bmatrix} \chi_1^T & \chi_2^T \end{bmatrix}^T$. Again we can split this large system as follows. First solve

$$\begin{bmatrix} \alpha^2 M_{11}^T + \alpha D_{11}^T + K_{11}^T & K_{21}^T \\ K_{12}^T & K_{22}^T \end{bmatrix} \begin{bmatrix} \chi_2 \\ \Gamma \end{bmatrix} = \begin{bmatrix} L_1^T c \\ L_2^T c \end{bmatrix}, \tag{11.25}$$

for χ_2. Then compute χ_1 from $\chi_1 = \alpha \chi_2$.

Applying this splitting idea to the linear systems (11.18) and (11.24), instead of solving a $2n_\xi + n_\varphi$ dimensional linear system, we can solve only a $n_\xi + n_\varphi$ dimensional linear system which ensures faster computation. The whole procedure of IRKA for computing ROM (11.7) for the second-order index 1 descriptor system (11.1) is summarized in Algorithm 39.

Algorithm 39: IRKA for second-order index 1 systems.

Input : M_{11}, D_{11}, K_{11}, K_{12}, K_{21}, K_{22}, H_1, H_2, L_1, L_2.
Output: $\hat{E}$, $\hat{A}$, $\hat{B}$, $\hat{C}$, $\hat{D}_a := L_2^T K_4^{-1} H_2$.

Form E_1, A_1, A_2, A_4, B_1, B_2, C_1, C_2 as in (11.4).
Make an initial selection of the interpolation points $\{\alpha_i\}_{i=1}^r$ and tangential directions $\{b_i\}_{i=1}^r$ and $\{c_i\}_{i=1}^r$.
while *(not converged)* **do**
    for $i = 1, 2, \cdots, r$ **do**
        solve $\begin{bmatrix} \alpha_i^2 M_{11} + \alpha_i D_{11} + K_{11} & K_{12} \\ K_{12}^T & K_{22} \end{bmatrix} \begin{bmatrix} \chi_2 \\ \Gamma \end{bmatrix} = \begin{bmatrix} H_1 b_i \\ H_2 b_i \end{bmatrix}$, for χ_2
        and $\begin{bmatrix} \alpha_i^2 M_{11}^T + \alpha_i D_{11}^T + K_{11}^T & K_{21}^T \\ K_{12}^T & K_{22}^T \end{bmatrix} \begin{bmatrix} \chi_2' \\ \Gamma \end{bmatrix} = \begin{bmatrix} L_1^T c_i \\ L_2^T c_i \end{bmatrix}$ for χ_2'.
        Construct $\mathrm{v}_i = \begin{bmatrix} \alpha_i \chi_2 \\ \chi_2 \end{bmatrix}$ and $\mathrm{w}_i = \begin{bmatrix} \alpha_i \chi_2' \\ \chi_2' \end{bmatrix}$,
        Form
        $V = \begin{bmatrix} \mathrm{v}_1, \mathrm{v}_2, \cdots, \mathrm{v}_r \end{bmatrix}$.
        $W = \begin{bmatrix} \mathrm{w}_1, \mathrm{w}_2, \cdots, \mathrm{w}_r \end{bmatrix}$.
    $\hat{E} = V^T E_1 V$, $\hat{A} = V^T A_1 V - (V^T A_2) A_4^{-1} (A_3 V)$
    $\hat{B} = V^T B_1 V - (V^T A_2) A_4^{-1} B_2$ and $\hat{C} = C_1 V - C_2 A_4^{-1} (A_3 V)$
    Compute $\hat{A} z_i = \hat{\lambda}_i \hat{E} z_i$, where z_i is the eigenvector associated with $\hat{\lambda}_i$.
    $\alpha_i \leftarrow -\hat{\lambda}_i$, $b_i \leftarrow -\hat{B}^T z_i$ and $c_i \leftarrow \hat{C} z_i$, for $i = 1, \cdots, r$.
Form the reduced matrices $\hat{E}$, $\hat{A}$, $\hat{B}$ and $\hat{C}$ as in (11.13).

11.3 Second-order-to-second-order MOR techniques

Second-order-to-second-order (i.e., structure preserving) MOR techniques of standard second-order systems were introduced in Chapter 5. We can apply the techniques of Chapter 5 to the second-order index 1 system (11.1) by converting the system into its standard form. In the following subsection, we will show how to convert the second-order index 1 system into an equivalent form of ODE system. Then, we will discuss how to apply the balanced truncation and the PDEG techniques onto the ODE system efficiently. Note that, although we apply the model reduction onto the converted ODE system, we never construct this explicitly because such a conversion is infeasible for the large-scale dynamical system.

11.3.1 Conversion into equivalent form of ODE system

This subsection discusses how to eliminate the algebraic equations in system (11.1) to convert into its equivalent form of ODE system. For this purpose, rewrite the index 1 system in (11.1) as

$$M_{11}\ddot{\xi}(t) + D_{11}\dot{\xi}(t) + K_{11}\xi(t) + K_{12}\varphi(t) = H_1 u(t) \tag{11.26a}$$

$$K_{21}\xi(t) + K_{22}\varphi(t) = H_2 u(t) \tag{11.26b}$$

$$y(t) = L_1\xi(t) + L_2\varphi(t) \tag{11.26c}$$

Since the block matrix K_{22} is invertible, from (11.26b) we obtain

$$\varphi(t) = -K_{22}^{-1}K_{21}\xi(t) + K_{22}^{-1}H_2 u(t).$$

Insert this identity into (11.26a) and simplify the resulting equation which yields

$$M_{11}\ddot{\xi}(t) + D_{11}\dot{\xi}(t) + (K_{11} - K_{12}K_{22}^{-1}K_{21})\xi(t) = (H_1 - K_{12}K_{22}^{-1}H_2)u(t). \tag{11.27}$$

From the output equation (11.26c) we also obtain

$$y(t) = L_1\xi(t) + L_2(-K_{22}^{-1}K_{21})\xi(t) + L_2K_{22}^{-1}H_2u(t)),$$

which gives

$$y(t) = (L_1 - L_2K_{22}^{-1}K_{21})\xi(t) + L_2K_{22}^{-1}H_2u(t)). \tag{11.28}$$

Combining (11.27) and (11.28), the index 1 system (11.1) is transformed into an equivalent ODE system

$$\begin{aligned} M\ddot{\xi}(t) + D\dot{\xi}(t) + K\xi(t) &= Hu(t), \\ y(t) &= Lx(t) + D_a u(t), \end{aligned} \tag{11.29}$$

where

$$\begin{aligned} &M = M_{11}, \;\; D = D_{11}, \\ &K = K_{11} - K_{12}K_{22}^{-1}K_{21}, \;\; H = H_1 - K_{12}K_{22}^{-1}H_2, \\ &L = L_1 - L_2K_{22}^{-1}K_{12}, \;\; D_a = L_2^T K_{22}^{-1}H_2. \end{aligned} \tag{11.30}$$

Descriptor system (11.1) and standard system (11.29) are equivalent. Because the transfer function of the system (11.1) which is defined by

$$\bar{G}(s) = \bar{L}(s^2\bar{M} + s\bar{D} + \bar{K})^{-1}\bar{H}, \tag{11.31}$$

and the transfer function of the system (11.29) which is defined by

$$G(s) = L(s^2M + sD + K)^{-1}H + L_2^T K_{22}^{-1}H_2, \tag{11.32}$$

are the same. Moreover, the finite spectra of (11.1) and (11.29) are also the same. These two statements are left to be tested in the exercise section.

In principle, we can apply the structure preserving model reduction techniques to system (11.29) by following the approaches discussed in Chapter 5. In this case, the matrix K is usually dense which causes enormous computational complexity. Moreover, for a large-scale system with a large K_{22} block (e.g., the system that we consider for the numerical experiments), due to memory restriction, forming (11.29) is simply infeasible. In the following subsections we discuss how to perform the model reduction for the DAE system (11.1) without forming the ODE system (11.29) explicitly.

11.3.2 Balancing-based method

Convert the system (11.29) into first-order form

$$\mathcal{E}\dot{z}(t) = \mathcal{A}z(t) + \mathcal{B}u(t), \quad y(t) = \mathcal{C}z(t) + D_a u(t), \tag{11.33}$$

where

$$\begin{aligned} \mathcal{E} &:= \begin{bmatrix} 0 & M \\ M & D \end{bmatrix}, \quad \mathcal{A} := \begin{bmatrix} M & 0 \\ 0 & -K \end{bmatrix}, \quad \mathcal{B} := \begin{bmatrix} 0 \\ H \end{bmatrix}, \\ \mathcal{C} &:= \begin{bmatrix} 0 \\ L \end{bmatrix}, \quad z(t) := \begin{bmatrix} \dot{\xi}(t) \\ \xi(t)) \end{bmatrix}. \end{aligned} \tag{11.34}$$

This is now like a generalized state-space system (4.1). In fact $(\mathcal{E}, \mathcal{A}, \mathcal{B}, \mathcal{C})$ in (11.34) and (E, A, B, C) in (11.6) are exactly the same. Therefore, the corresponding Lyapunov equations for the system (11.34) are (11.8) and (11.9), where P and Q are the controllability and observability Gramians of the second-order system (11.29). Let us recall the Gramians of the standard second-order systems as defined in Chapter 5. Following the discussion of Chapter 5, due to the structure of the system, the Gramians P and Q can be partitioned as

$$P = \begin{bmatrix} P_v & P_0 \\ P_0^T & P_p \end{bmatrix}, \quad Q = \begin{bmatrix} Q_v & Q_0 \\ Q_0^T & Q_p \end{bmatrix}, \tag{11.35}$$

where P_v and P_p respectively denote the controllability velocity and position Gramians; Q_v and Q_p respectively denote the observability velocity and position Gramians. These Gramians have rank deficiency and can be approximated by their low-rank factors. Note that we will discuss how to compute the low-rank Gramian factors by solving the corresponding Lyapunov equations later. We have already considered in Subsection 11.2.1 that R and L are the low-rank controllability and observability Gramian factors of P and Q, respectively. Due to the structure of system (11.33), they can be partitioned as $R = \begin{bmatrix} R_v^T & R_p^T \end{bmatrix}^T$ such that

$$P \approx RR^T = \begin{bmatrix} R_v \\ R_p \end{bmatrix} \begin{bmatrix} R_v^T & R_p^T \end{bmatrix} = \begin{bmatrix} R_v R_v^T & R_v R_p^T \\ R_p R_v^T & R_p R_p^T \end{bmatrix}, \tag{11.36}$$

where R_v is called the low-rank factor of the controllability velocity Gramian, and R_p is called the low-rank factor of the controllability position Gramian. Therefore we obtain the following relations

$$P_v \approx R_v R_v^T \quad \text{and} \quad P_p \approx R_p R_p^T. \tag{11.37}$$

Once the low-rank controllability Gramian factor R is computed by solving the Lyapunov equation (11.8), R_v and R_p can be obtained by taking upper n_ξ and lower n_ξ rows of R. Similarly, by partitioning L as $L = \begin{bmatrix} L_v^T & L_p^T \end{bmatrix}^T$ such that

$$Q \approx LL^T = \begin{bmatrix} L_v \\ L_p \end{bmatrix} \begin{bmatrix} L_v^T & L_p^T \end{bmatrix} = \begin{bmatrix} L_v L_v^T & L_v L_p^T \\ L_p L_v^T & L_p L_p^T \end{bmatrix}, \tag{11.38}$$

we can define the low-rank observability velocity Gramian factor L_v and position Gramian factor L_p. Once the low-rank controllability Gramian factor L is computed by solving the Lyapunov equation (11.9), L_v and L_p can be obtained by taking upper n_ξ and lower n_ξ rows of L. Using R_v and L_v construct the balancing and truncating transformations as follows.

- Perform the SVD

$$L_v^T M R_v = U_{vv} \Sigma_{vv} V_{vv}^T = \begin{bmatrix} U_{vv,1} & U_{vv,2} \end{bmatrix} \begin{bmatrix} \Sigma_{vv,1} & \\ & \Sigma_{vv,2} \end{bmatrix} \begin{bmatrix} V_{vv,1}^T \\ V_{vv,2}^T \end{bmatrix}. \tag{11.39}$$

- Define the balancing and truncating transformations as

$$W_s := L_v U_{vv,1} \Sigma_{vv,1}^{-\frac{1}{2}}, \quad V_s := R_v V_{vv,1} \Sigma_{vv,1}^{-\frac{1}{2}}, \tag{11.40}$$

  where $U_{vv,1}$ and $V_{vv,1}$ are composed of the leading k columns of U_{vv} and V_{vv}, respectively, and $\Sigma_{vv,1}$ is the first $k \times k$ block of the matrix Σ_{vv}.

By applying W_s and V_s to system (11.29), we obtain the reduced model

$$\begin{aligned} \hat{M}\ddot{\hat{\xi}}(t) + \hat{D}\dot{\hat{\xi}}(t) + \hat{K}\hat{\xi}(t) &= \hat{H}u(t), \\ \hat{y}(t) &= \hat{L}\hat{\xi}(t) + \hat{D}_a u(t), \end{aligned} \tag{11.41}$$

where the reduced coefficient matrices are formed as

$$\begin{aligned} &\hat{M} = W_s^T M_{11} V_s, \quad \hat{D} = W_s^T D_{11} V_s, \quad \hat{K} = \hat{K}_{11} - \hat{K}_{12} K_{22}^{-1} \hat{K}_{21} \\ &\hat{H} = \hat{H}_1 - \hat{K}_{12} K_{22}^{-1} H_2, \quad \hat{L} = \hat{L}_1 - L_2 K_{22}^{-1} \hat{K}_{21}, \quad \hat{D}_a = D_a, \end{aligned} \tag{11.42}$$

where

$$\begin{aligned} \hat{K}_{11} = W_s^T K_{11} V_s, \quad \hat{K}_{12} = W_s^T K_{12}, \quad \hat{K}_{21} = K_{21} V_s, \\ \hat{H}_1 = W_s^T H_1, \quad \hat{L}_1 = L_1 V_s. \end{aligned}$$

type	SVD	left proj. W_s	right proj. V_s
VV	$L_v^T M_{11} R_v = U_{vv}\Sigma_{vv}U_{vv}^T$	$L_v U_{vv,1}\Sigma_{vv,1}^{-\frac{1}{2}}$	$R_v U_{vv,1}\Sigma_{vv,1}^{-\frac{1}{2}}$
PP	$L_p^T M_{11} R_p = U_{pp}\Sigma_{pp}U_{pp}^T$	$L_p U_{pp,1}\Sigma_{pp,1}^{-\frac{1}{2}}$	$R_p U_{pp,1}\Sigma_{pp,1}^{-\frac{1}{2}}$
VP	$L_v^T M_{11} R_p = U_{vp}\Sigma_{vp}V_{vp}^T$	$L_p U_{vp,1}\Sigma_{vp,1}^{-\frac{1}{2}}$	$R_v V_{vp,1}\Sigma_{vp,1}^{-\frac{1}{2}}$
PV	$L_p^T M_{11} R_v = U_{pv}\Sigma_{pv}V_{pv}^T$	$L_v U_{pv,1}\Sigma_{pv,1}^{-\frac{1}{2}}$	$R_p V_{pv,1}\Sigma_{pv,1}^{-\frac{1}{2}}$

Table 11.1: Balancing transformations for the second-order index 1 descriptor systems.

Algorithm 40: SOLR-SRM for second-order index 1 system.

Input : $M_{11}, D_{11}, K_{11}, K_{12}, K_{21}, K_{22}\ H_1, H_2, L_1, L_2$.
Output: $\hat{M}$, $\hat{D}$, $\hat{K}$, $\hat{H}$, $\hat{L}$, $\hat{D}_a := D_a$.

1 Solve the Lyapunov equations (11.8) and (11.9) to compute R_v, R_p, L_v and L_p.
2 Compute one of the four types of transformations by following Table 11.1.
3 Construct $\hat{M}$, $\hat{D}$, $\hat{K}$, $\hat{H}$ and $\hat{L}$ by following (11.42) for each pair of balancing and truncating transformations (W_s, V_s) in Table 11.1.

When we use the pair $(R_v,\ L_v)$ to construct the balancing and truncating transformation, the balancing criterion is called *velocity-velocity* (VV) balancing. Analogously, the balancing criteria are called *position-position* (PP), *velocity-position* (VP) and *position-velocity* (PV) balancing if we use the low-rank Gramian factor pairs $(R_p,\ L_p)$, $(R_v,\ L_p)$ and $(R_p,\ L_v)$, respectively. Following (11.39-11.40), we can compute four types of balancing and truncating transformations which are summarized in Table 11.1. Algorithm 40 summarizes the above procedure to construct the structure preserving ROMs from the second-order index 1 system (11.1).

11.3.3 PDEG-based method

Structure preserving methods for second-order standard systems via projecting the system onto the dominant eigenspace of the system Gramian (PDEG) was introduced in Chapter 5. In this subsection, we extend the idea for the structured second-order index 1 systems. In the above subsection, we already have discussed how to convert the second-order index 1 system (11.1) into the second-order standard system (11.29). We have also discussed the Gramians of the system, i.e., controllability velocity, position Gramians (P_v, P_p) and observability velocity, position Gramians (Q_v, Q_p) and their low-rank factors in the above subsection. In the following text, we discuss how to construct the projectors efficiently from the low-rank Gramian factor to project the system onto the dominant eigenspace of the system's Gramians.

Let us first consider the controllability velocity Gramian P_v. Since it is symmetric positive definite (spd), it has a symmetric decomposition, that is to say,

$$P_v = R_v R_v^T. \tag{11.43}$$

The SVD of R_v is

$$R_v = U_v \Sigma_v V_v^T, \tag{11.44}$$

where the diagonal matrix Σ_v consists of the decreasingly ordered singular values σ_{v_i}, $i = 1, 2, \ldots, n_\xi$, of R_v. Using this SVD we obviously have

$$P_v = (U_v \Sigma_v V_v^T)(V_v \Sigma_v U_v^T) = U_v \Sigma_v^2 U_v^T. \tag{11.45}$$

This is also an eigenvalue decomposition where Σ_v^2 is a diagonal matrix whose entries are the decreasingly ordered eigenvalues of P_v and U_v is the orthogonal matrix consisting of the eigenvectors corresponding to the eigenvalues. We observe that U_v is the left singular vector matrix of R_v. Hence U_v is obtained by the SVD of R_v. Now we identify the k largest eigenvalues of P_v and construct

$$U_k = \left[u_1, u_2, \ldots, u_k\right], \tag{11.46}$$

where u_i, $i = 1, 2, \ldots, k$ are the eigenvectors corresponding to the eigenvalues σ_i^2. Then we construct the k dimensional reduced order model, as in (11.41), by forming the reduced dimensional matrices as in (11.42), where $W_s = V_s = U_k$.

Now if we consider R_v as a low-rank factor of the controllability velocity Gramian such that $P_v \approx R_v R_v^T$ then we can compute U_k in (11.46) by identifying the k largest left singular vectors of the SVD of R_v. The procedure that constructs a k dimensional ROM (11.41) via projecting the system onto the dominant eigenspaces of the controllability velocity Gramian P_v is summarized in Algorithm 41. This algorithm can also be used to obtain a k

Algorithm 41: PDEG for second-order index 1 system.

Input : M_{11}, D_{11}, $K_{11}, K_{12}, K_{21}, K_{22}$ H_1, H_2, L_1, L_2.
Output: $\hat{M}$, $\hat{D}$, $\hat{K}$, $\hat{H}$, $\hat{L}$, $\hat{D}_a := D_a$.

1 Compute the low-rank factor of the controllability velocity Gramian, i.e., R_v by solving the Lyapunov equation (11.8).
2 Construct U_k as in (11.46) by using the *thin* SVD of R_v.
3 Form the reduced dimensional matrices $\hat{M}$, $\hat{D}$, $\hat{K}$, $\hat{H}$ and $\hat{L}$ by following (11.42), where $W_s = V_s = U_k$.

dimensional ROM via projecting the system onto the eigenspace of the controllability position Gramian P_p. In that case, in Step 2, instead of R_v we use the low-rank controllability position Gramian factor R_p, where $P_p \approx R_p R_p^T$ to construct the transformation matrix U_k. Likewise, we can construct the ROMs via projecting the system onto the dominant eigenspace of the observability velocity and position Gramian factors by using the low-rank Gramian factors of the low-rank observability velocity and position Gramian factors L_v and L_p, respectively, which are defined in the above subsection.

Section 11.4 discusses how to find the low-rank factors of the second-order Gramians by solving the Lyapunov equations.

Note that the pre-assigned order k of the reduced-order model should satisfy the inequality

$$k \leq \dim(\text{low-rank factors of the Gramians}).$$

It can be shown that the ROMs obtained by the PDEG method preserve the symmetry and definiteness of the original system. According to [178, Theorem 4] it can be guaranteed that the reduced model preserves the stability also.

11.4 Solution of Lyapunov equations using LR-ADI iteration

In the previous sections we have seen, in order to carry out the BT and PDEG methods (i.e., Algorithms 38, 40 and 41), the main tools are the low-rank Gramian factors R and L which can be obtained by solving the Lyapunov equations (11.8) and (11.9), respectively. This section concentrates on how to solve these Lyapunov equations efficiently using the LR-ADI iteration that was introduced in Chapter 3. To ensure fast convergence of the LRCF-ADI method, we also discuss techniques for selecting the shift parameters.

11.4.1 Computation of low-rank controllability and observability Gramian factors

To compute the low-rank controllability Gramian factor, we need to solve the Lyapunov equation (11.8). We can apply Algorithm 15 by using the input matrices E, A and B defined in (11.6). Here we will show how to avoid the explicit formulation of matrices E, A and B. Recalling Algorithm 3, the initial guess of the residual is

$$W_0 = B = \begin{bmatrix} 0 \\ H_1 - K_{12}K_{22}^{-1}H_2 \end{bmatrix},$$

which can be partitioned as

$$W_0^{(1)} = 0 \quad \text{and} \quad W_0^{(2)} = H_1 - K_{12}K_{22}^{-1}H_2. \tag{11.47}$$

At the i-th step of the LRCF-ADI iteration (see, e.g., Algorithm 15), we need to compute $V_i = (A + \mu_i E)^{-1}W_{i-1}$ by solving the linear system

$$(A + \mu_i E)V_i = W_{i-1}. \tag{11.48}$$

Inserting E and A from (11.2) we obtain

$$\left(\begin{bmatrix} M_{11} & 0 \\ 0 & -(K_{11} - K_{12}K_{22}^{-1}K_{21}) \end{bmatrix} + \mu_i \begin{bmatrix} 0 & M_{11} \\ M_{11} & D_{11} \end{bmatrix}\right) \begin{bmatrix} V_i^{(1)} \\ V_i^{(2)} \end{bmatrix} = \begin{bmatrix} W_{i-1}^{(1)} \\ W_{i-1}^{(2)} \end{bmatrix}, \tag{11.49}$$

i.e.,

$$\begin{bmatrix} M_{11} & \mu_i M_{11} \\ \mu_i M_{11} & (\mu_i D_{11} - K_{11}) + K_{12}K_{22}^{-1}K_{21} \end{bmatrix} \begin{bmatrix} V_i^{(1)} \\ V_i^{(2)} \end{bmatrix} = \begin{bmatrix} W_{i-1}^{(1)} \\ W_{i-1}^{(2)} \end{bmatrix}. \tag{11.50}$$

It can be easily shown that by reversing the Schur complement, instead of solving the linear system (11.50) we can solve the linear system

$$\begin{bmatrix} M_{11} & \mu_i M_{11} & 0 \\ \mu_i M_{11} & (\mu_i D_{11} - K_{11}) & -K_{12} \\ 0 & -K_{21} & -K_{22} \end{bmatrix} \begin{bmatrix} V_i^{(1)} \\ V_i^{(2)} \\ \Gamma \end{bmatrix} = \begin{bmatrix} W_{i-1}^{(1)} \\ W_{i-1}^{(2)} \\ 0 \end{bmatrix}, \tag{11.51}$$

for $\begin{bmatrix} V_i^{(1)^T} & V_i^{(2)^T} \end{bmatrix}^T$. Although the dimension of the matrices in (11.51) is higher than that of (11.50), it is sparse and therefore, it can be solved by using a sparse direct solver, e.g., [58, Ch. 5] or any suitable iterative solver [143]. To ensure fast solution, we can partition the linear system (11.51) as follows.

Rewrite system (11.51) as

$$M_{11}V_i^{(1)} + \mu_i M_{11}V_i^{(2)} = W_{i-1}^{(1)} \tag{11.52}$$

$$\mu_i M_{11}V_i^{(1)} + (\mu_i D_{11} - K_{11})V_i^{(2)} - K_{12}\Gamma = W_{i-1}^{(2)} \tag{11.53}$$

$$-K_{21}V_i^{(2)} - K_{22}\Gamma = 0 \tag{11.54}$$

From (11.52) we obtain

$$V_i^{(1)} = M_{11}^{-1}W_{i-1}^{(1)} - \mu_i V_i^{(2)}. \tag{11.55}$$

Inserting this into (11.53) and some algebraic manipulations leads to

$$(\mu_i^2 M_{11} - \mu_i D_{11} + K_{11})V_i^{(2)} + K_{12}\Gamma = \mu_i W_{i-1}^{(1)} - W_{i-1}^{(2)}. \tag{11.56}$$

Combining (11.56) and (11.54) will result

$$\begin{bmatrix} \mu_i^2 M_{11} - \mu_i D_{11} + K_{11} & K_{12} \\ K_{21} & K_{22} \end{bmatrix} \begin{bmatrix} V_i^{(2)} \\ \Gamma \end{bmatrix} = \begin{bmatrix} \mu_i W_{i-1}^{(1)} - W_{i-1}^{(2)} \\ 0 \end{bmatrix}. \tag{11.57}$$

We have seen that instead of solving a large linear system (11.51) for $\begin{bmatrix} V_i^{(1)T} & V_i^{(2)T} \end{bmatrix}^T$, one can solve subsystem (11.57) for $V_i^{(2)}$ and then compute $V_i^{(1)}$ from (11.55). That means, the splitting idea reduces the dimension of the linear system from $2n_\xi + n_\varphi$ to $n_\xi + n_\varphi$ which ensures faster computation. Here, $W_{i-1}^{(1)}$ and $W_{i-1}^{(2)}$ are already computed from the previous step. At the i-th step, the ADI residual can be obtained by

$$W_i = W_{i-1} - 2\,\mathrm{Re}\,(\mu_i) E V_i,$$

which implies

$$\begin{bmatrix} W_i^{(1)} \\ W_i^{(2)} \end{bmatrix} = \begin{bmatrix} W_{i-1}^{(1)} \\ W_{i-1}^{(2)} \end{bmatrix} - 2\,\mathrm{Re}\,(\mu_i) \begin{bmatrix} 0 & M_{11} \\ M_{11} & D_{11} \end{bmatrix} \begin{bmatrix} V_i^{(1)} \\ V_i^{(2)} \end{bmatrix}$$
$$= \begin{bmatrix} W_{i-1}^{(1)} - 2\,\mathrm{Re}\,(\mu_i) M_{11} V_i^{(2)} \\ W_{i-1}^{(2)} - 2\,\mathrm{Re}\,(\mu_i)(M_{11} V_i^{(1)} + D_{11} V_i^{(2)}) \end{bmatrix}.$$

From this we get

$$\begin{aligned} W_i^{(1)} &= W_{i-1}^{(1)} - 2\,\mathrm{Re}\,(\mu_i) M_{11} V_i^{(2)}, \\ W_i^{(2)} &= W_{i-1}^{(2)} - 2\,\mathrm{Re}\,(\mu_i)(M_{11} V_i^{(1)} + D_{11} V_i^{(2)}). \end{aligned} \tag{11.58}$$

If the two consecutive shift parameters are complex conjugates of each other, i.e., $\{\mu_i, \mu_{i+1} := \overline{\mu_i}\}$, recalling (3.44) we have

$$W_{i+1} = W_{i-1} - 4\,\mathrm{Re}\,(\mu_i) E \left(\mathrm{Re}\,(V_i) + \delta\, \mathrm{Im}\,(V_i) \right),$$

where $\delta = \frac{\mathrm{Re}\,(\mu_i)}{\mathrm{Im}\,(\mu_i)}$ which gives

$$\begin{bmatrix} W_{i+1}^{(1)} \\ W_{i+1}^{(2)} \end{bmatrix} = \begin{bmatrix} W_{i-1}^{(1)} \\ W_{i-1}^{(2)} \end{bmatrix} - 4\,\mathrm{Re}\,(\mu_i) \begin{bmatrix} 0 & M_{11} \\ M_{11} & D_{11} \end{bmatrix} \begin{bmatrix} \chi_1 \\ \chi_2 \end{bmatrix}$$
$$= \begin{bmatrix} W_{i-1}^{(1)} - 4\,\mathrm{Re}\,(\mu_i) M_{11} \chi_2 \\ W_{i-1}^{(2)} - 4\,\mathrm{Re}\,(\mu_i)(M_{11} \xi_1 + D_{11} \xi_2) \end{bmatrix},$$

where $\chi_1 = \left(\mathrm{Re}\,(V_i^{(1)}) + \delta\, \mathrm{Im}\,(V_i^{(1)}) \right)$, $\chi_2 = \left(\mathrm{Re}\,(V_i^{(2)}) + \delta\, \mathrm{Im}\,(V_i^{(2)}) \right)$. This results in

$$\begin{aligned} W_{i+1}^{(1)} &= W_{i-1}^{(1)} - 4\,\mathrm{Re}\,(\mu_i) M_{11} \chi_2, \\ W_{i+1}^{(2)} &= W_{i-1}^{(2)} - 4\,\mathrm{Re}\,(\mu_i)(M_{11} \chi_1 + D_{11} \chi_2). \end{aligned} \tag{11.59}$$

Algorithm 42: LR-ADI for the second-order index 1 systems.

Input : $M_{11},\ D_{11},\ K_{11},\ K_{12},\ K_{21},\ K_{22},\ H_1,\ H_2,\ \{\mu_i\}_{i=1}^{J}$.

Output: $R = Z_i$, such that $P \approx RR^T$.

Set $Z_0 = [\,]$, $i = 1$.

$W_0^{(1)} = 0$ and $W_0^{(2)} = H_1 - K_{12}K_{22}^{-1}H_2$.

while $\|{W_{i-1}^{(1)}}^T W_{i-1}^{(1)} + {W_{i-1}^{(2)}}^T W_{i-1}^{(2)}\| \geq tol$ *and* $i \leq i_{max}$ **do**

  Solve $\begin{bmatrix} \mu_i^2 M_{11} - \mu_i D_{11} + K_{11} & K_{12} \\ K_{21} & K_{22} \end{bmatrix} \begin{bmatrix} V_i^{(2)} \\ \Gamma \end{bmatrix} = \begin{bmatrix} \mu_i W_{i-1}^{(1)} - W_{i-1}^{(2)} \\ 0 \end{bmatrix}$

  for $V_i^{(2)}$.

  Compute $V_i^{(1)} = M_{11}^{-1} W_{i-1}^{(1)} - \mu_i V_i^{(2)}$, $\quad V_i = \begin{bmatrix} {V_i^{(1)}}^T & {V_i^{(2)}}^T \end{bmatrix}^T$.

  if $\mathrm{Im}\,(\mu_i) = 0$ **then**

    $Z_i = \begin{bmatrix} Z_{i-1} & \sqrt{2\mu_i}\,\mathrm{Re}\,(V_i) \end{bmatrix}$,

    $W_i^{(1)} = W_{i-1}^{(1)} - 2\mu_i M_{11} V_i^{(2)}$,

    $W_i^{(2)} = W_{i-1}^{(2)} - 2\mu_i (M_{11} V_i^{(1)} + D_{11} V_i^{(2)})$.

  else

    $\gamma = -2\,\mathrm{Re}\,(\mu_i), \quad \delta = \frac{\mathrm{Re}\,(\mu_i)}{\mathrm{Im}\,(\mu_i)}$,

    $Z_{i+1} = \begin{bmatrix} Z_{i-1} & \sqrt{2\gamma}\,(\mathrm{Re}\,(V_i) + \delta\,\mathrm{Im}\,(V_i)) & \sqrt{2\gamma}\sqrt{(\delta^2+1)}.\,\mathrm{Im}\,(V_i) \end{bmatrix}$,

    $W_{i+1}^{(1)} = W_{i-1}^{(1)} + 2\gamma M_{11}\chi_2, \quad W_{i+1}^{(2)} = W_{i-1}^{(2)} + 2\gamma(M_{11}\chi_1 + D_1\chi_2)$,

    where $\chi_1 = \mathrm{Re}\,(V_i^{(1)}) + \delta\,\mathrm{Im}\,(V_i^{(1)})$, $\chi_2 = \mathrm{Re}\,(V_i^{(2)}) + \delta\,\mathrm{Im}\,(V_i^{(2)})$.

    $i = i + 1$

  $i = i + 1$

The procedure to compute the low-rank controllability Gramian factor (R) by solving the Lyapunov equation (11.8) is outlined in Algorithm 42. This algorithm can also be applied for computing the low-rank observability Gramian factor (L) by solving the Lyapunov equation (11.9). In this case the inputs M_{11}, D_{11}, K_{11}, K_{12}, K_{22}, H_1 and H_2 are changed by M_{11}^T, D_{11}^T, K_{11}^T, K_{21}^T, K_{12}^T K_{22}^T, L_1^T and L_2^T, respectively.

> Once the low-rank controllability Gramian factor R is computed, then R_v (low-rank controllability velocity Gramian factor) and R_p (low-rank controllability position Gramian factor) can be respectively obtained by taking upper n_ξ and lower n_ξ rows of R. Similarly, by partitioning the computed low-rank observability Gramian factor L, we consider upper n_ξ and lower n_ξ rows of L to respectively obtain the low-rank observability velocity Gramian factor (L_v) and low-rank observability position Gramian factor (L_p).

11.4.2 ADI shift parameter selection

It is known that the LR-ADI iteration relays on ADI shift parameters. For the fast convergence of Algorithm 42, a set of proper ADI shift parameters selection is necessary. In Chapter 3, we have mentioned that among the different kinds of ADI shift parameters proposed in the literature, Penzl's heuristic shifts [133] are more commonly used for large-scale dynamical systems. Another prominent shift selection criterion is the adaptive approach. In this subsection, we investigate both the approaches for computing the shift parameters of second-order index 1 DAE system.

Penzl's heuristic shifts. Recalling the discussion in Section 3.6 for computing Penzl's heuristic shifts, first we need to compute

- k_+ large magnitude ritzvalues with respect to the matrix pair (A, E),
- k_- small magnitude ritzvalues which are equivalent to the reciprocal of k_- large magnitude Ritz values with respect to the matrix pair (E, A),

where A and E are defined in (11.33). In both cases, to compute the ritzvalues, we can follow the Arnoldi procedure as summarized in Algorithm 5 (in Chapter 1).

Adaptive shifts. Besides Penzl's heuristic shifts, we can also use the adaptive approach to generate the shift parameters inside the LR-ADI iteration. See Section 3.6 for the motivation of this class of ADI shift parameters. Following the discussion in Section 3.6, the shifts are initialized by the eigenvalues of the matrix pencil $\lambda E - A$ projected to the span of B where E, A and B are defined in (11.2). Then, whenever all the shifts in the set have been used, the pencil is projected to the span of the current V_i and the eigenvalues are used as the next set of shifts. Here we use the same initialization. For the update step, however, we extend the subspace to all the V_i generated with the current set of shifts. Let us assume that U is the basis of the extended subspace. Now, from the eigenvalues of $\lambda U^T EU - U^T AU$, select some desired number of optimal shifts by solving the ADI min-max problem (see Chapter 3) like in the heuristic procedure. This approach is repeated until the algorithm has converged to the given tolerance. Note that our system is dissipative, i.e., all the eigenvalues of $\lambda(E + E^T) - (A + A^T)$ lie in the left complex plane. Therefore, Bendixon's theorem [120] ensures that all the eigenvalues of the projected pencil $\lambda U^T EU - U^T AU$ are stable.

11.5 Symmetric second-order index 1 system

If the block matrices of (11.1) satisfy the following conditions

$$\begin{aligned} &M_{11} = M_{11}^T, \quad D_{11} = D_{11}^T, \quad K_{11} = K_{11}^T, \quad K_{21} = K_{12}^T, \\ &K_{22} = K_{22}^T, \quad L_1 = H_1^T \quad \text{and} \quad L_2 = H_2^T, \end{aligned} \tag{11.60}$$

then we say this system is symmetric. Then the first-order form (11.6) is also symmetric, i.e.,

$$E = E^T, \quad A = A^T, \quad C = B^T.$$

Under these circumstances, we have an advantage; the two Lyapunov equations (11.8) and (11.9) are coincided and their solutions P and Q are the same. Therefore, we need only solve one Lyapunov equation, i.e.,

$$AG_mE + EG_mA = -BB^T, \tag{11.61}$$

where $P = Q = G_m$. We can compute the low-rank Gramian factor Z by using Algorithm 42 which satisfies $ZZ^T \approx G_m$. Considering

$$R = L = Z$$

one can find balancing and PDEG-based model reduction as discussed above. We should remember that in the balancing-based model reduction (in both second-to-first and second-to-second cases), the left- and right-balancing and truncating transformations are the same. Note that the data that we have used for the numerical test has this symmetric structure.

11.6 Numerical results

In this section, we present numerical results to assess the accuracy and efficiency of our proposed techniques. The techniques are applied to a set of data for the finite element discretization of an adaptive spindle support (ASS) [101]. Details of this data is also available in Appendix A.5.1. The dimension of the original model is $n = 290\,137$ which consists of $n_\xi = 282\,699$ differential equations and $n_\varphi = 7\,438$ algebraic equations. The number of the collocated inputs and outputs is 9.

All results were obtained using MATLAB 7.11.0 (R2012a) on a board with 4 Intel® Xeon® E7-8837 CPUs with a 2.67-GHz clock speed, 8 Cores each and 1TB of total RAM.

no. of iterations	normalized residual norm	
	heuristic shifts	adaptive shifts
100	9.88×10^{-1}	1.85×10^{-2}
200	9.99×10^{-1}	8.85×10^{-3}
300	9.78×10^{-1}	5.04×10^{-3}
400	9.69×10^{-1}	3.99×10^{-3}

Table 11.2: Comparison of the normalized residual norms using heuristic and adaptive shifts at different iteration steps in Algorithm 42.

This ASS model is symmetric. Moreover, the input and the output are the transpose of each other. Therefore, to implement the BT- and PDEG-based reduced order models, we solve only one Lyapunov equation (see Section 11.5). Suppose by applying Algorithm 42, the computed low-rank Gramian factor is Z. Both the *heuristic* and the *adaptive* shifts are compared to carry out this algorithm. In the case of the heuristic approach, we select 40 optimal shift parameters out of 60 large and 50 small magnitude approximate eigenvalues (see, e.g., [40] for details on the computation of heuristic ADI shift parameters for the ASS model). The algorithm is stopped by the maximum number of iteration steps $i_{\max} = 400$. Next, we apply the adaptive shift computation approach to implement this algorithm. In this case, the algorithm is also stopped by $i_{\max} = 400$. The convergence is compared in Table 11.2 in different iteration steps for both types of shift parameters. As we can see from this table, the performance of the adaptive shifts is better than that of the heuristic shifts. Before using the computed low-rank Gramian factor for the model reduction algorithm, we compress the columns of Z applying the technique proposed in [132].

11.6.1 Second-order-to-first-order reduction

Balancing-based methods. Algorithm 38 generated a 152 dimensional reduced-order model for the tolerance 10^{-5}. However, the dimension of the reduced order model can be reduced further by increasing the error tolerance which is shown in Table 11.3. All the ROMs obtained by different truncation tolerances have a good match with the original model keeping the relative error below the error which is reflected in Figure 11.1. We can also compute even lower dimensional ROMs if they are required for the controller design. In Figure 11.2 we see that although the approximation quality of the 5 dimensional ROM gets worse, order 50-10 dimensional models are satisfactory if an error of no more than 5% is desired. Even an order 10 model captures the important features of the original system.

Comparison with IRKA. To compare the balancing-based method with IRKA, we also compute different dimensional (e.g., 50 and 10) ROMs with Algorithm 39. Figure 11.3 shows the accuracy of the 50 and 10 dimensional

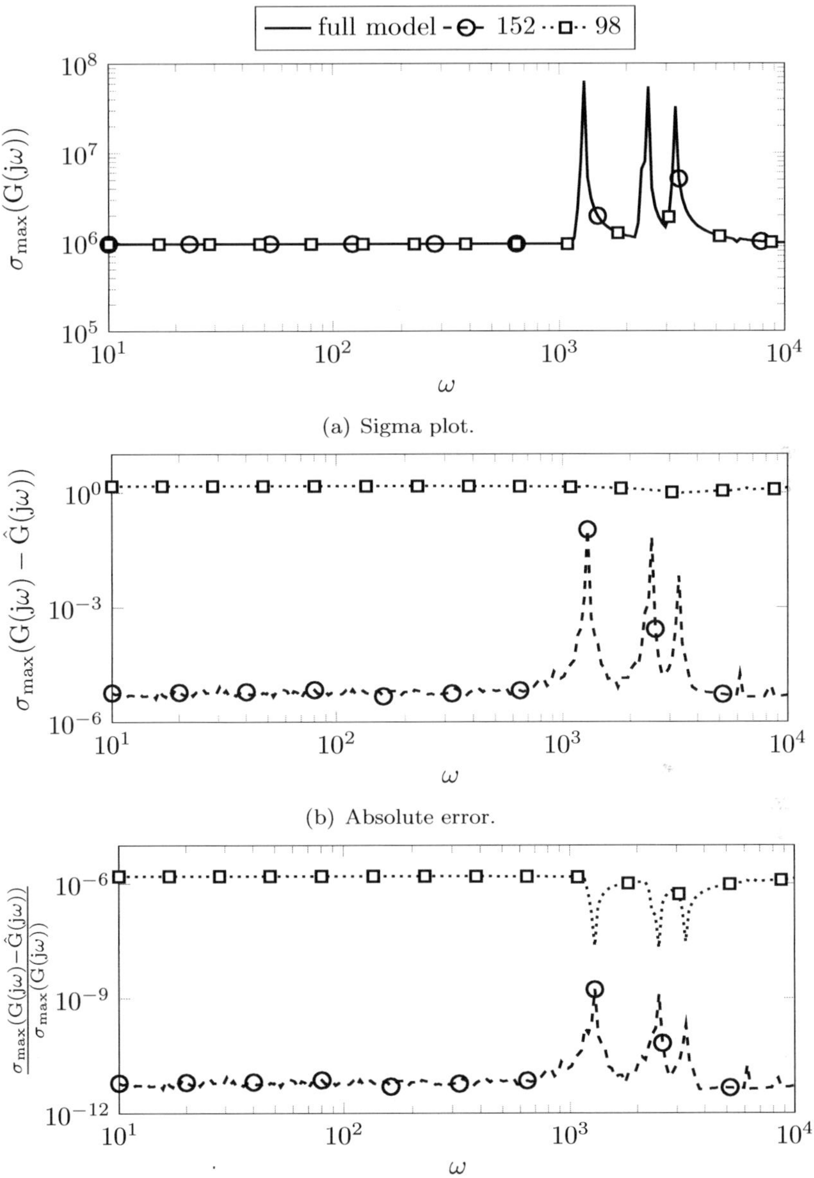

(a) Sigma plot.

(b) Absolute error.

(c) Relative error.

Figure 11.1: Comparison of different dimensional reduced systems obtained by different tolerances.

MOR tolerance	ROM dimension
10^{-4}	146
10^{-3}	140
10^{-2}	132
10^{-1}	123
10^{0}	98

Table 11.3: Dimension of reduced models obtained by using different truncation tolerances.

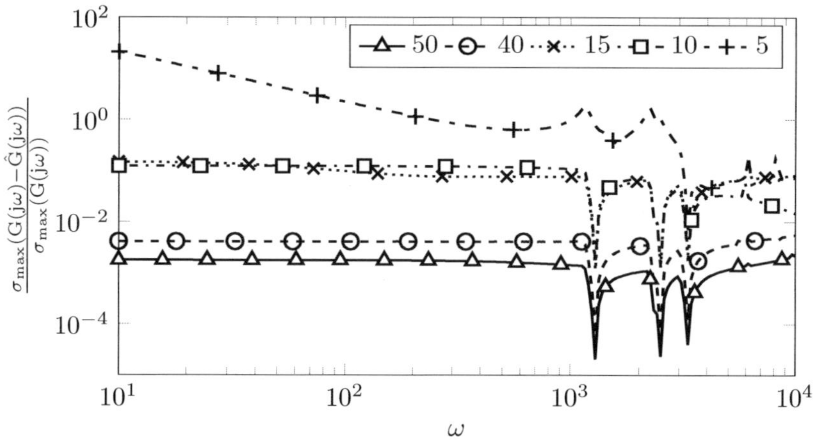

Figure 11.2: Relative errors between the original and different dimensional reduced systems obtained by balanced truncation and IRKA.

BT and IRKA-based reduced models. Here IRKA-based reduced models show a higher relative error. Note that Algorithm 39 is stopped after 50 cycles. That means we have updated the interpolation points and tangential directions 50 times; this number is still large. Perhaps, the quality of the ROMs can be improved further by considering even more cycles. In that case, the computation would be more expensive.

11.6.2 Second-order-to-second-order reduction

Balancing-based methods. For computing second-order-to-second-order ROMs using balanced truncation, we first partition the computed Z as Z_v and Z_p by taking upper and lower n_ξ rows of Z and then applying Algorithm 40. This algorithm computes different dimensional reduced systems for the truncation tolerance 10^{-5} by using different types of balancing labels as shown in Table 11.1. The comparisons of the full and different dimensional reduced systems are shown in Figure 11.4. The absolute error and the relative error of the frequency responses of full and reduced systems that are exhibited

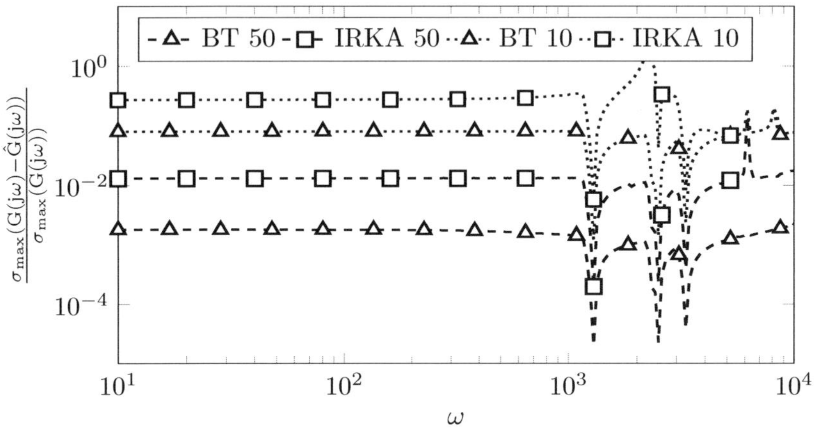

Figure 11.3: Relative error for different dimensional ROMs computed by Algorithms 38 and 39.

system dimension	execution time (sec)	speedup
290 137	90.00	1
50	0.0014	64 285
40	0.0012	75 000
30	0.0009	100 000
20	0.0007	128 571
10	0.0003	300 000

Table 11.4: Average execution time and speedup against full order model for computing the maximum Hankel singular value at a given sampling frequency.

in Figure 11.4(b) and Figure 11.4(b) respectively show very good accuracy. As we can see in Figure 11.4(c), the relative errors for all the reduced systems are far below the truncation tolerance (10^{-5}).

We further compute the 40, 30, 20 and 10 dimensional reduced order models using the same algorithm via balancing the system on the position-position levels. Figure 11.5 discusses the SISO relation of full and different dimensional reduced order models computed by position-position balancing. Since in the SISO case, we know that the transfer function matrix is just a scalar rational function, we have computed the absolute values of the transfer function in different frequencies. The relative error between the original and reduced order models of the respective SISO relation are also shown in the same figure. Table 11.4 shows the possible execution time gains that can be expected from the reduced order modeling. Here, the computation at one sampling frequency (out of 200 used in the figures) is used as a representative for one evaluation of the model. This roughly corresponds to the most expensive step in a time step for a simple integrator that is applied in the transient simulation of the system. Therefore similar speedups can be expected in those simulations.

PDEG-based methods. Algorithm 41 is applied again on the ASS model to obtain the reduced systems via projecting the system onto the dominant eigenspace of the Gramians. We know, due to the symmetric property, the controllability and observability Gramians are the same. By applying the low-rank ADI iteration, we compute the low-rank Gramian factor Z which is partitioned as velocity Gramian factor (Z_v) and the position Gramian factor (Z_p) as we have done for implementing the balancing-based method. By predefining the dimension of the ROM, we compute 40, 30, 20 and 10 dimensional models by projecting the system onto the dominant eigenspaces of both velocity and position Gramians. Figure 11.6 shows the relative error between the original and the different dimensional reduced models when we project the system onto the dominant eigenspace of the velocity Gramian (VG) (Figure 11.6(a)) and position Gramian (PG) (Figure 11.6(b)). We observe that the constructed reduced systems of the ASS model by PDEG methods are asymptotically stable which is shown in Figure 11.7. This figure shows that all the eigenvalues of the reduced systems, which are obtained via projecting the system onto the dominant eigenspace of the position Gramian, lie in the left complex half plane. From this figure, one can also see that the reduced system of successively decreasing dimension contains the eigenvalues closer to the imaginary axis.

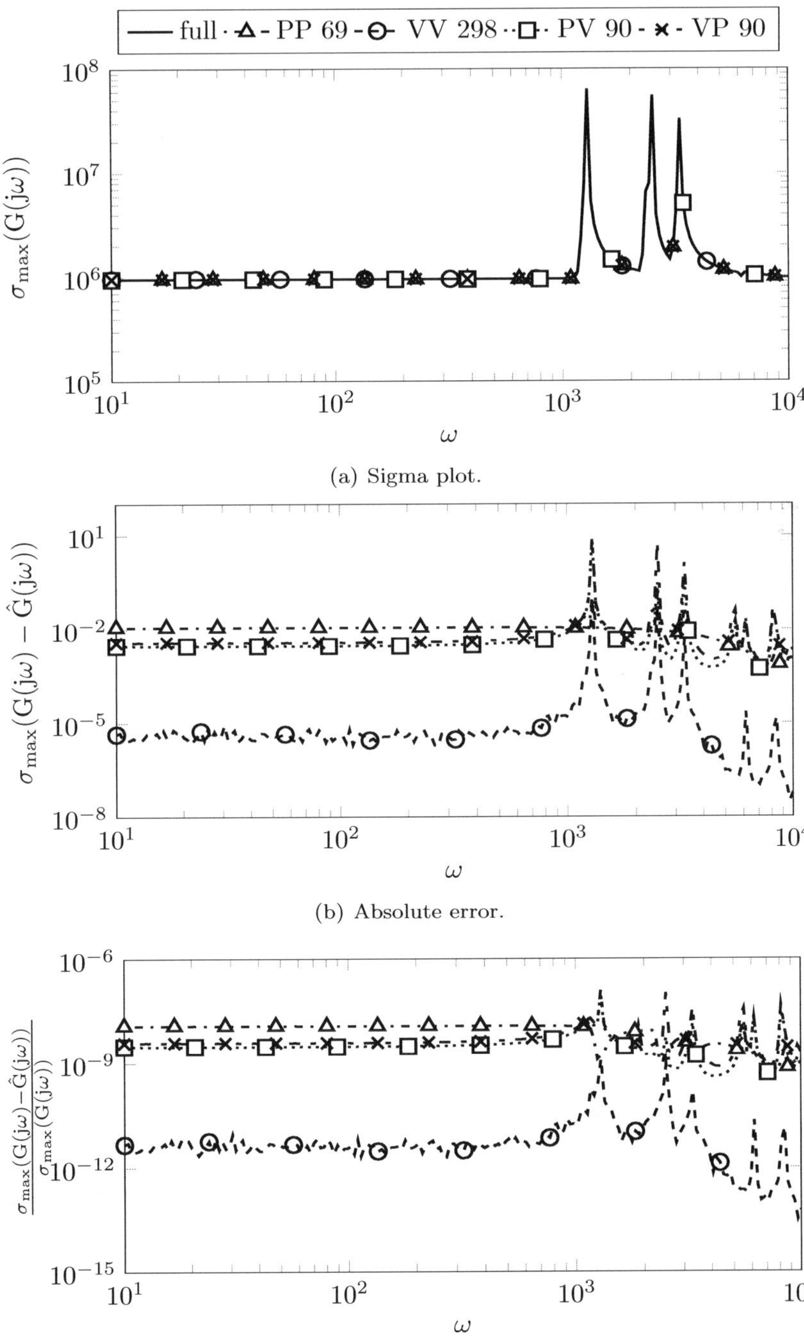

(a) Sigma plot.

(b) Absolute error.

(c) Relative error.

Figure 11.4: Comparison of different dimensional reduced systems obtained by different balancing levels using a truncation tolerance of 10^{-5}.

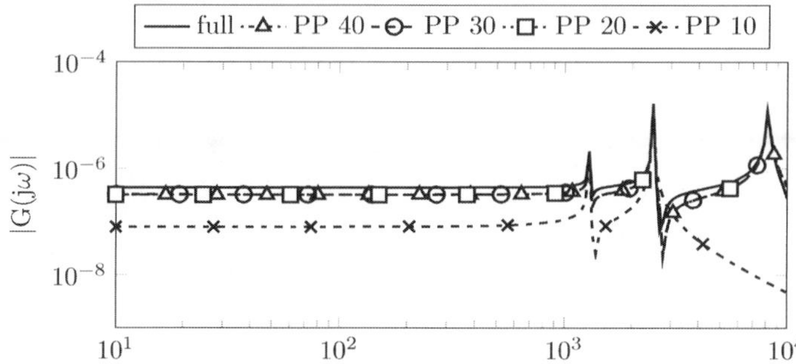

(a) 1st input to 1st output relations.

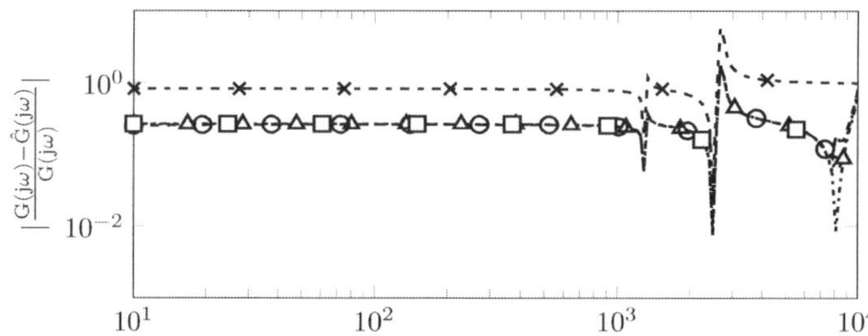

(b) Relative error of 1st input to 1st output relations.

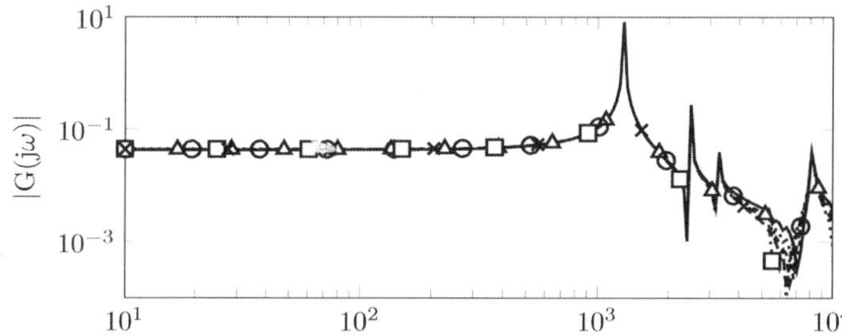

(c) 9th input to 1st output relations.

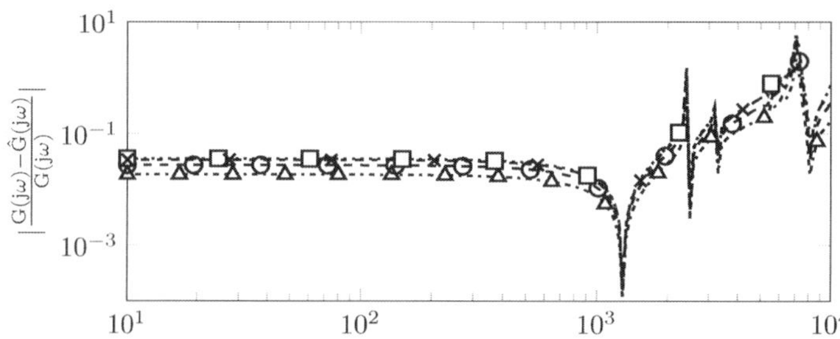

(d) Relative error of 9th input to 1st output relations.

Figure 11.5: Single-input single-output relations.

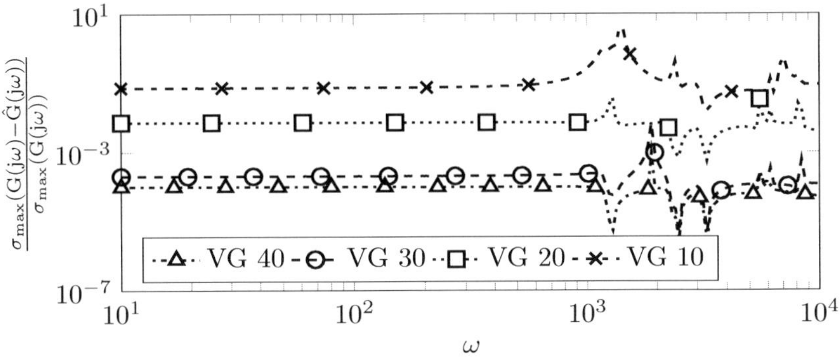

(a) Velocity Gramian.

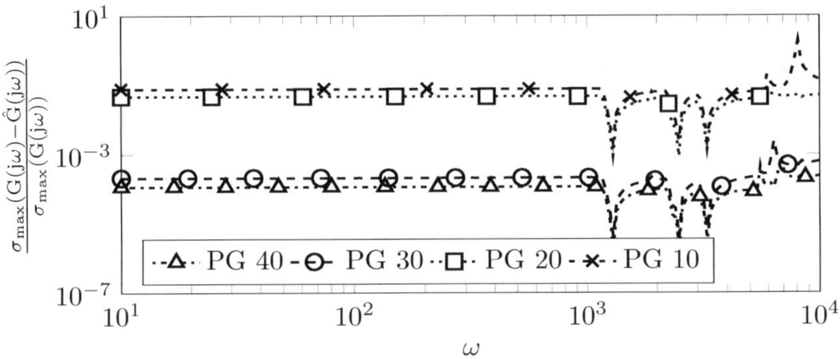

(b) Position Gramian.

Figure 11.6: Relative error between the full and different dimensional reduced models via projecting onto the dominant eigenspace of the Gramians.

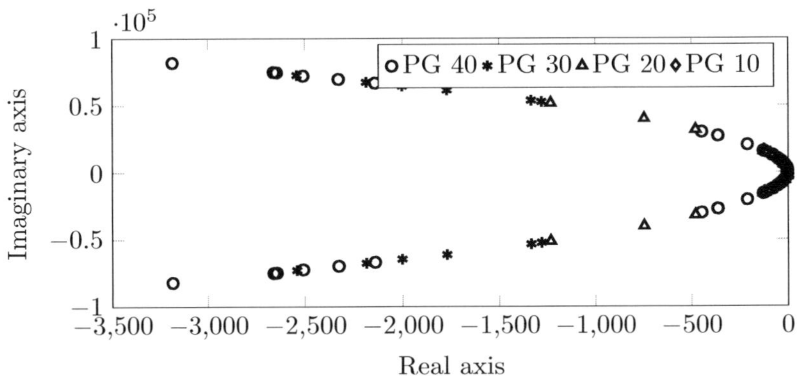

Figure 11.7: Eigenvalues of the ROM via projecting onto the dominant eigenspace of the position Gramian.

■ Exercises:

11.1 Considering a second-order index 1 descriptor system

$$\begin{aligned} M_{11}\ddot{\xi}(t) + D_{11}\dot{\xi}(t) + K_{11}\xi(t) + K_{12}\varphi(t) &= H_1 u(t), \\ K_{12}^T\xi(t) + K_{22}\varphi(t) &= H_2 u(t) \\ y(t) &= H_1^T\xi + H_2^T\varphi(t), \end{aligned}$$

show the different types of first-order transformations. Which one is the efficient transformations in your opinion?

11.2 For the second-order index 1 descriptor system in Exercise 11.1, construct the block matrices[1] using MATLAB as follows

```
n=500; %number of differential variables
l=50; %number of algebraic variables
io=1; %(number of input/output)
den=0.01; I=speye(n);
M11=.5*I+spdiags(-0.2*ones(n,1),2,n,n)+
     spdiags(-0.2*ones(n,1),-2,n,n)+
     spdiags(0.2*ones(n,1),4,n,n)+
     spdiags(0.2*ones(n,1),-4,n,n);
K11=spdiags(5*ones(n,1),0,n,n)+
    spdiags(-1*ones(n,1),2,n,n)+
    spdiags(-1*ones(n,1),-2,n,n)+
    spdiags(2*ones(n,1),4,n,n)+
    spdiags(2*ones(n,1),-4,n,n);
mu=0.005; nu = .1;
D11=mu*M1+nu*K11;
K22=spdiags(5*ones(l,1),0,l,l)+
    spdiags(-1*ones(l,1),2,l,l)+
    spdiags(-1*ones(l,1),-2,l,l)+
    spdiags(2*ones(n,1),4,l,l)+
    spdiags(2*ones(n,1),-4,l,l);
K12= sprand(n,l,den);
H1=spdiags(ones(n,1),0,n,io);
H2 = spdiags(ones(l,1),0,l,io);
```

Using this data, construct the different first-order transformed systems. Now show that the frequency responses and step responses are the same for all the different first-order transformed systems.

[1]Note, M11 denotes M_{11}, similarly other block matrices

11.3 For the data in Exercise 11.2, find the eigenvalues using MATLAB function `polyeig`. Show that the number of infinite eigenvalues is equal to the number of algebraic equations.

11.4 For the data in Exercise 11.2, you can change the dimension of the model by changing n and l. Consider different dimensional models and for each model, find suitable ROMs using both balanced truncation and IRKA. Show the comparison of their performance.

11.5 The second-order index 1 system (11.26) can be converted into second-order index 0 system (11.29); show that the transfer functions for both systems are equal. Also verify this using the data in Exercise 11.2.

11.6 For the second-order index 1 system in Exercise 1, develop an efficient LR-ADI algorithm where the controllability and the observability Gramians can be computed by solving only one Lyapunov equation. Show the efficiency of the algorithm by applying it to the data of the previous exercise.

11.7 For the second-order index 1 system in Exercise 11.1, develop a structure preserving MOR algorithm using

   (i) Balanced truncation.

   (ii) PDEG method.

   Verify its efficiency using the data in Exercise 11.2.

Chapter 12

Model Reduction of Second-Order Index 3 Descriptor Systems

12.1 Introduction

This chapter presents model reduction methods for a class of structured second-order index 3 descriptor systems [167] of the form

$$\begin{aligned}
\underbrace{\begin{bmatrix} M_1 & 0 \\ 0 & 0 \end{bmatrix}}_{M} \begin{bmatrix} \ddot{\xi}(t) \\ \ddot{\varphi}(t) \end{bmatrix} + \underbrace{\begin{bmatrix} D_1 & 0 \\ 0 & 0 \end{bmatrix}}_{D} \begin{bmatrix} \dot{\xi}(t) \\ \dot{\varphi}(t) \end{bmatrix} + \underbrace{\begin{bmatrix} K_1 & G_1^T \\ G_1 & 0 \end{bmatrix}}_{K} \begin{bmatrix} \xi(t) \\ \varphi(t) \end{bmatrix} &= \underbrace{\begin{bmatrix} H_1 \\ 0 \end{bmatrix}}_{H} u(t), \\
\underbrace{\begin{bmatrix} L_1 & 0 \end{bmatrix}}_{L} \begin{bmatrix} \xi(t) \\ \varphi(t) \end{bmatrix} &= y(t),
\end{aligned} \tag{12.1}$$

where $\xi(t) \in \mathbb{R}^{n_\xi}$ and $\varphi(t) \in \mathbb{R}^{n_\varphi}$ are the states with $n_\xi > n_\varphi$ and $u(t) \in \mathbb{R}^m$ are the inputs, $y(t) \in \mathbb{R}^p$ are the outputs and M, D, K, H, L are all sparse matrices with appropriate dimensions. Such structured systems arise in many applications like in constraint multibody system dynamics [64, 151] and mechanical systems with *holonomic constraints* [122, 178]. The system (12.1) is called an index 3 system due to its analogy to first-order index 3 (see, e.g., Chapter 10) linear time invariant (LTI) systems.

If the model is very large, performing simulations either have a prohibitively expensive computational effort associated with them or are simply impossible due to the limited computer memory. Therefore, reducing the size of the system is essential for fast simulation. Hence, we want to apply model reduction onto the large-scale descriptor system and replace it by a substantially lower dimensional system.

In general, the algebraic part of the system can be eliminated by projecting it onto the subspace where the solution of the descriptor system exists. It can be shown that the projected and the original systems are equivalent in the sense that they have the same finite spectrum and transfer function. Then both the second-order-to-first-order and second-order-to-second-order reduction techniques can be applied to the projected systems. However, to overcome the computational complexity, explicit formulation of the projected system has to be avoided.

This chapter discusses both second-order-to-first-order and second-order-to-second-order reduction techniques of second-order index 3 systems (12.1). In the case of second-order-to-first-order reduction, we discuss both balanced truncation and the interpolatory technique with IRKA. To implement the methods, the second-order projected system is converted into its first-order form. These first-order projected systems are very similar to the projected systems considered in [89, 93]. Following their strategies (see also, e.g., Chapter 10), we show a technique to avoid the explicit computation of the projector for implementing the BT and interpolatory methods. On the other hand, for the second-order-to-second-order reduction method, besides balanced truncation, we also discuss the PDEG method that was introduced in the previous chapter. In this case, we also discuss the issues that are associated with avoiding the projector.

The BT and PDEG methods rely on the controllability and observability Gramian factors. To compute the Gramian factors, we need to solve two projected continuous-time algebraic Lyapunov equations. Following the discussions from Chapter 10, here we show an efficient technique to solve the projected Lyapunov equations by handling the projector implicitly. Moreover, we discuss the difficulties of computing the ADI shift parameters by using both the heuristic and adaptive approaches and suggest how to overcome them.

The proposed techniques are applied to several test examples and numerical results are discussed to show the efficiency of the techniques.

12.2 Reformulation of the dynamical systems

Let us rewrite the second-order index 3 system (12.1) as

$$M_1\ddot{\xi}(t) = -D_1\dot{\xi}(t) - K_1\xi(t) - G_1^T\varphi(t) + H_1u(t), \tag{12.2a}$$
$$G_1\xi(t) = 0, \tag{12.2b}$$
$$y(t) = L_1\xi(t). \tag{12.2c}$$

From (12.2b) we obtain $G_1\ddot{\xi}(t) = 0$. Inserting this identity after multiplying both sides of (12.2a) by $G_1M_1^{-1}$, we find

$$0 = -G_1M_1^{-1}D_1\dot{\xi}(t) - G_1M_1^{-1}K_1\xi(t) - G_1M_1^{-1}G_1^T\varphi(t) + G_1M_1^{-1}H_1u(t), \tag{12.3}$$

which implies

$$\varphi(t) = -(G_1M_1^{-1}G_1^T)^{-1}G_1M_1^{-1}D_1\dot{\xi}(t) - (G_1M_1^{-1}G_1^T)^{-1}G_1M_1^{-1}K_1\xi(t)+ (G_1M_1^{-1}G_1^T)^{-1}G_1M_1^{-1}H_1u(t). \tag{12.4}$$

Inserting $\varphi(t)$ into (12.2a) we obtain

$$M_1\ddot{\xi}(t) = -\Pi D_1\dot{\xi}(t) - \Pi K_1\xi(t) + \Pi H_1u(t), \tag{12.5}$$

where

$$\Pi := I_{n_\xi} - G_1^T(G_1M_1^{-1}G_1^T)^{-1}G_1M_1^{-1}, \tag{12.6}$$

in which I_{n_ξ} is an identity matrix of size n_ξ. In fact, Π is a projector since

$$\begin{aligned}\Pi^2 &= (I_{n_\xi} - G_1^T(G_1M_1^{-1}G_1^T)^{-1}G_1M_1^{-1})(I_{n_\xi} - G_1^T(G_1M_1^{-1}G_1^T)^{-1}G_1M_1^{-1})\\ &= I_{n_\xi} - 2G_1^T(G_1M_1^{-1}G_1^T)^{-1}G_1M_1^{-1} + G_1^T(G_1M_1^{-1}G_1^T)^{-1}G_1M_1^{-1}\\ &= I_{n_\xi} - G_1^T(G_1M_1^{-1}G_1^T)^{-1}G_1M_1^{-1} = \Pi.\end{aligned}$$

The projector Π satisfies the following properties.

Proposition 4. *Let Π be the projector defined above. The following conditions must hold.*

1. $\Pi M_1 = M_1\Pi^T$.
2. $\operatorname{Null}(\Pi) = \operatorname{Range}(G_1^T)$.
3. $\operatorname{Range}(\Pi) = \operatorname{Null}(G_1M_1^{-1})$.

Proof. For the proofs see, e.g., Proposition 4 in Chapter 8. □

Theorem 20. *The vector a is in the null space of G_1, i.e., $G_1 a = 0$ iff $\Pi^T a = a$ where Π is defined in (12.6).*

Proof. See, e.g., Theorem 18 in Chapter 10. □

Following Theorem 20, equation (12.2b) implies

$$\Pi^T \xi(t) = \xi(t). \tag{12.7}$$

Inserting this identity into (12.5) and multiplying the resulting equation by Π, we obtain

$$\Pi M_1 \Pi^T \ddot{\xi}(t) = -\Pi K_1 \Pi^T \xi(t) - \Pi D_1 \Pi^T \dot{\xi}(t) + \Pi H_1 u(t). \tag{12.8}$$

Moreover, applying (12.7) into the output equation (12.2c), we find the system in (12.2) is equivalent to

$$\Pi M_1 \Pi^T \ddot{\xi}(t) = -\Pi D_1 \Pi^T \dot{\xi}(t) - \Pi K_1 \Pi^T \xi(t) + \Pi H_1 u(t), \tag{12.9a}$$
$$y(t) = L_1 \Pi^T \xi(t). \tag{12.9b}$$

The system dynamics of (12.9) are projected onto the $n_m := n_\xi - n_\varphi$ dimensional subspace Range $\left(\Pi^T\right)$. This subspace is, however, still represented in the coordinates of the n_ξ dimensional space. The n_m dimensional representation can be made explicit by employing the *thin* singular value decomposition (SVD)

$$\Pi = U\Sigma V^T = \begin{bmatrix} U_1 & U_2 \end{bmatrix} \begin{bmatrix} \Sigma_1 & 0 \\ 0 & 0 \end{bmatrix} \begin{bmatrix} V_1^T \\ V_2^T \end{bmatrix} = U_1 \Sigma_1 V_1^T = \Theta_l \Theta_r^T, \tag{12.10}$$

where $\Theta_l = U_1 \Sigma_1^{\frac{1}{2}}$, $\Theta_r = V_1 \Sigma_1^{\frac{1}{2}}$ and $U_1,\ V_1 \in \mathbb{R}^{n_\xi \times n_m}$ consist of the corresponding leading n_m columns of U, $V \in \mathbb{R}^{n_\xi \times n_\xi}$. Moreover, Θ_l and Θ_r satisfy

$$\Theta_l^T \Theta_r = I_{n_m}. \tag{12.11}$$

Inserting the decomposition of Π from (12.10) into (12.9) and considering $\tilde{\xi}(t) = {\Theta_l}^T \xi(t)$, the resulting dynamical system leads to

$$\Theta_r^T M_1 \Theta_r \ddot{\tilde{\xi}}(t) = -\Theta_r^T D_1 \Theta_r \dot{\tilde{\xi}}(t) - \Theta_r^T K_1 \Theta_r \tilde{\xi}(t) + \Theta_r^T H_1 u(t), \tag{12.12a}$$
$$y(t) = L_1 \Theta_r \tilde{\xi}(t). \tag{12.12b}$$

System (12.12) is now a standard second-order system like (5.1). This system is practically system (12.9) with the redundant equations removed by the Θ_r projection. The dynamical systems (12.2), (12.9) and (12.12) are equivalent in the sense that they are different realizations of the same transfer function. Moreover, their finite spectrum is the same which we prove in the next section.

12.3 Equivalent finite spectra

The quadratic matrix polynomial [11, 142, 161, 162] associated with the index 3 DAE system (12.1) is

$$Q(\lambda) = \lambda^2 \underbrace{\begin{bmatrix} M_1 & 0 \\ 0 & 0 \end{bmatrix}}_{M} + \lambda \underbrace{\begin{bmatrix} D_1 & 0 \\ 0 & 0 \end{bmatrix}}_{D} + \underbrace{\begin{bmatrix} K_1 & G_1^T \\ G_1 & 0 \end{bmatrix}}_{K}, \tag{12.13}$$

where $\lambda \in \mathbb{C}$. Although $Q(\lambda)$ is regular, due to the singularity of M, it contains some infinite eigenvalues as well. If the degree of $\det(Q(\lambda))$ is $r < 2\tilde{n}$, where $\tilde{n} = n_\xi + n_\varphi$, then $Q(\lambda)$ has r finite and $2\tilde{n} - r$ infinite eigenvalues [162]. Again, the quadratic matrix polynomials corresponding to systems (12.9) and (12.12), respectively, are

$$\tilde{Q}(\lambda) = \lambda^2 \Pi M_1 \Pi^T + \lambda \Pi D_1 \Pi^T + \Pi K_1 \Pi^T \tag{12.14}$$

and

$$\bar{Q}(\lambda) = \lambda^2 \Theta_r^T M_1 \Theta_r + \lambda \Theta_r^T D_1 \Theta_r + \Theta_r^T K_1 \Theta_r. \tag{12.15}$$

We know that $\Theta_r^T M_1 \Theta_r \in \mathbb{R}^{n_\xi \times n_\xi}$ is nonsingular and $\bar{Q}$ is regular. Therefore, all of the eigenvalues of the polynomial $\bar{Q}(\lambda)$ are finite [162]. The degree of $\det\left(\bar{Q}(\lambda)\right)$ is $2(n_\xi - n_\varphi)$. Hence the number of finite eigenvalues of $\bar{Q}(\lambda)$ is exactly $2(n_\xi - n_\varphi)$ and the number of infinite eigenvalues is $2\tilde{n} - 2(n_\xi - n_\varphi) = 2n_\xi + 2n_\varphi - 2n_\xi + 2n_\varphi = 4n_\varphi$ infinite eigenvalues. By applying the appropriate projectors onto the index 3 DAE system, we can preserve all the finite eigenvalues of the system. The following theorem proves that both the original and the projected systems have the same finite eigenvalues.

Theorem 21. *Let us consider the matrix polynomials $Q(\lambda)$ and $\bar{Q}(\lambda)$ respectively defined in (12.13) and (12.14). An eigenvalue λ_1 is a finite eigenvalue of $Q(\lambda)$ with corresponding eigenvector $\begin{bmatrix} v_1^T & v_2^T \end{bmatrix}^T$ if and only if λ_1 is an eigenvalue of $\bar{Q}(\lambda)$ with corresponding eigenvector $\tilde{v}_1$ where $\tilde{v}_1 = \Theta_l^T v_1$ with Θ_l defined in (12.11).*

Proof. Suppose λ_1 is a finite eigenvalue of $Q(\lambda)$ corresponding to the eigenvector $\begin{bmatrix} v_1^T & v_2^T \end{bmatrix}^T$. Then the quadratic eigenvalue problem of the matrix polynomial (12.13) is

$$\left(\lambda_1^2 \begin{bmatrix} M_1 & 0 \\ 0 & 0 \end{bmatrix} + \lambda_1 \begin{bmatrix} D_1 & 0 \\ 0 & 0 \end{bmatrix} + \begin{bmatrix} K_1 & G_1^T \\ G_1 & 0 \end{bmatrix}\right) \begin{bmatrix} v_1 \\ v_2 \end{bmatrix} = \begin{bmatrix} 0 \\ 0 \end{bmatrix}. \tag{12.16}$$

The last line of (12.16) gives $G_1 v_1 = 0$ which implies v_1 is in the null space of G_1. By applying Theorem 20, we obtain $\Pi^T v_1 = v_1$. Plug in $\Pi^T v_1 = v_1$ into the first equation of (12.16) and then project the resulting equation from the left by Π. Since $\Pi G_1^T = 0$, by Proposition 4, it leads to

$$(\lambda_1^2 \Pi M_1 \Pi^T + \lambda_1 \Pi D_1 \Pi^T + \Pi K_1 \Pi^T) v_1 = 0, \tag{12.17}$$

which is the eigenvalue problem for the matrix polynomial $\tilde{Q}(\lambda)$. Applying the decompositions of Π as defined above to (12.10) and using $\tilde{v}_1 = \Theta_l^T v_1$ we obtain

$$\Theta_l(\lambda_1^2 \Theta_r^T M_1 \Theta_r + \lambda_1 \Theta_r^T D_1 \Theta_r + \Theta_r^T K_1 \Theta_r)\tilde{v}_1 = 0.$$

Multiplying by Θ_r from the left and using (12.11) yields

$$(\lambda_1^2 \Theta_r^T M_1 \Theta_r + \lambda_1 \Theta_r^T D_1 \Theta_r + \Theta_r^T K_1 \Theta_r)\tilde{v}_1 = 0, \tag{12.18}$$

which is the eigenvalue problem of the matrix polynomial (12.14) where λ_1 is an eigenvalue of the polynomial. Conversely, we want to demonstrate that if $\tilde{v}_1$ is an eigenvector of $\bar{Q}(\lambda)$ to the corresponding eigenvalue, λ_1, i.e., equation (12.18) holds, then $\begin{bmatrix} v_1^T & v_2^T \end{bmatrix}^T$ is an eigenvector of $Q(\lambda)$ with the same eigenvalue. Again, plugging $\tilde{v}_1 = \Theta_l^T v_1$ in (12.18) and multiplying the resulting equation by Θ_l from the left we obtain

$$(\lambda_1^2 \Pi M_1 \Pi^T + \lambda_1 \Pi D_1 \Pi^T + \Pi K_1 \Pi^T) v_1 = 0, \tag{12.19}$$

Since the projector Π satisfies $\Pi^T v_1 = v_1$, (12.19) gives

$$\Pi(\lambda_1^2 M_1 v_1 + \lambda_1 D_1 v_1 + K_1 v_1) = 0.$$

which means that $\lambda_1^2 M_1 v_1 + \lambda_1 D_1 v_1 + K_1 v_1$ is in the null space of Π. We know that $\text{Null}\,(\Pi) = \text{Range}\left(G_1^T\right)$ (see Proposition 4 (2)). Therefore, there exists a vector v_2 such that

$$\lambda_1^2 M_1 v_1 + \lambda_1 D_1 v_1 + K_1 v_1 = -G_1^T v_2,$$

which implies

$$\lambda_1^2 M_1 v_1 + \lambda_1 D_1 v_1 + K_1 v_1 + G_1^T v_2 = 0. \tag{12.20}$$

Again, if $\Pi^T v_1 = v_1$, using Theorem 20 yields

$$G_1 v_1 = 0. \tag{12.21}$$

Equations (12.20) and (12.21) yield (12.16). □

Example 1. *In order to show the equivalent finite spectra of the second-order index 3 system (12.1) numerically, we consider the damped spring-mass system (DSMS) form [122]. See Appendix A.5.2 for details. Here we consider* $n_\xi = 10$. *As a result, the dimension of the second-order index 3 model is 11. Using the MATLAB* `polyeig` *command, we compute the eigenvalues of* $Q(\lambda)$, $\tilde{Q}(\lambda)$ *and* $\bar{Q}(\lambda)$. *As we can see in Table 12.1, by applying appropriate projectors to the index 3 system, we can preserve all the finite eigenvalues.*

$Q(\lambda)$	$\tilde{Q}(\lambda)$	$\bar{Q}(\lambda)$
∞	0	
∞	0	
-0.1220 ± 0.2876i	-0.1220 ± 0.2876i	-0.1220 ± 0.2876i
-0.1171 ± 0.2827i	-0.1171 ± 0.2827i	-0.1171 ± 0.2827i
-0.1000 ± 0.2646i	-0.1000 ± 0.2646i	-0.1000 ± 0.2646i
-0.0958 ± 0.2597i	-0.0958 ± 0.2597i	-0.0958 ± 0.2597i
-0.0270 ± 0.1445i	-0.0270 ± 0.1445i	-0.0270 ± 0.1445i
-0.0367 ± 0.1674i	-0.0367 ± 0.1674i	-0.0367 ± 0.1674i
-0.0423 ± 0.1789i	-0.0423 ± 0.1789i	-0.0423 ± 0.1789i
-0.0679 ± 0.2229i	-0.0679 ± 0.2229i	-0.0679 ± 0.2229i
-0.0663 ± 0.2206i	-0.0663 ± 0.2206i	-0.0663 ± 0.2206i
∞		
∞		

Table 12.1: The equivalent finite spectrum of original and projected systems.

12.4 Second-order-to-first-order reduction

Let us consider the second-order index 3 system (12.1). In the preceding section, it was shown that this system can be converted into the equivalent form of the projected second-order system (12.9). The first-order transformed form of this second-order projected system can be written as

$$\begin{aligned} \tilde{\Pi}E_1\tilde{\Pi}^T\dot{x}_1(t) &= \tilde{\Pi}A_1\tilde{\Pi}^T x_1(t) + \tilde{\Pi}B_s u(t), \\ y(t) &= C_s\tilde{\Pi}^T x_1(t), \end{aligned} \tag{12.22}$$

where

$$\begin{aligned} \tilde{\Pi} &= \begin{bmatrix} \Pi & \\ & \Pi \end{bmatrix}, \quad E_1 = \begin{bmatrix} M_1 & 0 \\ 0 & M_1 \end{bmatrix}, \quad A_1 = \begin{bmatrix} 0 & M_1 \\ -K_1 & -D_1 \end{bmatrix}, \\ B_s &= \begin{bmatrix} 0 \\ H_1 \end{bmatrix}, \qquad C_s = \begin{bmatrix} L_1 & 0 \end{bmatrix} \quad \text{and} \quad x_1(t) = \begin{bmatrix} \xi(t) \\ \dot{\xi}(t) \end{bmatrix}. \end{aligned} \tag{12.23}$$

In system (12.22), all the coefficient matrices are singular since $\tilde{\Pi}$ has rank deficiency (due to the singularity of Π). This means the system contains redundant elements. To remove the redundant elements, let us decompose $\tilde{\Pi}$ as

$$\tilde{\Pi} = \bar{\Theta}_l\bar{\Theta}_r^T, \quad \text{with} \quad \bar{\Theta}_l^T\bar{\Theta}_r = I_k, \tag{12.24}$$

where $\bar{\Theta}_l$, $\bar{\Theta}_r \in \mathbb{R}^{2n_\xi \times k}$ and $k = \text{rank}\left(\tilde{\Pi}\right)$. By applying the decomposition of $\tilde{\Pi}$ from (12.24) to (12.22) and defining $\tilde{x}_1(t) := \bar{\Theta}_l^T x_1(t)$, we obtain

$$\begin{aligned} \bar{\Theta}_r^T E_1 \bar{\Theta}_r \dot{\tilde{x}}_1(t) &= \bar{\Theta}_r^T A_1 \bar{\Theta}_r \tilde{x}_1(t) + \bar{\Theta}_r^T B_s u(t), \\ y(t) &= C_s \bar{\Theta}_r \tilde{x}_1(t). \end{aligned} \tag{12.25}$$

This system can be compared with the generalized state space system (4.1) and hence one can directly apply a naive approach of balanced truncation or IRKA-based model reduction methods. Unfortunately, considering computational costs, forming (12.25) is prohibitive for a large-scale system since the actual computation of $\bar{\Theta}_l$ and $\bar{\Theta}_r$ by decomposing $\tilde{\Pi}$ is rather expensive. Moreover, the coefficient matrices in system (12.25) are typically dense. Therefore, by following the approaches discussed in [89, 93] for first-order index 2 systems, we apply balanced truncation and IRKA to the system (12.22) and compute the substantially reduced dimensional model

$$\begin{aligned} \hat{E}\dot{\hat{x}}(t) &= \hat{A}\hat{x}(t) + \hat{B}u(t), \\ \hat{y}(t) &= \hat{C}x(t). \end{aligned} \tag{12.26}$$

In the following subsections, we will show how to achieve this goal efficiently.

12.4.1 Balancing-based technique

Let us assume we want to apply the balanced truncation method to system (12.24). Thus, we need to solve the Lyapunov equations

$$\bar{\Theta}_r^T A_1 \bar{\Theta}_r \bar{P} \bar{\Theta}_r^T E_1^T \bar{\Theta}_r + \bar{\Theta}_r^T E_1 \bar{\Theta}_r \bar{P} \bar{\Theta}_r^T A_1^T \bar{\Theta}_r = -\bar{\Theta}_r^T B_s B_s^T \bar{\Theta}_r, \tag{12.27a}$$
$$\bar{\Theta}_r^T A_1^T \bar{\Theta}_r \bar{Q} \bar{\Theta}_r^T E_1 \bar{\Theta}_r + \bar{\Theta}_r^T E_1^T \bar{\Theta}_r \bar{Q} \bar{\Theta}_r^T A_1 \bar{\Theta}_r = -\bar{\Theta}_r^T C_s^T C_s \bar{\Theta}_r, \tag{12.27b}$$

where $\bar{P}, \bar{Q} \in \mathbb{R}^{k \times k}$ are, respectively, the controllability and observability Gramians of system (12.25). The solutions, $\bar{P}$ and $\bar{Q}$, of the Lyapunov equations are unique because (according to Theorem 21) the corresponding system is asymptotically stable and symmetric positive (semi-)definite since the right-hand side is semi-definite. Multiplying both equations in (12.27) by $\bar{\Theta}_l$ from the left and $\bar{\Theta}_l^T$ from the right and exploiting the property in (12.24), we obtain

$$\tilde{\Pi} A_1 \tilde{\Pi}^T \tilde{P} \tilde{\Pi} E_1^T \tilde{\Pi}^T + \tilde{\Pi} E_1 \tilde{\Pi}^T \tilde{P} \tilde{\Pi} A_1^T \tilde{\Pi}^T = -\tilde{\Pi} B_s B_s^T \tilde{\Pi}^T, \tag{12.28a}$$
$$\tilde{\Pi} A_1^T \tilde{\Pi}^T \tilde{Q} \tilde{\Pi} E_1 \tilde{\Pi}^T + \tilde{\Pi} E_1^T \tilde{\Pi}^T \tilde{Q} \tilde{\Pi} A_1 \tilde{\Pi}^T = -\tilde{\Pi} C_s^T C_s \tilde{\Pi}^T, \tag{12.28b}$$

where

$$\tilde{P} = \bar{\Theta}_r \bar{P} \bar{\Theta}_r^T, \quad \tilde{Q} = \bar{\Theta}_r \bar{Q} \bar{\Theta}_r^T. \tag{12.29}$$

The Lyapunov equations in (12.28) are nothing but the Lyapunov equations of the projected system (12.22) where $\tilde{P}$, $\tilde{Q} \in \mathbb{R}^{2n_\xi \times 2n_\xi}$ are the system's controllability and observability Gramians, respectively. Under condition (12.29), it can be shown that $\tilde{P}$ and $\tilde{Q}$ satisfy

$$\tilde{P} = \tilde{\Pi}\tilde{P}\tilde{\Pi}^T \quad \text{and} \quad \tilde{Q} = \tilde{\Pi}\tilde{Q}\tilde{\Pi}^T, \tag{12.30}$$

which ensures that the solutions are unique even though the equations in (12.28) are singular due to the singular projectors. The solution techniques of the projected Lyapunov equations (12.28) for computing the low-rank Gramian factors will be discussed in Section 12.6. Let $\tilde{R}$ and $\tilde{L}$ respectively be the low-rank factors of the controllability and observability Gramians of system (12.22) such that

$$\tilde{P} \approx \tilde{R}\tilde{R}^T, \quad \tilde{Q} \approx \tilde{L}\tilde{L}^T. \tag{12.31}$$

Also, let $\bar{R}$ and $\bar{L}$ respectively be the low-rank factors of the controllability and observability Gramians of system (12.25) such that

$$\bar{P} \approx \bar{R}\bar{R}^T, \quad \bar{Q} \approx \bar{L}\bar{L}^T.$$

Then the controllability Gramian factors and the observability Gramian factors of systems (12.22) and (12.25) are related by

$$\tilde{R} = \bar{\Theta}_r\bar{R} \quad \text{and} \quad \tilde{L} = \bar{\Theta}_r\bar{L}. \tag{12.32}$$

This relation can be easily obtained since

$$\begin{aligned} \tilde{R}\tilde{R}^T &\approx \tilde{P} = \bar{\Theta}_r\bar{P}\bar{\Theta}_r^T \approx \bar{\Theta}_r\bar{R}\bar{R}^T\bar{\Theta}_r^T, \quad \text{and} \\ \tilde{L}\tilde{L}^T &\approx \tilde{Q} = \bar{\Theta}_r\bar{Q}\bar{\Theta}_r^T \approx \bar{\Theta}_r\bar{L}\bar{L}^T\bar{\Theta}_r^T. \end{aligned}$$

Let us consider the singular value decomposition of $\bar{L}^T\bar{\Theta}_r^T E_1\bar{\Theta}_r\bar{R}$, i.e.,

$$\bar{L}^T\bar{\Theta}_r^T E_1\bar{\Theta}_r\bar{R} = U\Sigma V^T.$$

Now construct the left and right *balancing and truncating* transformations $\bar{W}$ and $\bar{V}$ as

$$\bar{W} = \bar{L}U_1\Sigma_1^{-\frac{1}{2}}, \quad \bar{V} = \bar{R}V_1\Sigma_1^{-\frac{1}{2}},$$

where U_1 and V_1 consist of the corresponding leading l ($l \ll k$) columns of U and V while Σ_1 is the first leading $l \times l$ block of Σ. Considering, again, the singular value decomposition (using the Gramian factors of system (12.22))

$$\tilde{L}^T E_1\tilde{R} = \bar{R}^T\bar{Q}_r^T E_1\bar{Q}_r\bar{L} = U\Sigma V^T, \tag{12.33}$$

we can construct the left and right balancing and truncating transformations as

$$\tilde{W} = \tilde{L}U_1\Sigma_1^{-\frac{1}{2}}, \quad \tilde{V} = \tilde{R}V_1\Sigma_1^{-\frac{1}{2}}. \tag{12.34}$$

Algorithm 43: LR-SRM for second-order index 3 systems.

Input : M_1, D_1, K_1, G_1, H_1, L_1.
Output: $\hat{E}$, $\hat{A}$, $\hat{B}$, $\hat{C}$.

1. Set up matrices E_1, A_1, B_s, C_s as in (12.22).
2. Compute the low-rank Gramian factors $\tilde{R}$ and $\tilde{L}$ by solving the projected Lyapunov equations (12.28).
3. Construct the balancing and truncating transformations $\tilde{W}$ and $\tilde{V}$ by performing (12.33)-(12.34).
4. Form the reduced matrices using (12.37).

We observe that

$$\begin{aligned} \tilde{W} &= \tilde{L}U_1\Sigma_1^{-\frac{1}{2}} = \bar{\Theta}_r\bar{L}U_1\Sigma_1^{-\frac{1}{2}} = \bar{\Theta}_r\bar{W} = \bar{\Theta}_r\bar{\Theta}_l^T\bar{\Theta}_r\tilde{W} = \tilde{\Pi}^T\tilde{W}, \\ \tilde{V} &= \tilde{R}U_1\Sigma_1^{-\frac{1}{2}} = \bar{\Theta}_r\bar{R}U_1\Sigma_1^{-\frac{1}{2}} = \bar{\Theta}_r\bar{V} = \bar{\Theta}_r\bar{\Theta}_l^T\bar{\Theta}_r\tilde{V} = \tilde{\Pi}^T\tilde{V}. \end{aligned} \tag{12.35}$$

By applying the balancing and truncating transformations $\bar{W}$ and $\bar{V}$ to system (12.25), we can construct the reduced-order model (12.26) where the coefficient matrices are formed by

$$\hat{E} = \bar{W}^T\bar{\theta}_r^T E_1\bar{\theta}_r\bar{V}, \quad \hat{A} = \bar{W}^T\bar{\theta}_r^T A_1\bar{\theta}_r\bar{V}, \quad \hat{B} = \bar{W}^T\bar{\theta}_r^T B_s \quad \text{and } \hat{C} = C_s\bar{\theta}_r\bar{V}.$$

Close observation reveals that by applying property (12.35), the above reduced matrices can be computed efficiently by

$$\begin{aligned} \hat{E} = \bar{W}^T\bar{E}\bar{V} &= \bar{W}^T\bar{\Theta}_r^T E_1\bar{\Theta}_r\bar{V} = \tilde{W}^T\Pi E_1\Pi^T\tilde{V} = \tilde{W}^T E_1\tilde{V}, \\ \hat{A} = \bar{W}^T\bar{A}\bar{V} &= \bar{W}^T\bar{\Theta}_r^T A_1\bar{\Theta}_r\bar{V} = \tilde{W}^T\Pi A_1\Pi^T\tilde{V} = \tilde{W}^T A_1\tilde{V}, \\ \hat{B} = \bar{W}^T\bar{B} &= \bar{W}^T\bar{\Theta}_r^T B_s = \tilde{W}^T\Pi B_s = \tilde{W}^T B_s, \\ \hat{C} = \bar{C}\bar{V} &= C_s\bar{\Theta}_r\bar{V} = C_s\Pi^T\tilde{V} = C_s\tilde{V}. \end{aligned} \tag{12.36}$$

Therefore, from the above discussion, it is clear that to obtain the reduced model (12.26), we need not to form system (12.22) or system (12.25). We just form the balancing and truncating transformations $\tilde{W}$ and $\tilde{V}$ as given in (12.34) and then construct the reduced matrices as

$$\hat{E} = \tilde{W}^T E_1\tilde{V}, \; \hat{A} = \tilde{W}^T A_1\tilde{V}, \; \hat{B} = \tilde{W}^T B_s \quad \text{and} \quad \hat{C} = C_s\tilde{V}. \tag{12.37}$$

The procedure to compute the reduced first-order system (12.26) from the second-order index 3 DAEs (12.2) is summarized in Algorithm 43.

12.4.2 Interpolatory method via IRKA

We concentrate on the interpolatory method via IRKA for model reduction of system (12.1). The method that we follow here was introduced in Chapters 8 and 10 for first-order index 2 and index 3 systems, respectively. The main focus here is to apply the IRKA to the projected system without creating the projector explicitly. Gugercin et al. [89] first showed this for first-order index 2 DAE system. We have extended that idea here for second-order index 3 systems.

Like the already discussed balanced truncation, we want to apply the IRKA onto system (12.25) which is an equivalent form of system (12.22). To construct a reduced system from the projected system (12.22) based on the interpolatory method, we need to construct the right and left transformations defined as

$$\tilde{V} = \left[(\alpha_1\tilde{E}-\tilde{A})^I\tilde{B}b_1,\cdots,(\alpha_r\tilde{E}-\tilde{A})^I\tilde{B}b_r\right], \quad \text{and} \tag{12.38a}$$

$$\tilde{W} = \left[(\alpha_1\tilde{E}^T-\tilde{A}^T)^I\tilde{C}^Tc_1,\cdots,(\alpha_r\tilde{E}^T-\tilde{A}^T)^I\tilde{C}^Tc_r\right], \tag{12.38b}$$

where $\tilde{E} := \tilde{\Pi}E_1\tilde{\Pi}^T$, $\tilde{A} := \tilde{\Pi}A_1\tilde{\Pi}^T$, $\tilde{B} := \tilde{\Pi}B_s$, $\tilde{C} := C_s\tilde{\Pi}^T$ and $(\alpha_i\tilde{E}-\tilde{A})^I = (\alpha_i\tilde{E}^T-\tilde{A}^T)^I := \bar{\Theta}_r(\bar{\Theta}_r^TE_1\bar{\Theta}_r - \alpha_i\bar{\Theta}_r^TA_1\bar{\Theta}_r)^{-1}\bar{\Theta}_r^T$ for $i = 1,\cdots,r$. If we can construct $\tilde{V}$ and $\tilde{W}$ as in (12.38), then the reduced model in (12.26) can be formed by computing the reduced matrices as in (12.37). Therefore, to construct the reduced model (12.26) we can avoid the projectors. However, in each term of $\tilde{V}$ and $\tilde{W}$ the projectors are implicitly hidden. The solution of this problem has been shown in [89] based on [93, Theorem 5.2]. In our case, each column of $\tilde{V}$ contains a vector like

$$v = (\alpha\tilde{E}-\tilde{A})^I\tilde{B}b,$$

which implies that we have to solve the linear system

$$(\alpha\tilde{E}-\tilde{A})v = \tilde{B}b, \tag{12.39}$$

for v. Equation (12.39) is equivalent to

$$\begin{bmatrix}\alpha\Pi M_1\Pi^T & -\Pi M_1\Pi^T\\ \Pi K_1\Pi^T & \Pi M_1\Pi^T+\alpha\Pi D_1\Pi^T\end{bmatrix}\begin{bmatrix}v_1\\v_2\end{bmatrix} = \begin{bmatrix}0\\ \Pi H_1 b\end{bmatrix}. \tag{12.40}$$

This linear system can be solved efficiently by following Theorem 19.

It can be shown that the vector $v = \begin{bmatrix}v_1^T & v_2^T\end{bmatrix}^T$ in (12.43) solves

$$\begin{bmatrix}\alpha M_1 & -M_1 & G_1^T & 0\\ K_1 & \alpha M_1+D_1 & 0 & G_1^T\\ G_1 & 0 & 0 & 0\\ 0 & G_1 & 0 & 0\end{bmatrix}\begin{bmatrix}v_1\\v_2\\\Gamma_1\\\Gamma_2\end{bmatrix} = \begin{bmatrix}0\\H_1b\\0\\0\end{bmatrix}. \tag{12.41}$$

Analogously, to construct $\tilde{W}$ in (12.38), at each iteration, we need to solve a linear system which implies solving the linear system

$$(\alpha\tilde{E}^T-\tilde{A}^T)w = \tilde{C}^Tc, \tag{12.42}$$

Algorithm 44: IRKA for second-order index 3 system.

Input : M_1, D_1, K_1, G_1, H_1, L_1.
Output: $\hat{E}$, $\hat{A}$, $\hat{B}$, $\hat{C}$.

Setup matrices E_1, A_1, B_s, C_s as in (12.22).
Select initial interpolation points $\{\sigma_i\}_{i=1}^r$ and tangent directions $\{b_i\}_{i=1}^r$ and $\{c_i\}_{i=1}^r$.
while *(not converged)* **do**
  for $i = 1, 2, \cdots, r$ **do**
    Solve $\begin{bmatrix} \alpha M_1 & -M_1 & G_1^T & 0 \\ K_1 & \alpha M_1 + D_1 & 0 & G_1^T \\ G_1 & 0 & 0 & 0 \\ 0 & G_1 & 0 & 0 \end{bmatrix} \begin{bmatrix} v_1 \\ v_2 \\ \Gamma_1 \\ \Gamma_2 \end{bmatrix} = \begin{bmatrix} 0 \\ H_1 b \\ 0 \\ 0 \end{bmatrix}$,
    for $\mathrm{v}_i = \begin{bmatrix} v_1^T & v_2^T \end{bmatrix}^T$.
    Solve $\begin{bmatrix} \alpha M_1^T & K_1^T & G_1^T & 0 \\ -M_1^T & \alpha M_1^T + D_1^T & 0 & G_1^T \\ G_1 & 0 & 0 & 0 \\ 0 & G_1 & 0 & 0 \end{bmatrix} \begin{bmatrix} w_1 \\ w_2 \\ \Gamma_1 \\ \Gamma_2 \end{bmatrix} = \begin{bmatrix} L_1^T c \\ 0 \\ 0 \\ 0 \end{bmatrix}$,
    for $\mathrm{w}_i = \begin{bmatrix} w_1^T & w_2^T \end{bmatrix}^T$.
    Construct $\tilde{V} = \begin{bmatrix} \mathrm{v}_1, \mathrm{v}_1, \cdots, \mathrm{v}_r \end{bmatrix}$, $\tilde{W} = \begin{bmatrix} \mathrm{w}_1, \mathrm{w}_1, \cdots, \mathrm{w}_r \end{bmatrix}$.
  $\hat{E} = \tilde{W}^T E_1 \tilde{V}$, $\hat{A} = \tilde{W}^T A_1 \tilde{V}$, $\hat{B} = \tilde{W}^T B_s$ and $\hat{C} = C_s \tilde{V}$
  Compute $\hat{A} z_i = \lambda_i \hat{E} z_i$ and $y^* \hat{A} = \lambda_i y^* \hat{E}$.
  $\alpha_i \leftarrow -\lambda_i$, $b_i^* \leftarrow y^* \hat{B}$ and $c_i \leftarrow \hat{C} z_i$.
Form $\hat{E}$, $\hat{A}$, $\hat{B}$ and $\hat{C}$ with (12.37).

for w, which is equivalent to

$$\begin{bmatrix} \alpha \Pi M_1^T \Pi^T & \Pi K_1^T \Pi^T \\ -\Pi M_1^T \Pi^T & \Pi D_1^T \Pi^T + \alpha \Pi M_1^T \Pi^T \end{bmatrix} \begin{bmatrix} w_1 \\ w_2 \end{bmatrix} = \begin{bmatrix} \Pi L_1^T c \\ 0 \end{bmatrix}. \tag{12.43}$$

This is equivalent to solving the linear system

$$\begin{bmatrix} \alpha M_1^T & K_1^T & G_1^T & 0 \\ -M_1^T & \alpha M_1^T + D_1^T & 0 & G_1^T \\ G_1 & 0 & 0 & 0 \\ 0 & G_1 & 0 & 0 \end{bmatrix} \begin{bmatrix} w_1 \\ w_2 \\ \Gamma_1 \\ \Gamma_2 \end{bmatrix} = \begin{bmatrix} L_1^T c \\ 0 \\ 0 \\ 0 \end{bmatrix}, \tag{12.44}$$

where $w = \begin{bmatrix} w_1^T & w_2^T \end{bmatrix}^T$. A complete procedure to compute the reduced model (12.26) from the second-order index 3 system (12.1) is presented in Algorithm 44.

12.5 Second-order-to-second-order reduction

In this section, we discuss second-order-to-second-order, i.e., structure preserving model reduction (SPMOR) of second-order index 3 DAE system (12.2). In Section 12.2, we have shown that index 3 system (12.2) can be converted into a standard ODE system (12.12). In principle, we can now apply that model reduction technique onto this converted system by following the approach in Chapter 5. However, considering the computational complexity, we should avoid constructing ODE system (12.12). In Section 12.2 it was shown that the systems in (12.2), (12.9) and (12.12) are equivalent. Therefore, instead of applying the MOR method on system (12.12), it is convenient to apply the method onto system (12.9). In this case, the main concern is how to avoid the explicit computation of the projected system to perform the model reduction by applying BT and PDEG methods. In both cases, we want to apply the model reduction techniques to system (12.9) and obtain a substantially reduced-order model

$$\begin{aligned} \hat{M}_1\ddot{\hat{\xi}}(t) + \hat{D}_1\dot{\hat{\xi}}(t) + \hat{K}_1\hat{\xi}(t) &= \hat{H}_1 u(t), \\ \hat{y}(t) &= \hat{L}_1\hat{\xi}(t). \end{aligned} \tag{12.45}$$

In the following subsections, we will show how to apply both methods to the projected system (12.37) while avoiding the projector (Π).

12.5.1 The BT method

Chapter 5 (see Subsection 5.4.1) discussed the second-order-to-second-order balancing criterion for standard second-order systems. For a second-order index 1 system (11.1), of a slightly different form than (12.1), a second-order-to-second-order balancing technique is shown in Chapter 11. The fundamental procedure here is the same as it is in Chapter 11. We can convert the index 3 system (12.2) to the ODE system (12.12); thus allowing the balancing idea from Chapter 5 to be employed. As mentioned already, forming (12.9) is infeasible for a large-scale system. Therefore, we want to apply the BT technique to a system that is equivalent to system (12.9). For this purpose, recalling the discussion in Chapter 5, we need to solve the projected Lyapunov equations (12.28). We already know that the controllability and observability Gramians, $\tilde{P}$ and $\tilde{Q}$, are the unique positive semi-definite solutions of the projected Lyapunov equations (12.28). By employing the block subdivision of the controllability and observability Gramians as in, e.g., [37], we get

$$\tilde{P} = \begin{bmatrix} \tilde{P}_{pp} & \tilde{P}_{pv} \\ \tilde{P}_{pv}^T & \tilde{P}_{vv} \end{bmatrix}, \qquad \tilde{Q} = \begin{bmatrix} \tilde{Q}_{pp} & \tilde{Q}_{pv} \\ \tilde{Q}_{pv}^T & \tilde{Q}_{vv} \end{bmatrix},$$

where $\tilde{P}_{pp}$, $\tilde{Q}_{pp} \in \mathbb{R}^{n_\xi \times n_\xi}$ and $\tilde{P}_{vv}$, $\tilde{Q}_{vv} \in \mathbb{R}^{n_\xi \times n_\xi}$ are called controllability, observability position and velocity Gramians, respectively. Using the LR-ADI

Algorithm 45: LR-SRM for second-order index 3 system.

Input : M_1, D_1, K_1, H_1, L_1.
Output: $\hat{M}_1$, $\hat{D}_1$, $\hat{K}_1$, $\hat{H}_1$, $\hat{L}_1$.
1 Solve the Lyapunov equations (12.28a) to compute $\tilde{R}_p$ and $\tilde{L}_p$.
2 Compute the balancing and truncating transformations as in (12.47).
3 Construct $\hat{M}_1$, $\hat{D}_1$, $\hat{K}_1$, $\hat{H}_1$, $\hat{L}_1$ following (12.48).

---

iterations (will be discussed in the next section) we can compute the low-rank controllability Gramian factor $\tilde{R}$ and observability Gramian factor $\tilde{L}$ defined in (12.31) by solving the Lyapunov equations (12.28). Due to the block structure of the $\tilde{P}$ and $\tilde{Q}$, the low-rank Gramian factors can be partitioned as

$$\tilde{R} = \begin{bmatrix} \tilde{R}_p \\ \tilde{R}_v \end{bmatrix}, \qquad \tilde{L} = \begin{bmatrix} \tilde{L}_p \\ \tilde{L}_v \end{bmatrix},$$

where $\tilde{R}_p$, $\tilde{L}_p$ and $\tilde{R}_v$, $\tilde{L}_v$ denote the low-rank position, velocity controllability and observability Gramian factors, respectively. Let us consider the Gramian factors $\tilde{R}_p$ and $\tilde{L}_p$ to compute the thin SVD

$$\tilde{L}_p^T M_1 \tilde{R}_p = \begin{bmatrix} U_{pp,1} & U_{pp,1} \end{bmatrix} \begin{bmatrix} \Sigma_{pp,1} & 0 \\ 0 & \Sigma_{pp,2} \end{bmatrix} \begin{bmatrix} V_{pp,1}^T & V_{pp,2}^T \end{bmatrix}, \tag{12.46}$$

and construct the balancing and truncating transformations

$$W_s = \tilde{L}_p U_{pp,1} \Sigma_{pp,1}^{-\frac{1}{2}}, \quad V_s = \tilde{R}_p U_{pp,1} \Sigma_{pp,1}^{-\frac{1}{2}}. \tag{12.47}$$

Now by applying W_s and V_s to system (12.9), we can construct the reduced model (12.45). Like the second-order-to-first-order balancing based reduction method (see the discussion above), we can also prove that the constructed balancing and truncating transformations are Π^T invariant, i.e., $\Pi^T V_s = V_s$ and $\Pi^T W_s = W_s$. Therefore, the coefficient matrices in (12.45) can be constructed as

$$\begin{aligned} &\hat{M}_1 = W_s^T M_1 V_s, && \hat{D}_1 = W_s^T D_1 V_s, && \\ &\hat{K}_1 = W_s^T K_1 V_s, && \hat{H}_1 = W_s^T H_1, && \hat{L}_1 = L_1 V_s, \end{aligned} \tag{12.48}$$

which prevent from constructing the projected system (12.9). The process of obtaining a reduced model by using a pair of low-rank controllability and observability position Gramian factors, i.e., $(\tilde{R}_p, \tilde{L}_p)$ is summarized in Algorithm 45. The constructed reduced model via $(\tilde{R}_p, \tilde{L}_p)$ is called position-position (PP) balancing. Likewise, the reduced models are called velocity-velocity (VV), velocity-position (VP) and position-velocity (PV) balancing if we use the pairs (R_v, L_v), (R_v, L_p) and (R_p, L_v), respectively.

12.5.2 The PDEG method

Here we want to construct the ROMs via projecting the system onto the dominant eigenspaces of the Gramians. The PDEG technique is introduced in Chapter 11 to obtain second-order-to-second-order reduced models for second-order index 1 systems; we follow the same procedure for the second-order index 3 systems. We first convert the second-order index 3 system (12.1) into its equivalent form of the second-order projected system (12.9). In the preceding subsection, we have already defined the Gramians for the second-order projected system. Let us first consider the controllability position Gramian $\tilde{P}_{pp}$. Since $\tilde{P}_{pp}$ is symmetric positive (semi-)definite, it has the singular value decomposition,

$$\tilde{P}_{pp} = \tilde{U}_{pp}\tilde{\Sigma}_{pp}\tilde{V}_{pp}^T. \tag{12.49}$$

If rank $\left(\tilde{P}_{pp}\right) = k$, where $k \ll n_\xi$, then the first k columns of $\tilde{U}_{pp}$ are the eigenvectors of $\tilde{P}_{pp}$. Now suppose $\tilde{R}_p$ is the low-rank factor (as defined previously) of the controllability position Gramian $\tilde{P}_{pp}$ such that $\tilde{P}_{pp} \approx \tilde{R}_p\tilde{R}_p^T$. Compute the *thin* SVD as

$$\tilde{R}_p = U_k\Sigma_k V_k^T, \tag{12.50}$$

where it can be proved that U_k consists of the first k columns of $\tilde{U}_{pp}$. Now forming

$$V_s = \begin{bmatrix} u_1, u_2, \cdots, u_r \end{bmatrix}, \tag{12.51}$$

where u_i, $i = 1, \cdots, r$, are the first r columns of U_k and applying V_s to system (12.9) we can construct the r dimensional reduced model (12.45), where the reduced coefficient matrices can be formed as

$$\begin{aligned} \hat{M}_1 = V_s^T\Pi M_1\Pi^T V_s,\ \hat{D}_1 = V_s^T\Pi D_1\Pi^T V_s,\ \hat{K}_1 = V_s^T\Pi K_1\Pi^T V_s, \\ \hat{H}_1 = V_s^T\Pi H_1,\ \hat{L}_1 = L_1\Pi^T V_s. \end{aligned} \tag{12.52}$$

By applying Theorem 20, we have $\Pi^T V_s = V_s$ which implies $V_s^T\Pi = V_s^T$. Therefore, the reduced matrices in (12.45) can be constructed as

$$\hat{M}_1 = V_s^T M_1 V_s,\ \hat{D}_1 = V_s^T D_1 V_s,\ \hat{K}_1 = V_s^T K_1 V_s, \hat{H}_1 = V_s^T H_1,\ \hat{L}_1 = L_1 V_s. \tag{12.53}$$

This procedure to compute the ROM via projecting the system onto the dominant eigenspace of the controllability position Gramian is summarized in Algorithm 46. A reduced model that is obtained via projecting the system onto the dominant eigenspace of the controllability position Gramian is called PDEG-CP. Similarly, the reduced models are called PDEG-CV, PDEG-OP or PDEG-OV if they are obtained via projecting the system onto the eigenspaces of the controllability position, observability position or observability velocity Gramians, respectively.

Algorithm 46: PDEG for second-order index 3 system.

Input : M_1, D_1, K_1, H_1, L_1.
Output: $\hat{M}_1$, $\hat{D}_1$, $\hat{K}_1$, $\hat{H}_1$, $\hat{L}_1$.

1 Solve the Lyapunov equations (12.28a) to compute $\tilde{R}_p$.
2 Compute V_s by performing (12.50-12.51).
3 Construct $\hat{M}_1$, $\hat{D}_1$, $\hat{K}_1$, $\hat{H}_1$, $\hat{L}_1$ following (12.53).

12.6 Solution of the projected Lyapunov equations

In previous sections, we noted that to implement the balancing and PDEG-based model reduction methods for the second-order index 3 descriptor system (12.1), we need to solve the projected Lyapunov equations (12.28) for computing the low-rank Gramian factors $\tilde{R}$ and $\tilde{L}$. This section discusses how to apply the LR-ADI iteration introduced in Chapter 3 to solve such projected Lyapunov equations efficiently. For convenience, we rewrite the projected Lyapunov equations from (12.28) as

$$\tilde{A}\tilde{P}\tilde{E}^T + \tilde{E}\tilde{P}\tilde{A}^T = -\tilde{B}\tilde{B}^T, \tag{12.54a}$$

$$\tilde{A}^T\tilde{Q}\tilde{E} + \tilde{E}^T\tilde{Q}\tilde{A} = -\tilde{C}^T\tilde{C}, \tag{12.54b}$$

where $\tilde{E} = \tilde{\Pi}E_1\tilde{\Pi}^T$, $\tilde{A} = \tilde{\Pi}A_1\tilde{\Pi}^T$, $\tilde{B} = \tilde{\Pi}B_s$ and $\tilde{C} = C_s\tilde{\Pi}^T$. These Lyapunov equations look like the projected Lyapunov equations from (10.16) (in Chapter 10) for the first-order index 3 DAEs. An efficient solution technique of such projected Lyapunov equations was discussed in Chapter 8 and also in Chapter 10 by using the LR-ADI method. Here we briefly review how to solve them without using the projector explicitly. Let us first concentrate on the solution of the controllability Lyapunov equation (12.54a). Recalling Algorithm 15, a close observation reveals that the initial residual factor $\tilde{W}_0$ can be partitioned as

$$\tilde{W}_0 = \tilde{B} = \tilde{\Pi}B_s = \begin{bmatrix} \Pi & \\ & \Pi \end{bmatrix} \begin{bmatrix} 0 \\ H_1 \end{bmatrix} = \begin{bmatrix} 0 \\ \Pi H_1 \end{bmatrix} = \begin{bmatrix} \tilde{W}_0^{(1)} \\ \tilde{W}_0^{(2)} \end{bmatrix}, \tag{12.55}$$

which gives $\tilde{W}_0^{(1)} = 0$ and $\tilde{W}_0^{(2)} = \Pi H_1$. We can compute $\tilde{W}_0^{(2)} = \Pi H_1$ efficiently by using the following observation.

Lemma 8. *The matrix Ξ satisfies $\Xi = \Pi^T\Xi$ and $M_1\Xi = \Pi H_1$, where Π is defined in (12.6) if and only if*

$$\begin{bmatrix} M_1 & G_1^T \\ G_1 & 0 \end{bmatrix} \begin{bmatrix} \Xi \\ \Lambda \end{bmatrix} = \begin{bmatrix} H_1 \\ 0 \end{bmatrix}. \tag{12.56}$$

Proof. For the proof, refer to Lemma 5. □

To solve the Lyapunov equation (12.54) by using Algorithm 15, at the i-th iteration step we need to solve the linear system

$$(\tilde{A} + \mu_i \tilde{E})V_i = \tilde{W}_{i-1}, \tag{12.57}$$

where $\tilde{W}_{i-1}$ is the ADI residual factor computed from the $(i-1)$-st iteration. Now partitioning $\tilde{W}_{i-1}$ as $\tilde{W}_{i-1} = \begin{bmatrix} \tilde{W}_{i-1}^{(1)} \\ \tilde{W}_{i-1}^{(2)} \end{bmatrix}$, equation (12.57) can be written as

$$\begin{bmatrix} \mu_i \Pi M_1 \Pi^T & \Pi M_1 \Pi^T \\ -\Pi K_1 \Pi^T & -\Pi D_1 \Pi^T + \mu_i \Pi M_1 \Pi^T \end{bmatrix} \begin{bmatrix} V_i^{(1)} \\ V_i^{(2)} \end{bmatrix} = \begin{bmatrix} \tilde{W}_{i-1}^{(1)} \\ \tilde{W}_{i-1}^{(2)} \end{bmatrix}. \tag{12.58}$$

According to Theorem 19, instead of solving the linear system (12.58) the linear system

$$\begin{bmatrix} \alpha_i M_1 & M_1 & G_1^T & 0 \\ -K_1 & -D_1 + \alpha_i M_1 & 0 & G_1^T \\ G_1 & 0 & 0 & 0 \\ 0 & G_1 & 0 & 0 \end{bmatrix} \begin{bmatrix} V_i^{(1)} \\ V_i^{(2)} \\ \Gamma_1 \\ \Gamma_2 \end{bmatrix} = \begin{bmatrix} \tilde{W}_{i-1}^{(1)} \\ \tilde{W}_{i-1}^{(2)} \\ 0 \\ 0 \end{bmatrix}, \tag{12.59}$$

can be solved for $\begin{bmatrix} {V_i^{(1)}}^T & {V_i^{(2)}}^T \end{bmatrix}^T$. Note that although the matrix in (12.59) has larger dimensions, it is highly sparse and can be solved efficiently using any sparse solver. The matrix/vector $\begin{bmatrix} {\tilde{W}_{i-1}^{(1)}}^T & {\tilde{W}_{i-1}^{(2)}}^T \end{bmatrix}^T$ is updated in each iteration which is computed from the ADI residual factor of the previous step. At each iteration, the ADI residual factor can be computed by (see Step 5 in Algorithm 15)

$$\tilde{W}_i = \tilde{W}_{i-1} - 2\,\mathrm{Re}\,(\alpha_i)\tilde{E}V_i, \tag{12.60}$$

which can be partitioned as

$$\begin{aligned} \begin{bmatrix} \tilde{W}_i^{(1)} \\ \tilde{W}_i^{(2)} \end{bmatrix} &= \begin{bmatrix} \tilde{W}_{i-1}^{(1)} \\ \tilde{W}_{i-1}^{(2)} \end{bmatrix} - 2\,\mathrm{Re}\,(\alpha_i) \begin{bmatrix} \Pi M_1 \Pi^T & 0 \\ 0 & \Pi M_1 \Pi^T \end{bmatrix} \begin{bmatrix} V_i^{(1)} \\ V_i^{(2)} \end{bmatrix}. \\ &= \begin{bmatrix} \tilde{W}_{i-1}^{(1)} - 2\,\mathrm{Re}\,(\alpha_i)\Pi M_1 \Pi^T V_i^{(1)} \\ \tilde{W}_{i-1}^{(2)} - 2\,\mathrm{Re}\,(\alpha_i)\Pi M_1 \Pi^T V_i^{(2)} \end{bmatrix}. \end{aligned}$$

By exploiting the properties of Π, that is to say, by using $\Pi M_1 = M_1 \Pi^T$, $\Pi^T V_i^{(1)} = V_i^{(1)}$ and $\Pi^T V_i^{(2)} = V_i^{(2)}$, the above equation results in

$$\begin{aligned} \tilde{W}_i^{(1)} &= \tilde{W}_{i-1}^{(1)} - 2\,\mathrm{Re}\,(\alpha_i) M_1 V_i^{(1)}, \\ \tilde{W}_i^{(2)} &= \tilde{W}_{i-1}^{(2)} - 2\,\mathrm{Re}\,(\alpha_i) M_1 V_i^{(2)}. \end{aligned}$$

Algorithm 47: LR-ADI for projected Lyapunov equation.

Input : M_1, D_1, K_1, G_1, H_1, $\{\alpha_i\}_{i=1}^{J}$, a tolerance $0 < \tau \ll 1$ for the normalized residual.

Output: $\tilde{R} = Z_i$ such that $\tilde{P} \approx \tilde{R}\tilde{R}^T$.

Set $Z_0 = [\,]$, $i = 1$ and $\tilde{W}_0 = \tilde{B}$ as in (12.55) following Lemma 8.

while $\frac{\|\tilde{W}_{i-1}^T \tilde{W}_{i-1}\|}{\|\tilde{B}\tilde{B}^T\|} \geq \tau$ **do**

  Solve the linear system (12.59) for $V_i = \begin{bmatrix} V_i^{(1)} & V_i^{(2)} \end{bmatrix}^T$

  if $\operatorname{Im}(\alpha_i) = 0$ **then**

    $Z_i = \begin{bmatrix} Z_{i-1} & \sqrt{-2\alpha_i}\operatorname{Re}(V_i) \end{bmatrix}$

    Update, $\tilde{W}_i$ as in (12.61)

  else

    $\gamma = -2\operatorname{Re}(\alpha_i)$, $\delta = \frac{\operatorname{Re}(\alpha_i)}{\operatorname{Im}(\alpha_i)}$ and $\beta = \sqrt{(\delta^2 + 1)}$

    $Z_i = \sqrt{2\gamma}\begin{bmatrix} \operatorname{Re}(V_i) + \delta\operatorname{Im}(V_i) & \beta\operatorname{Im}(V_i) \end{bmatrix}$

    $Z_{i+1} = \begin{bmatrix} Z_{i-1} & Z_i \end{bmatrix}$,

    Compute $V_{i+1} = \operatorname{Re}(V_i) + \delta\operatorname{Im}(V_i)$

    Update $\tilde{W}_{i+1}$ as in (12.62).

    $i = i + 1$

  $i = i + 1$

Therefore, (12.60) becomes

$$\tilde{W}_i = \tilde{W}_{i-1} - 2\operatorname{Re}(\alpha_i)E_1 V_i, \tag{12.61}$$

If the two consecutive shift parameters are complex conjugates of each other, i.e., $\{\alpha_i, \alpha_{i+1} := \overline{\alpha_i}\}$, then recalling Step 5 in Algorithm 15,

$$\tilde{W}_{i+1} = \tilde{W}_{i-1} - 4\operatorname{Re}(\alpha_i)\tilde{E}\left(\operatorname{Re}(V_i) + \delta\operatorname{Im}(V_i)\right),$$

where $\delta = \frac{\operatorname{Re}(\alpha_i)}{\operatorname{Im}(\alpha_i)}$. Some algebraic manipulations will enable this to be written as

$$\tilde{W}_{i+1} = \tilde{W}_{i-1} - 4\operatorname{Re}(\alpha_i)E_1\left(\operatorname{Re}(V_i) + \delta\operatorname{Im}(V_i)\right). \tag{12.62}$$

Based on previous discussions, Algorithm 47 is presented to compute the low-rank controllability Gramian factor by solving the controllability Lyapunov equation (12.28a). The same procedure can be applied to solve the observability Lyapunov equation (12.28b).

Algorithm 47 relies on certain shift parameters that are crucial for fast convergence of the method. We investigate two types of ADI shift parameters. Penzl's heuristic approach that was introduced in [133] is one of the most commonly used approaches to compute the ADI shift parameters for a large-scale system. Recently, another approach has been introduced in [26, 41] to compute the ADI shifts adaptively. Both approaches were discussed in Chapter 3. This section focuses on some technical problems that arise in both approaches for the system considered in this chapter.

Heuristic shifts. In this approach, we often require a set of approximate finite eigenvalues which consist of some large magnitude and small magnitude ritzvalues of the matrix pencil corresponding to the underlying system (see, e.g., Penzl's heuristic in [133]). For the second-order index 3 descriptor system (12.2) the corresponding matrix pencil is

$$\lambda \underbrace{\begin{bmatrix} I_{n_\xi} & 0 & 0 \\ 0 & M_1 & 0 \\ 0 & 0 & 0 \end{bmatrix}}_{\breve{E}} - \underbrace{\begin{bmatrix} 0 & I_{n_\xi} & 0 \\ -K_1 & -D_1 & -G_1^T \\ -G_1 & 0 & 0 \end{bmatrix}}_{\breve{A}_3}. \tag{12.63}$$

Due to the singularity of $\breve{E}$, the matrix pencil features some infinite eigenvalues that prevent the direct usage of Arnoldi method for the approximation of large magnitude eigenvalues. To overcome this problem, we can employ the strategy introduced in [52], looking at the modified eigenvalue problem of the first-order structured index 2 DAE system such as (9.1). Following the strategy, we modify the matrix pencil (12.63) as

$$\lambda \underbrace{\begin{bmatrix} I_{n_\xi} & 0 & 0 \\ 0 & M_1 & 0 \\ 0 & 0 & 0 \end{bmatrix}}_{\breve{E}} - \underbrace{\begin{bmatrix} 0 & I_{n_\xi} & 0 \\ -K_1 & -D_1 & -G_1^T \\ 0 & -G_1 & 0 \end{bmatrix}}_{\breve{A}_2}. \tag{12.64}$$

The matrix pencil (12.64) has the same structure as the matrix pencil corresponding to system (9.1). Moreover, according to [64, Theorem 2.7.3], the matrix pencils in (12.64) and (12.63) share the same non-zero finite spectrum. Now the modified matrix pencil

$$\lambda \underbrace{\begin{bmatrix} I_{n_\xi} & 0 & 0 \\ 0 & M_1 & -\alpha G_1^T \\ 0 & -\alpha G_1 & 0 \end{bmatrix}}_{\breve{E}_\alpha} - \underbrace{\begin{bmatrix} 0 & I_{n_\xi} & 0 \\ -K_1 & -D_1 & -G_1^T \\ 0 & -G_1 & 0 \end{bmatrix}}_{\breve{A}_2} \tag{12.65}$$

moves all infinite eigenvalues to $\frac{1}{\alpha}$ ($\alpha \in \mathbb{R}$) (see, e.g., Chapter 9) without altering the finite eigenvalues. The parameter α can be chosen such that $\frac{1}{\alpha}$ is close to the smallest magnitude eigenvalues after those have been determined with respect to the original matrices. Note that the matrix $\bar{A}_2$ in (12.65) will always be singular such that the small eigenvalue approximations cannot be computed since the inversion of $\bar{A}_2$ would be required. Therefore, small magnitude ritzvalues should be always computed from the matrix pencil (12.63).

Adaptive shifts. A second shift computation strategy, which is rather simple and more efficient, is already discussed in Chapter 3. Here the computed shift parameters are associated with the projected system (12.22) in which the corresponding matrix pencil is

$$\lambda\tilde{\Pi}E_1\tilde{\Pi}^T - \tilde{\Pi}A_1\tilde{\Pi}^T. \tag{12.66}$$

From the deliberation of Section 12.3, we already know the matrix pencil (12.66) incorporates all of the finite eigenvalues of the index 3 system (12.2). For the initialization of the shifts, we can proceed with the same procedure given in [26] as long as the system has sufficiently many inputs and outputs. If the input or output matrix consists of only a few columns, particularly for SISO systems, sometimes we may not achieve any stable eigenvalue from the projected pencil of (12.66). To overcome this problem, we propose a different initialization technique. Instead of using $\tilde{W}_0$ to project the pencil (12.66), we want to use a random thin rectangular matrix $\breve{B} \in \mathbb{R}^{2n_\xi \times k}$ where $k \ll n_\xi$. For the updated shifts, we follow the same procedure as discussed in Chapter 3.

12.7 Numerical results

To assess the accuracy and efficiency of the proposed model reduction methods, we present numerical results for two model examples. The first example is a holonomically constrained damped mass-spring (CDMS) system [122] and details of this model are given in Appendix A.5.2. The second example is a constrained triple chain oscillator (CTCO) model which is introduced in Appendix A.5.3. For the numerical test, the constraint matrix $G_1 \in \mathbb{R}^{n_\varphi \times n_\xi}$ is chosen as a random sparse matrix. In this particular test example, there are 2 000 masses and 5 000 constraints. Therefore, G_1 becomes a $5\,000 \times 6\,001$ matrix. Here we consider the 2 000-th off-diagonal and 4 000-th diagonal elements of G_1 all to be 1 and -1, respectively. The dimension of the second-order index 3 system is 11 001. The input and output matrices $H_1 \in \mathbb{R}^{n_\xi \times 1}$ and $L_1 \in \mathbb{R}^{1 \times n_\xi}$ are chosen randomly.

All the results were obtained using MATLAB 7.11.0 (R2012a) on a board with 2 Intel® Xeon® X5650 CPUs with a 2.67 GHz clock speed, 6 Cores each and 48 GB of total RAM.

12.7.1 LR-ADI iteration

In order to perform the BT- and PDEG-based techniques, we must compute the low-rank controllability and observability Gramian factors $\tilde{R}$ and $\tilde{L}$. These Gramian factors are computed by applying Algorithm 47. The algorithm is experimented by using both heuristic and adaptive ADI shift parameters. The comparison of the performance of heuristic and adaptive shifts

models	sizes	tolerance	no. of iterations			
			heuristic shifts		adaptive shifts	
			$\tilde{R}$	$\tilde{L}$	$\tilde{R}$	$\tilde{L}$
CDMS	10001	10^{-8}	80	92	26	31
CTCO	11001	10^{-8}	270	282	153	240

Table 12.2: The performance of the heuristic and adaptive shifts in the LR-ADI in terms of iteration number.

model size	CPU time (sec)				
	heuristic shift			adaptive shift	
	$\tilde{R}$	$\tilde{L}$	μ	$\tilde{R}$	$\tilde{L}$
CDMS	2.27	3.84	119	1.16	1.56
CTCO	2.41	3.55	35	2.19	3.45

Table 12.3: The performances of the heuristic and adaptive shifts in the LR-ADI in terms of computational time.

for both model examples is shown in Tables 12.2 and 12.3. Table 12.2 shows the comparison in terms of number of iterations while Table 12.3 compares the performance of both types of shifts in terms of computational time. From both tables, we can conclude that in both cases (number of iterations taken to converge within the given tolerance and execution time), the adaptive shifts perform better than the heuristic shifts. Note that, for the CDMS model, we selected 15 optimal heuristic shifts out of 30 large and 25 small magnitude ritz values. On the other hand, for the CTCO model, from 50 large and 180 small magnitude ritzvalues, 100 heuristic shifts were selected. For the adaptive shifts, in each cycle, we were restricted to 10 proper shift parameters for the CDMS model. In case of the CTCO model, the number of adaptive shifts was 70.

12.7.2 Second-order-to-first-order reduction

First we apply Algorithm 43 to the CDMS model which generates an 11 dimensional standard state space model with a truncation tolerance of 10^{-4}. Figure 12.1(a) shows the sigma plot, i.e., the maximum singular values of the transfer function matrix of the full and reduced models on a wide frequency domain, more precisely, from 10^{-4}Hz to 10^{4}Hz. The corresponding absolute error between the full and reduced dimensional models is shown in Figure 12.1(b). We observe that the error is below the MOR tolerance for a very low dimensional model. When the same algorithm is applied to the system in the CTCO model, we obtain a 55 dimensional reduced system with a truncation tolerance of 10^{-3}. However, the dimension of the ROM can be

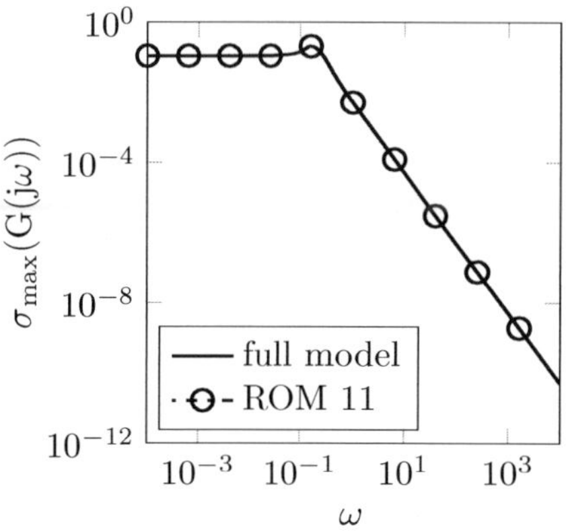

(a) Sigma plot.

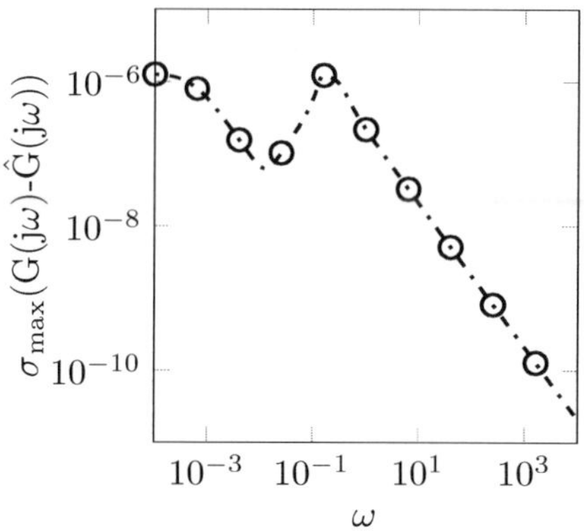

(b) Absolute error.

Figure 12.1: Comparison of full and first-order reduced model of the CDMS system.

dimension of ROM	H_∞ norm			
	absolute		relative	
	BT	IRKA	BT	IRKA
50	2.88×10^{-4}	8.00×10^{-2}	3.77×10^{-9}	1.25×10^{-6}
40	9.00×10^{-3}	1.33×10^{0}	1.70×10^{-7}	2.42×10^{-5}
30	6.19×10^{-1}	6.29×10^{2}	1.24×10^{-5}	1.00×10^{-2}
20	1.17×10^{1}	9.59×10^{2}	2.35×10^{-4}	1.50×10^{-2}

Table 12.4: Comparisons of balancing and IRKA-based methods for different dimensional ROMs with the CTCO model.

reduced further by using higher truncation tolerances. For instance, 10^{-4} and 10^{-3} truncation tolerances respectively generate 65 and 55 dimensional reduced systems. The comparison of the full and 55 dimensional reduced system is shown in Figure 12.2. This figure depicts the frequency response of the full and different dimensional reduced systems; it is observed that the frequency response of the original and the reduced systems have a good match and both errors indicate good accuracy. To compare the balancing-based method with IRKA, we compute 60, 50, 40, 30 and 20 dimensional reduced models using both Algorithms 43 and 44. The absolute and relative deviations between the full and 60 dimensional reduced models are shown in Figure 12.3. On the other hand, Table 12.4 lists the absolute and relative H_∞ norm of the error systems for the 50, 40, 30 and 20 dimensional ROMs. From the figures and tables, one can observe that the balancing-based methods generate more accurate ROMs.

12.7.3 Second-order-to-second-order reduction

We consider an 11001 dimensional second-order index 3 model for the CTCO example. To compute the Gramian factors, we follow the same strategy discussed previously. Applying Algorithm 45, we compute a 44 dimensional reduced order model by balancing the system on the position-position level. The same algorithm generates 41, 44 and 38 dimensional reduced systems via balancing the system onto the velocity-velocity, position-velocity and velocity-position levels. In all cases, the truncation tolerance is set to 10^{-3}. The frequency response of the full and the reduced models and their absolute and relative errors are shown in Figure 12.4. Although the accuracy is not satisfactory for the approximated models on the position-position and the velocity-position levels, for the other balancing levels the accuracy is good enough. We also apply the PDEG method to the CTCO model. In this case, we construct 45 dimensional reduced models by projection of the systems onto the dominant eigenspaces of the different Gramians. Figure 12.5 shows very good accuracy of all the ROMs computed by the PDEG method. Moreover, the ROMs constructed by this method preserve the stability of the original model which is reflected in Figure 12.6.

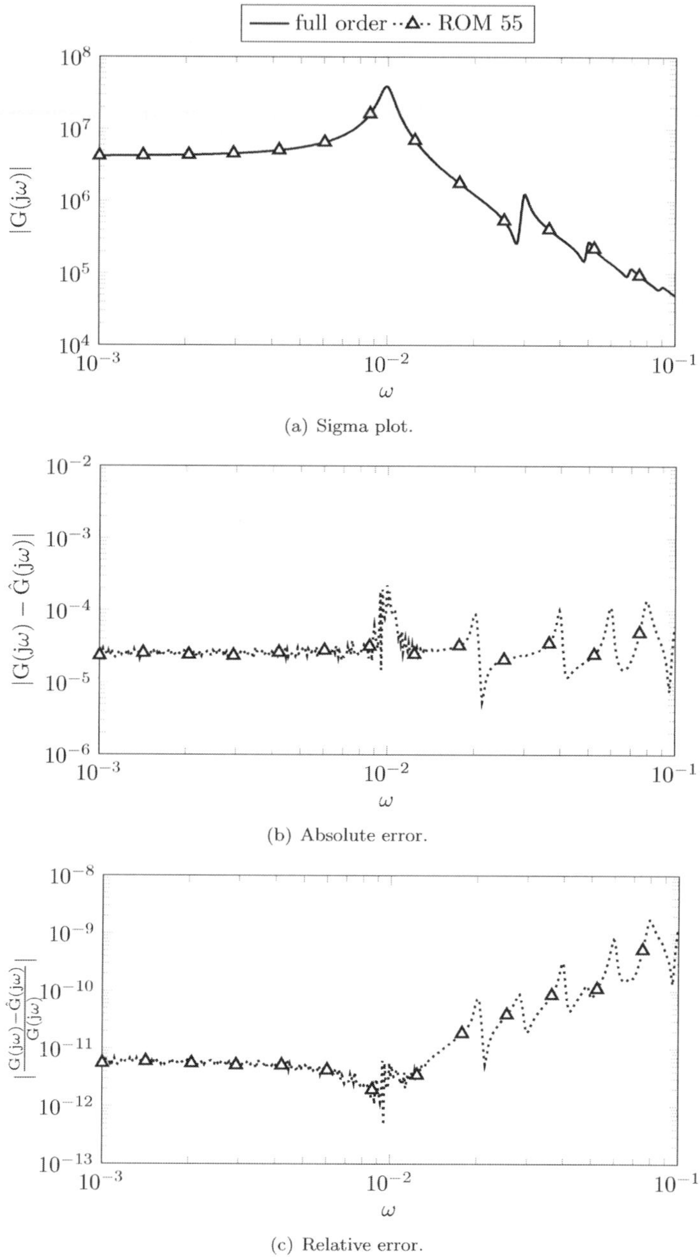

Figure 12.2: Comparison between the original and the different dimensional first-order reduced models in the frequency domain for the CTCO model.

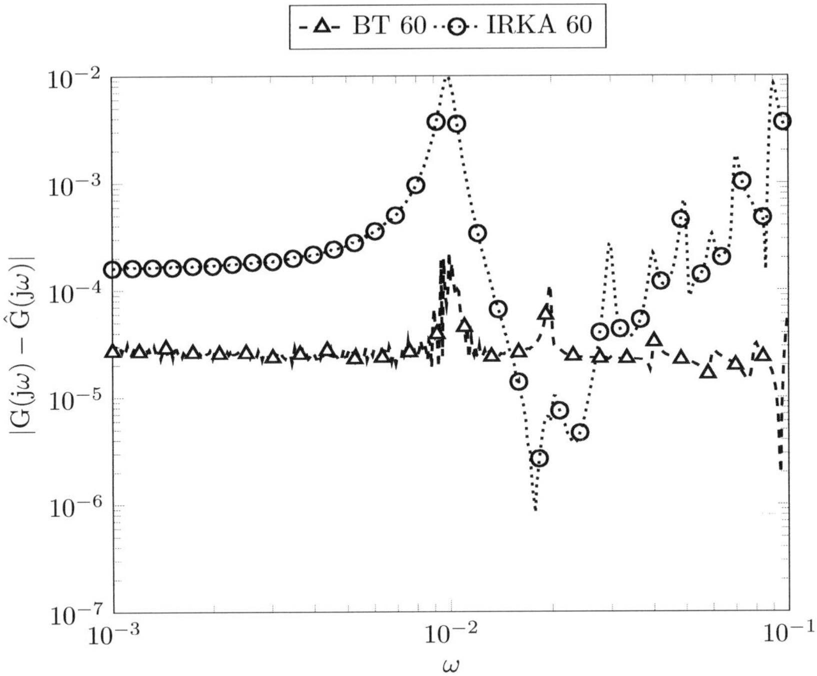

(a) Sigma plot

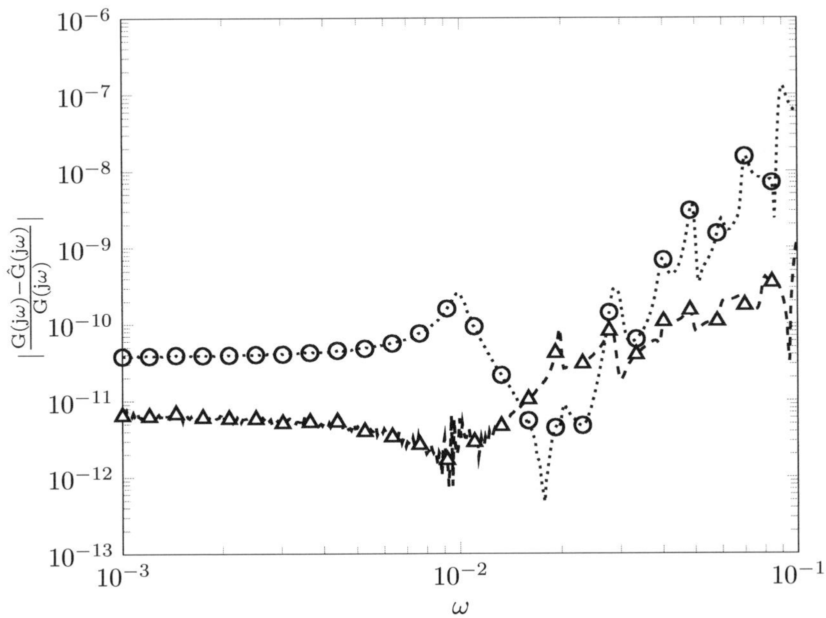

(b) Absolute error

Figure 12.3: Comparison of balanced truncation and IRKA with a 60 dimensional reduced model of the CTCO model.

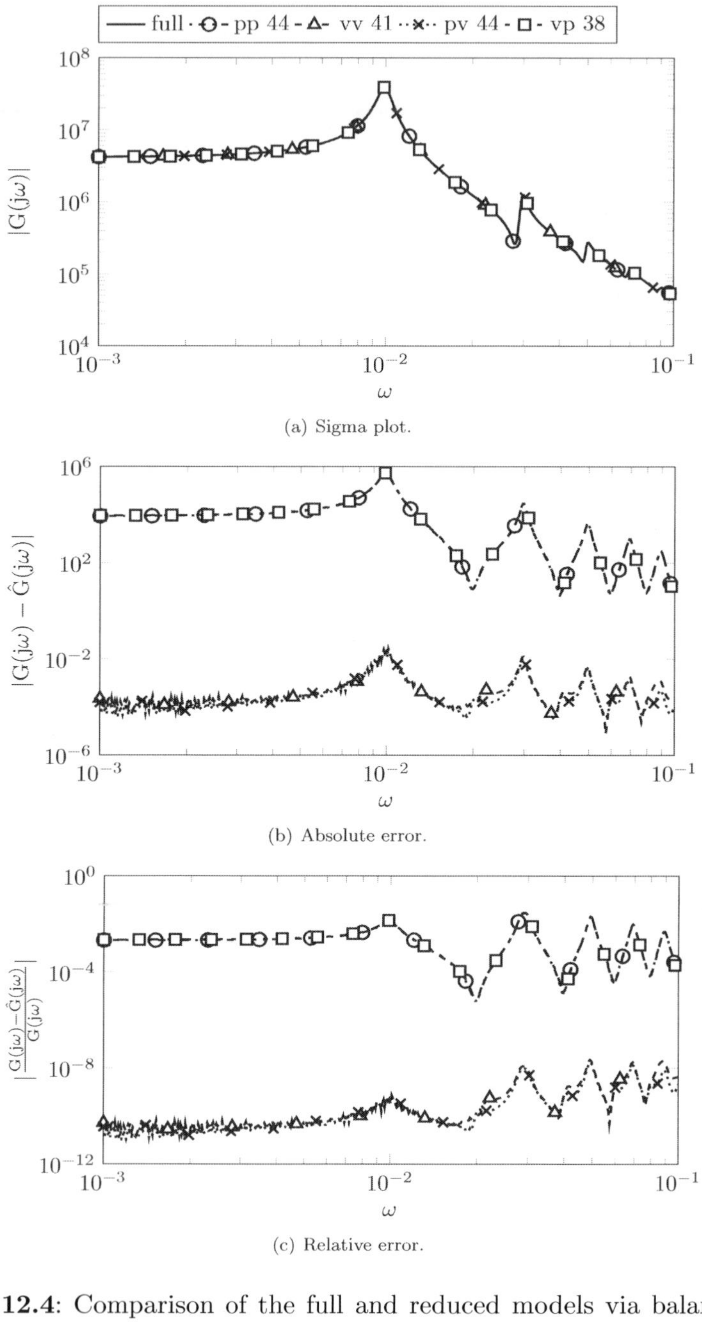

Figure 12.4: Comparison of the full and reduced models via balancing on different levels for the CTCO model.

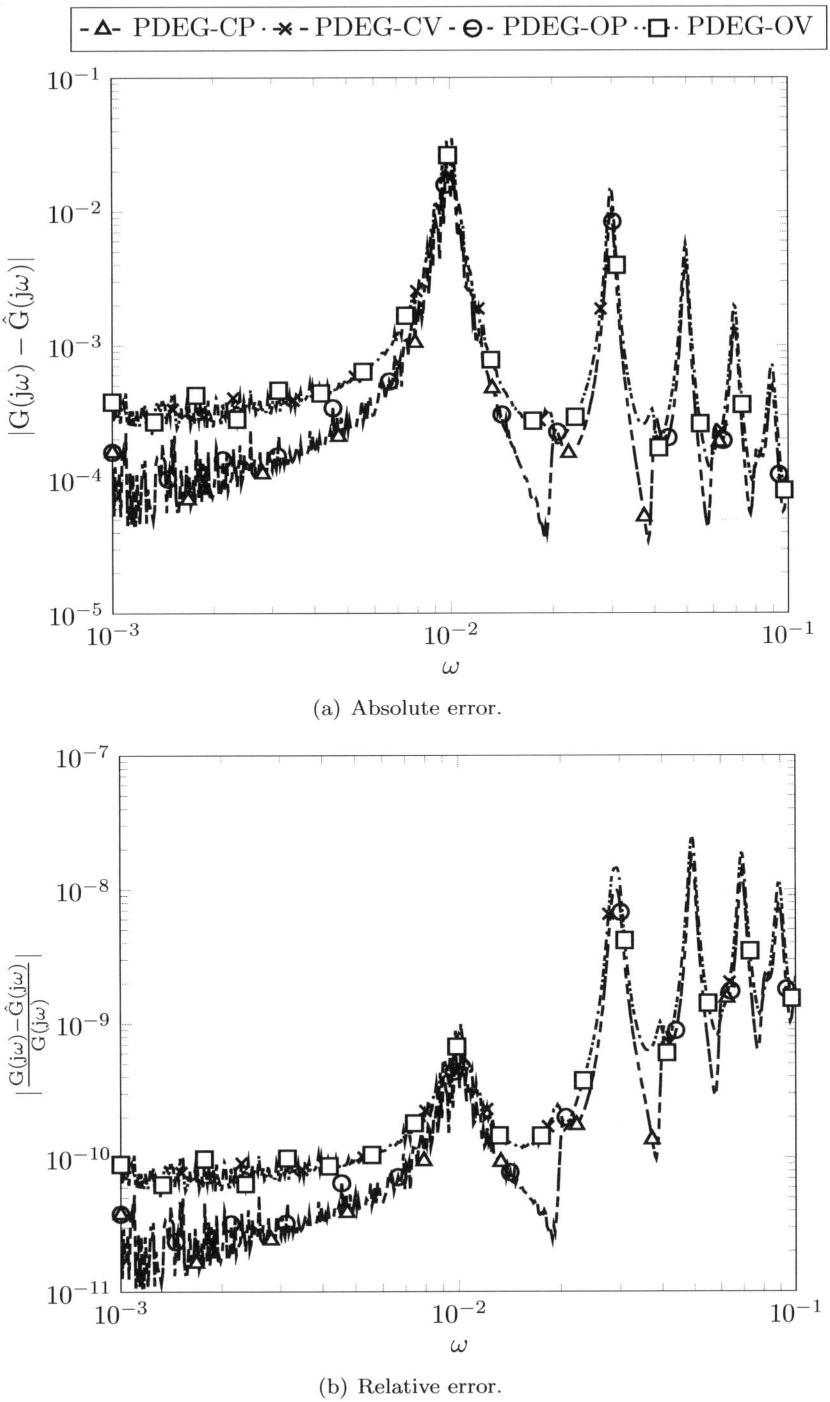

Figure 12.5: PDEG-based 45 dimensional ROMs for the CTCO model.

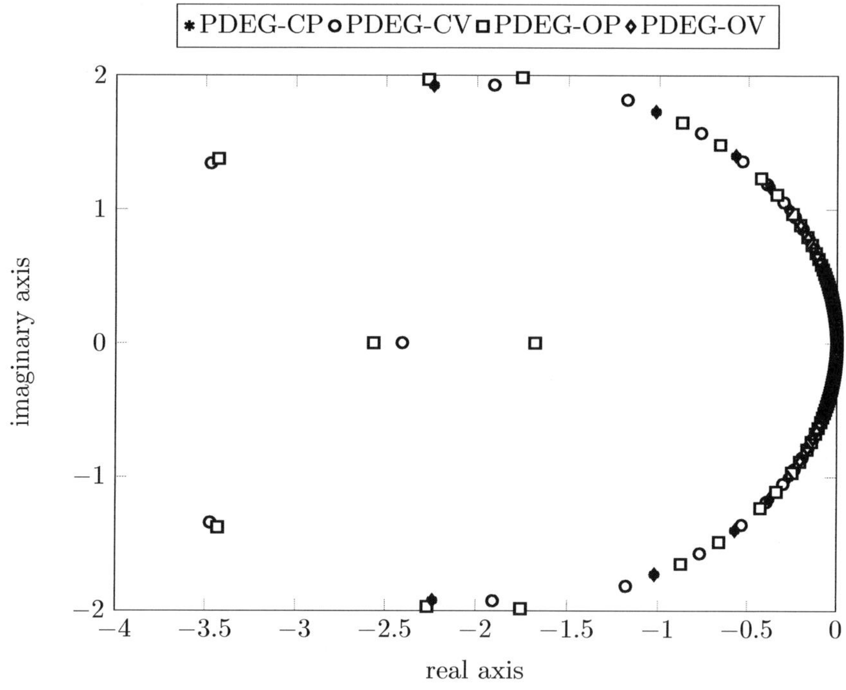

Figure 12.6: Eigenvalues of the ROMs obtained by the PDEG method.

■ Exercises:

12.1 Consider the following second-order index 3 descriptor system:

$$\begin{aligned} M\ddot{\xi}(t) + D\dot{\xi}(t) + K\xi(t) - G_1\lambda = H_1u(t), \\ G_1^T\xi(t) = 0 \\ y(t) = H_1^T\xi(t) \end{aligned}$$

Convert the system into its first-order form such that the system is symmetric and the input and output matrices are transpose of each other. Also show that the transfer functions of the second-order and the first-order systems are equal.

12.2 For the second-order index 3 descriptor system in Exercise 12.1, construct the coefficient matrices using the following MATLAB code snippet:

```
n=2;
I=speye(n);   k=100*(1:n)';
M =spdiags(k,0,n,n);
K =spdiags(-5*ones(n,1),0,n,n)+
     spdiags(2*ones(n,1),1,n,n)+
     spdiags(2*ones(n,1),-1,n,n);
mu=.001;nu=5;
D=mu*M+nu*K;
G1=sparse(1,n); G1(1,1)=1; G1(1,n)=-1;
G1(1,1)=1;G1(1,n-1)=-1;
H1=spdiags(ones(n,1),0,n,1);
```

Now use this data to construct the different first-order transformed systems. Show that the frequency response and step response of the original and the transformed systems are the same.

12.3 For the second-order index 3 system given in Exercise 12.1, the projector Π is defined by $\Pi = I - G_1^T(G_1M^{-1}G_1^T)^{-1}G_1M^{-1}$. Now, for the given data in Exercise 11.2, construct Π and show that it satisfies (i) $\Pi^2 = \Pi$ and (ii) $\Pi M = M\Pi^T$.

12.4 Verify Theorem 21 for the data given in Exercise 11.2.

12.5 If the second equation of the system in Exercise 12.1 is replaced by

$$G_1^T\dot{\xi}(t) = 0$$

then it becomes a second-order index 2 system. For such a system, develop algorithms for

(a) second-order-to-first-order reduction (BT and IRKA),

(b) second-order-to-second-order reduction (BT and PDEG).

You can assess the efficiency of your algorithms by using the data in Exercise 11.2 and considering a large n (e.g. $n = 10^5$).

12.6 Consider the following second-order index 3 descriptor system:

$$\begin{aligned} M\ddot{\xi}(t) + D\dot{\xi}(t) + K\xi(t) - G_1\lambda &= H_1 u(t), \\ G_1^T\xi(t) &= H_2, \\ y(t) &= H_1^T\xi(t). \end{aligned}$$

Based on the discussion in Section 12.2, can you convert the system into its equivalent standard second-order form? If possible, find the reduced first-order (using BT and IRKA) and second-order (using BT and PDEG) systems. Verify the methods by constructing an artificial data set as in Exercise 12.2.

Part III

APPENDICES

Appendix A

Data of Benchmark Model Examples

A.1 Introduction

This book has discussed numerical techniques for model reduction of large-scale dynamical systems. We have introduced different techniques for different types of models. To show the validity and capability of the methods, we have also discussed the numerical results that were obtained by applying the data of some Benchmark Model Examples. We have collected the experimental data from previous literature. In this appendix, we briefly discuss all the models including their sources, structures, applications and a few other properties. For the sake of convenience, we discuss the model examples by dividing them into four categories as,

(1) First-order LTI continuous-time (standard or generalized) systems,

(2) First-order LTI continuous-time descriptor systems,

(3) Second-order LTI continuous-time standard systems,

(4) Second-order LTI continuous-time descriptor systems.

A.2 First-order LTI continuous-time systems

This section considers the model examples of the form

$$\begin{aligned} E\dot{x}(t) &= Ax(t) + Bu(t); \quad x(t_0) = x_0,\, t \geq t_0, \\ y(t) &= Cx(t), \end{aligned} \tag{A.1}$$

where $E, A \in \mathbb{R}^{n \times n}$, $B \in \mathbb{R}^{n \times p}$ and $C \in \mathbb{R}^{m \times n}$. The matrix E is either invertible or an identity matrix. In the following subsections we consider examples of such a model.

A.2.1 CD player

This is the classical CD player (CDP) model which has been frequently used in various literature over the years to test the efficiency of the MOR methods of LTI dynamical systems. In this book we have also applied this model to test the efficiency of our proposed algorithms. Following the description in [103], the task is to construct a controller that ensures that the laser points to the track of pits on a rotating CD. The mechanism that is considered here consists of a swing arm on which a lens is mounted by means of two horizontal leaf springs. The rotation of the arm in the horizontal plane enables reading of the spiral-shaped disc tracks and the suspended lens is used to focus the laser on the disc. Due to the fact that the disc is not perfectly flat, and due to irregularities in the spiral of pits on the disc, the challenge is to find a low-cost controller that can make the servo-system faster and less sensitive to external shocks [61, 187].

dimension (n)	**input (m)**	**output (p)**
120	2	2

A.2.2 FOM

This model originated in [134]. The structure of the matrices is as follows.

$$A = \begin{bmatrix} A_1 & & & \\ & A_2 & & \\ & & A_3 & \\ & & & A_4 \end{bmatrix},$$

where

$$A_1 = \begin{bmatrix} -1 & 100 \\ -100 & -1 \end{bmatrix}, \quad A_2 = \begin{bmatrix} -1 & 200 \\ -200 & -1 \end{bmatrix}, \quad A_3 = \begin{bmatrix} -1 & 400 \\ -400 & -1 \end{bmatrix},$$

$$A_4 = \begin{bmatrix} -1 & & & \\ & -2 & & \\ & & \ddots & \\ & & & -1000 \end{bmatrix} \quad \text{and} \quad B^T = C = \begin{bmatrix} \underbrace{10 \cdots 10}_{6}, & \underbrace{1 \cdots 1}_{1000} \end{bmatrix}$$

dimension (n)	input (m)	output (p)
1006	1	1

A.3 Second-order LTI continuous-time systems

This section considers the model examples of the form

$$\begin{aligned} M\ddot{\xi}(t) + D\dot{\xi}(t) + K\xi(t) &= Hu(t), \\ y(t) &= L\xi(t), \end{aligned} \tag{A.2}$$

where $M, D, K \in \mathbb{R}^{n_\xi \times n_\xi}$ are invertible matrices; input matrix $H \in \mathbb{R}^{n_\xi \times p}$ and output matrix $L \in \mathbb{R}^{m \times n_\xi}$. Examples of such models are discussed in the subsections that follow.

A.3.1 International Space Station

The International Space Station (ISS) model is composed of a complex structure containing several parts. The structural part (part 1R of the Russian Service Module) of the international space station has been modeled in the second-order form of dimension 135 with 3 inputs and 3 outputs, see [86] for detailed description of the model.

dimension (n_ξ)	input (m)	output (p)
135	3	3

A.3.2 Clamped beam model

The clamped beam model (CBM) is a structural model with 348 states. The model is obtained by spatial discretization of an appropriate partial differential equation (see [6]). The input represents the force applied to the structure at the free end, and the output is the resulting displacement.

dimension (n_ξ)	input (m)	output (p)
348	1	1

A.3.3 Triple chain oscillator model

This example originates in [164] with the setup described in [145] which results in system (5.1). The triple chain oscillator (TCO) model contains three chains with each of them being coupled to a fixed mounting by an additional damper on one end and fixed rigidly to a large mass that couples all three of them. The large mass is bound to a fixed mount by a single spring element. Each of the chains consists of n_1 equal masses and spring elements of equal stiffness. Therefore, the model parameters are the masses m_1, m_2, m_3 and the corresponding stiffnesses k_1, k_2, k_3 for the three oscillator chains, the mass m_0 with its spring stiffness k_0 for the coupling mass, the viscosities ϑ of the additional wall-mount-dampers and the length n_1 of each of the oscillator chains. The resulting system is of order $n_\xi = 3n_1 + 1$. The mass matrix $M = \operatorname{diag}(m_1 I_{n_1}, m_2 I_{n_1}, m_3 I_{n_1}, m_0)$. The stiffness matrix K and damping matrix D consist of a leading block diagonal matrix (consisting of the three stiffness matrices for the three oscillator chains) and coupling terms in the last row and column at positions n_1, $2n_1$ and $3n_1$ in the diagonal elements. For the numerical experiment, we consider the values of the variables as follows:

$$\begin{aligned} m_1 = 1,\ m_2 = 2,\ m_3 = 3,\ m_0 &= 10, \\ k_1 = 10,\ k_2 = 20,\ k_3 = 1,\ k_0 &= 50, \\ \vartheta &= 5. \end{aligned}$$

Input matrix $H \in \mathbb{R}^{n_\xi \times 1}$ consists of all elements with one and the output matrix $L = H^T$. The default dimension of the model is $n_\xi = 6001$, i.e., $n_1 = 2000$.

dimension (n_ξ)	**input** (m)	**output** (p)
6001	1	1

A.3.4 Butterfly Gyro

The Butterfly Gyro [44, 100] is a vibrating micro-mechanical gyro that has sufficient theoretical performance characteristics to make it a promising candidate for use in inertial navigation applications. The gyro chip consists of a three-layer silicon wafer stack, in which the middle layer contains the sensor element. The sensor consists of two wing pairs that are connected to a common frame by a set of beam elements; this is the reason the gyro is called the Butterfly. The original model consists of 17 361 degrees of freedom which results in an order n_ξ= 17 361 second-order system. The system has a single input and 12 outputs. This model is available in oberwolfach Model Reduction Benchmark Collection[1].

dimension (n_ξ)	**input** (m)	**output** (p)
17 361	1	12

[1] http://www.imtek.de/simulation/benchmark/wb/35889/

data from	model name	n_1	n_2	m/p
BIPS98	BIPSM-606	606	1142	4/4
	BIPSM-1142	1142	8593	
	BIPSM-1450	1450	9855	
	BIPSM-1693	1693	11582	
BIPS07	BIPSM-1998	1998	12,520	
	BIPSM-2476	2476	13,837	
	BIPSM-3078	3078	18,050	

Table A.1: Brazilian interconnected power system models (BIPSM).

A.4 First-order LTI continuous-time descriptor systems

In this section we consider model examples of the form:

$$E\dot{x}(t) = Ax(t) + Bu(t); \quad x(t_0) = x_0,\, t \geq t_0, \tag{A.3a}$$
$$y(t) = Cx(t) + Du(t), \tag{A.3b}$$

where E is singular; $E, A \in \mathbb{R}^{n \times n}$, $B \in \mathbb{R}^{n \times p}$ and $C \in \mathbb{R}^{m \times n}$. This system consists of differential and algebraic equation parts. Therefore, the dimension of the system can be partitioned as $n = n_1 + n_2$, where n_1 is the number of differential variables and n_2 is the number of algebraic variables. In the following subsections we present some structured first-order descriptor systems.

A.4.1 Power system model

We consider two practical Brazilian Interconnected Power System (BIPS) models from [71]. The first one, BIPS98 has 2380 buses, 2536 nonlinear loads, 3450 ac branches, 8 HVDC converters, up to 124 synchronous machines and 6 FACTS devices. The second model, BIPS07 has 191 synchronous machines, 3647 buses, 5175 branches, 3639 nonlinear loads, 8 HVDC converters and 8 FACTS

$$E_1\dot{x}_1(t) = A_1x_1(t) + A_2x_2(t) + B_1u(t), \tag{A.4a}$$
$$0 = A_3x_1(t) + A_4x_2(t) + B_2u(t), \tag{A.4b}$$
$$y(t) = C_1x_1(t) + C_2x_2(t) + D_au(t), \tag{A.4c}$$

where $x_1(t) \in \mathbb{R}^{n_1}$ and $x_2(t) \in \mathbb{R}^{n_2}$ are the states, $u(t) \in \mathbb{R}^p$ is the control input to the system and the measurement output is $y(t) \in \mathbb{R}^m$. The data is available *here*[2]. Table A.1 summarizes the data of BIPS models.

[2] `https://sites.google.com/site/rommes/software#owners.`

A.4.2 Supersonic engine inlet

Supersonic engine inlet (SEI) considers the unsteady flow through a supersonic diffuser at a nominal Mach number of 2.2 as described in [181]. Linearization of the two-dimensional integral Euler equations about the steady-state solution and spatial discretization using a finite volume method leads to a semi-explicit descriptor system:

$$E_1\dot{x}_1(t) + E_2\dot{x}_2(t) = A_1x_1(t) + A_2x_2(t) + B_1u(t), \tag{A.5a}$$

$$0 = A_3x_1(t) + A_4x_2(t) + B_2u(t), \tag{A.5b}$$

$$y(t) = C_1x_1(t) + C_2x_2(t), \tag{A.5c}$$

where $x_1(t) \in \mathbb{R}^{n_1}$ and $x_2(t) \in \mathbb{R}^{n_2}$ are the states, the input is $u(t) \in \mathbb{R}^p$, the output is $y(t) \in \mathbb{R}^m$ and $E_1, A_1 \in \mathbb{R}^{n_1 \times n_1}$, $E_2, A_2 \in \mathbb{R}^{n_1 \times n_2}$, $A_3 \in \mathbb{R}^{n_2 \times n_1}$, $A_4 \in \mathbb{R}^{n_2 \times n_2}$, $B_1 \in \mathbb{R}^{n_1 \times p}$, $B_2 \in \mathbb{R}^{n_2 \times p}$, $C_1 \in \mathbb{R}^{m \times n_1}$, $C_2 \in \mathbb{R}^{m \times n_2}$. In the literature, such a system is called index 1 since the matrices A_4 and $E_1 - E_2A_4^{-1}A_3$ are assumed to be invertible. This model has 11,730 unknowns in which the number of differential and algebraic variables are 3280 and 8450, respectively. There are two inputs: the bleed actuation mass-flow input and the incoming density disturbance input. The system has a single output: the average Mach number at the inlet throat.

differential var. (n_1)	**algebraic var.** (n_2)	**input** (m)	**output** (p)
3280	8450	2	1

A.4.3 Semi-discretized linearized Navier-Stokes model

Let us consider the linearized Navier-Stokes equations:

$$\begin{aligned} \frac{\partial}{\partial t}\vec{v} - \frac{1}{Re}\Delta\vec{v} + (\vec{w}.\nabla)\vec{v} + (\vec{v}.\nabla)\vec{w} + \nabla p = 0, \\ \nabla.\vec{v} = 0, \end{aligned} \tag{A.6}$$

where $\vec{v}$ and $\vec{w}$ denote velocity vectors, p is the pressure and Re is the Reynolds number. The vector $\vec{w}$ represents the stationary solution of the incompressible nonlinear Navier-Stokes equations and $\vec{v}$ is the deviation of the original state from the stationary solution. The boundary and initial conditions as well as the derivation of this model are given in [15]. The authors apply a mixed finite element method based on the well-known Taylor-Hood finite elements [95] to discretize equation (A.6). This yields the differential-algebraic equations

$$E_1\dot{v}(t) = A_1v(t) + A_2p(t) + B_1u(t), \tag{A.7a}$$

$$A_2^Tv(t) = 0, \tag{A.7b}$$

where $v(t) \in \mathbb{R}^{n_1}$ denotes the nodal vector of discretized velocity deviations, $p(t) \in \mathbb{R}^{n_2}$ is the discretized pressure, $u(t) \in \mathbb{R}^m$ are the inputs, and $E_1, A_1 \in \mathbb{R}^{n_1 \times n_1}$, $A_2 \in \mathbb{R}^{n_1 \times n_2}$, $B_1 \in \mathbb{R}^{n_1 \times m}$ are all sparse matrices.

model name	n_1	n_2	m/p
SLNSM-1	3 142	453	
SLNSM-2	8 268	123	
SLNSM-3	19 770	615	2/7
SLNSM-5	44 744	783	
SLNSM-5	98 054	566	

Table A.2: Number of differential variables (n_1), algebraic variables (n_2), inputs (m) and outputs (p) of different discretization levels of semi-discretized linearized Navier stokes model.

Additionally, the vertical velocities in the observation nodes in the domain are modeled by the output equation

$$y(t) = C_1 v(t), \tag{A.8}$$

with the output $y(t) \in \mathbb{R}^p$ and the output matrix $C_1 \in \mathbb{R}^{p \times n_1}$. Such a structured dynamical system is known as index 2 system (see, e.g., [9, 106]). The semi-discretized linearized Navier-Stokes model (SLNSM) remains stable, i.e., the finite spectrum of the *matrix pencil*

$$\mathcal{P}(\lambda) = \lambda \begin{bmatrix} E_1 & 0 \\ 0 & 0 \end{bmatrix} - \begin{bmatrix} A_1 & A_2 \\ A_2^T & 0 \end{bmatrix}, \tag{A.9}$$

is located in the negative half plane $\mathbb{C}^-$ as long as the Reynolds number Re is small. However, for moderate Reynolds numbers (e.g., $Re \geq 400$) a few finite eigenvalues move to the positive half plane, $\mathbb{C}^+$ [4]. Bansch et al. [15] generate different-sized models using the Reynolds number $R_e = 500$. Table A.2 shows their dimension (n), number of algebraic/differential variables (n_1/n_2) and the number of inputs/outputs (m/p).

A.4.4 Semi-discretized linearized Stokes model

Following [160], the flow of an incompressible fluid can be described by the instationary Stokes equations:

$$\begin{aligned} \frac{\partial}{\partial t}\vec{v} - \Delta \vec{v} + \nabla p = \vec{f}, \qquad & (\xi, t) \in \vec{\Omega} \times (0, t_f), \\ \nabla . \vec{v} = 0, \qquad & (\xi, t) \in \vec{\Omega} \times (0, t_f), \end{aligned} \tag{A.10}$$

where the initial and boundary conditions are

$$\begin{aligned} \vec{v}(\xi, t) &= g(\xi, t), \quad (\xi, t) \in \partial\Omega \times (0, t_f), \\ \vec{v}(\xi, 0) &= \vec{v}_0(\xi), \quad \xi \in \Omega. \end{aligned}$$

Here $\vec{v}(\xi, t) \in \mathbb{R}^d$ is the velocity vector ($d = 2$ or 3 is the dimension of the spatial domain), $p(\xi, t) \in \mathbb{R}$ is the pressure, $\vec{f}(\xi, t) \in \mathbb{R}^d$ is the vector of external forces, $\vec{\Omega} \in \mathbb{R}^d$ is a bounded open domain with boundary $\partial\vec{\Omega}$ and $t_f > 0$ is the endpoint of the time interval. The spatial discretization of the Stokes equation by the finite difference or the finite element method produces a system in the generalized state space (or descriptor) form

$$E_1\dot{v}(t) = A_1v(t) - A_2p(t) + B_1u(t), \tag{A.11}$$

$$0 = -A_2^Tv(t) + B_2u(t), \tag{A.12}$$

where $v \in \mathbb{R}_1^n$ and $p \in \mathbb{R}_2^n$ are the semi-discretized vectors of velocity and pressure, respectively; $E_1 \in \mathbb{R}^{n_1\times n_1}$ is the symmetric positive definite mass matrix and $A_1 \in \mathbb{R}^{n_1\times n_1}$ is the discrete Laplace operator. The matrices $A_2 \in \mathbb{R}^{n_1\times n_2}$ and $A_2^T \in \mathbb{R}^{n_2\times n_1}$ are the discrete gradient and divergence operators. Due to the non-uniqueness of the pressure, the matrix A_2 has a rank defect which, in most spatial discretization methods, is equal to one. In this case, instead of A_2^T we can take a full row rank matrix obtained from A_2^T by discarding the last row. Therefore, in the following subsection we will assume, without loss of generality, that A_2^T has full row rank. The matrices $B_1 \in \mathbb{R}^{n_1\times p}$, $B_2 \in \mathbb{R}^{n_2\times p}$ and the control input $u(t) \in \mathbb{R}^m$ are resulting from the boundary conditions and external forces. Moreover, the output $y(t) \in \mathbb{R}^p$ can be measured by

$$y(t) = C_1v + C_2p,$$

where $C_1 \in \mathbb{R}^{p\times n_1}$ and $C_2 \in \mathbb{R}^{p\times n_2}$. The dimension of the model (in the provided MATLAB data) can be varied by choosing the number of discretization points. We can change the number of inputs and outputs as well. If we consider 50 discretization points along both x and y directions, then we obtain $n_1 = 4900$, $n_2 = 2500$. We also choose $m = 4$ and $p = 4$.

differential var. (n_1)	**algebraic var.** (n_2)	**input** (m)	**output** (p)
4 900	2 500	4	4

A.4.5 Constrained damped mass-spring system

We consider a holonomically constrained damped mass-spring (CDMS) system from [122]. The i-th mass of weight m_i is connected to the $(i-1)$-st mass by a spring and a damper with constants k_i and d_i, respectively, and also to ground by a spring and damper with constants κ_i and δ_i, respectively. Moreover, the first mass is connected to the last one by a rigid bar and influenced by the control $u(t)$. The dynamics of this system can be described by first-order index 3 descriptor system as follows,

$$\begin{aligned}\dot{p}(t) &= v(t)\\ M\dot{v}(t) &= Kp(t) + Dv(t) - G^T\lambda(t) + B_2u(t),\\ 0 &= Gp(t),\\ y(t) &= C_1p(t),\end{aligned} \tag{A.13}$$

where $p(t) \in \mathbb{R}^{n_g}$ is the position vector, $v(t) \in \mathbb{R}^{n_g}$ is the velocity vector, and $\lambda(t) \in \mathbb{R}$ is the Lagrange multiplier. The matrix $M \in \mathbb{R}^{n_g \times n_g}$ is a diagonal mass matrix, $K \in \mathbb{R}^{n_g \times n_g}$ and $D \in \mathbb{R}^{n_g \times n_g}$ are tridiagonal stiffness and damping matrices, respectively. $G = [1, 0, \cdots, 0, -1] \in \mathbb{R}^{1 \times n_g}$ is the constraint matrix. The input matrix $B_2 = e_1$ and the output matrix $C_1 = [e_1, e_2, e_{n_g-1}]^T$, where e_i denotes the i-th column of the identity matrix I_{n_ξ}, lead the system to one input and three outputs. In the experiments, matrices M, K and D are considered as $M = 100\, I_{n_g}$,

$$K = \begin{bmatrix} -6 & 2 & & & \\ 2 & -6 & 2 & & \\ & 2 & -6 & \ddots & \\ & & \ddots & \ddots & 2 \\ & & & 2 & -6 \end{bmatrix} \text{ and } D = \begin{bmatrix} -15 & 5 & & & \\ 5 & -15 & 5 & & \\ & 5 & -15 & \ddots & \\ & & \ddots & \ddots & 5 \\ & & & 5 & -15 \end{bmatrix}.$$

Now considering $n_g = 10\,000$ masses, we obtain $20\,000$ differential equations and 1 algebraic equation.

differential var. (n_1)	**algebraic var.** (n_2)	**input** (m)	**output** (p)
20 000	1	1	3

A.5 Second-order LTI continuous-time descriptor systems

This section considers the model examples of the form (A.14), that is to say

$$M\ddot{\xi}(t) + D\dot{\xi}(t) + K\xi(t) = Hu(t), \tag{A.14a}$$

$$y(t) = L_1\xi(t) + L_2\dot{\xi}(t), \tag{A.14b}$$

where the matrices M and D are singular. In the following subsections, we present some model examples of such types.

A.5.1 Piezo-actuator based adaptive spindle support

Piezo-actuator based adaptive spindle support (ASS) [127, 129, 130] is an important component of a mechatronic (electro-mechanical) [126, 128] system. Applying the finite element method (FEM) to the ASS, the following mathematical model is formed.

$$\begin{aligned} M_{11}\ddot{\xi}(t) + D_{11}\dot{\xi}(t) + K_{11}\xi(t) + K_{12}\varphi(t) &= H_1u(t), \\ K_{12}^T\xi(t) + K_{22}\varphi(t) &= H_2u(t), \\ L_1\xi(t) + L_2\varphi(t) &= y(t), \end{aligned} \tag{A.15}$$

where $\xi(t) \in \mathbb{R}^{n_\xi}$ is a vector of the mechanical displacements, $\varphi(t) \in \mathbb{R}^{n_\varphi}$ is a vector of electric potentials. The block matrices M_{11}, D_{11} and K_{11} are the mechanical mass, damping and stiffness matrices. The matrix K is composed of the mechanical (K_{11}), electrical (K_{22}) and coupling (K_{12}) terms. Selected general force quantities (mechanical forces and electrical charges) are chosen as the input quantities u, and the corresponding general displacements (mechanical displacements and electrical potential) are the output quantities y. The total mass matrix contains zeros at the locations of the electrical potential. More precisely, the electrical potential of the system (degrees of freedom (DoF) for the electrical part) is not associated with an inertia. The equation of motion of the system in (A.15) can be found in [101]. This equation results from a finite element discretization of the balanced equations. For piezo-mechanical systems, these are the mechanical balance of momentum (with the inertia term) and the electro-static balance. From this, the electrical potential without the inertia term is obtained. Thus, for the whole system (mechanical and electrical DoFs), the mass matrix has rank deficiency. In the literature, see, e.g., [165, 168, 169], this system is known as second-order index 1 system.

differential var. (n_ξ)	**algebraic var.** (n_φ)	**input** (m)	**output** (p)
282 699	7 438	9	9

A.5.2 Constrained damped mass-spring (second-order) system

This example has already been discussed in Subsection A.4.5. The first-order system A.13 can be written in second-order form

$$\begin{aligned} M\dot{v}(t) &= Kp(t) + Dv(t) - G^T\lambda(t) + B_2u(t), \\ 0 &= Gp(t), \\ y(t) &= C_1p(t). \end{aligned}$$

This is a second-order index 3 system.

differential var. (n_ξ)	algebraic var. (n_φ)	input (m)	output (p)
10 000	1	1	1

A.5.3 Constrained triple chain oscillator model

The triple chain oscillator (TCO) model has been discussed in Subsection A.3.3 which results in ODEs. To make it into a descriptor system, we add algebraic constrains. Then the constrained damped mass-spring system [167] is

$$\begin{aligned} M\ddot{\xi}(t) + D\dot{\xi}(t) + K\xi(t) - G^T\lambda(t) &= Hu(t), \\ 0 &= G\xi(t), \\ y(t) &= L\xi(t), \end{aligned} \tag{A.16}$$

where the matrices M, D, K, H and L are defined in Subsection A.3.3. The constraint matrix $G \in \mathbb{R}^{n_\xi \times n_\lambda}$ is chosen as a random sparse matrix. In this particular test example, there are 2 000 masses and 5 000 constraints. Therefore, G becomes a $5\,000 \times 6\,001$ matrix.

differential var. (n_ξ)	algebraic var. (n_φ)	input (m)	output (p)
6 001	5 000	1	1

Appendix B

MATLAB Codes

B.1 Algorithm 1

```
function [Q,R]=mgs(A)
[m,n]=size(A);
V=A;
Q=zeros(m,n);
R=zeros(n,n);
for i=1:n
  R(i,i)=norm(V(:,i));
  Q(:,i)=V(:,i)/R(i,i);
   for j=i+1:n
     R(i,j)=Q(:,i)'*V(:,j);
     V(:,j)=V(:,j)-R(i,j)*Q(:,i);
  end
end
```

B.2 Algorithm 2

```
function [Q, R] = update_mgs(A)
```

```
[m, n] = size(A);
Asave = A;
for j = 1:n
  for k = 1:j-1
    mult = (A(:, j)'*A(:, k)) / (A(:, k)'*A(:, k));
    A(:, j) = A(:, j) - mult*A(:, k);
  end
end
for j = 1:n
  if norm(A(:, j)) < sqrt(eps)
    error('Columns of A are linearly dependent.')
  end
  Q(:, j) = A(:, j) / norm(A(:, j));
end
R = Q'*Asave;
```

B.3 Algorithm 3

```
function [H,V] = arn_p(A, k,rv)
na = nargin;
n = size(A,1);
if k >= n-1, error('k must be smaller than the order
of A!');
end
if na<3, rv = randn(n,1);
end

V = zeros(n,k+1);
H = zeros(k+1,k);

V(:,1) = (1.0/norm(rv))*rv;

beta = 0;

for j = 1:k


  if j > 1
    H(j,j-1) = beta;
    V(:,j) = (1.0/beta)*rv;
  end

  if (mod(j,5)==0)
    V(:,1:j) = mgs(V(:,1:j));
  end
```

```
  w = A*V(:,j);

  rv = w;

  for i = 1:j
    H(i,j) = V(:,i)'*w;
    rv = rv-H(i,j)*V(:,i);
  end

  beta = norm(rv);
  H(j+1,j) = beta;

end

V(:,k+1) = (1.0/beta)*rv;
```

B.4 Algorithm 6

```
y = my_step(E,A,B,C,Tmax,N);
h=(Tmax-0)/N;
 t = linspace(0,Tmax,N);
 yout = zeros(size(C,1),n);
 [L,U,P,Q] = lu(E-h*A);
  for i=1:n
    x=Q*(U\(L\(P*(E*x+h*B))));
    y(:,i)=C*x;
  end
   plot(t,y)
```

B.5 Algorithm 7

```
function [s,SV] =singularvalues(E,A,B,C,D,Wmin,Wup,N)
   s=logspace(Wmin,Wup, N);
     for k=1:N
      G=C*((1j*s(k)*E-A)\B)+D;
      SV(k)=max(svds(G));
     end
loglog(s,SV)
end
```

B.6 Algorithm 8

```
function sigma =hankelsingularvalues(E,A,B,C)
    R=lyapchol(A,B,E);
       L=lyap(A',C',E');
       sigma=svd(R'*L);
      semilogy(sigma)
end
```

B.7 Algorithm 9

```
function h2 =h2norm(E,A,B,C)
    P=lyap(A,B*B',[],E);
    h2=sqrt(trace(C*P*C'));
end
```

B.8 Algorithm 10

```
function hinfinity =hinfinitynorm(E,A,B,C,D,Wmin,Wup,N)
    s=logspace(Wmin,Wup, N);
      for k=1:N
       G=C*((1j*s(k)*E-A)\B)+D;
       SV(k)=max(svds(G));
      end
   hinfinity=max(SV);
end
```

B.9 Algorithm 15

```
function [Zc,res1]=lr_adi(E,A,B,l,maxiter1,restol)
i=1;
Zc=[];ip=0;
res1=zeros(1,maxiter1+2);
m3=size(B,2);
W=B;
bnorm=norm(W'*W,'fro');
%% Initial shift computation
  Bn=sprand(size(A,1),100,.8);
```

```
   [Q,~]=qr(full(Bn),0);
   p = adaptive_shift(E,A,Q,l);
   l=length(p);
while i<=maxiter1
  if ip<l
     ip=ip+1;
  else
     m2=size(Zc,2);
     Zcn=Zc(:,m2+1-l*m3:end);
    [Q,~]=qr(full(Zcn),0);
     p = adaptive_shift(E,A,Q,l);
     l=length(p);
     ip =1;
  end %
     pc=p(ip);
     V=(A+pc*E)\W;
   if isreal(pc)
     Zc=[Zc sqrt(-2*real(pc))*real(V)];
    W = W-2*pc*E*V;
   else
     beta=real(pc)/imag(pc);
     gam=2*sqrt(-real(pc));
     Zc=[Zc,...
         gam*(real(V)+beta*imag(V)),...
         gam*sqrt(beta^2+1)*imag(V)];
         V=real(V)+beta*imag(V);
         pc=conj(pc);
        W = W-4*real(pc)*E*V;
         i=i+1;
         res1(i)=norm(W'*W,'fro')/bnorm;
    fprintf(1,'step: %4d  normalized residual: %d\n',i,
    res1(i))
    if res1(i)<restol
    break;
    end
    end
    i=i+1;
end
return
%% adaptive shifts computation
function p = adaptive_shift(E,A,Q,l)
    An=Q'*A*Q;
    En=Q'*E*Q;
    rw=eig(full(An),full(En));
    rw0 = rw;
```

```
    rw = [];
   for j = 1:length(rw0)
     if real(rw0(j))<0
       rw = [rw;rw0(j)];
     end
   end
     p=lp_mnmx(rw,l);
%% solve the minimax problem
function p = lp_mnmx(rw,l0)
if length(rw)<l0
  error('length(rw) must be at least l0.');
end
max_rr = +Inf;
for i = 1:length(rw)
  max_r = lp_s(rw(i),rw);
  if max_r < max_rr
    p0 = rw(i);
    max_rr = max_r;
  end
end

if imag(p0)
  p = [ p0; conj(p0) ];
else
  p = p0;
end
[max_r,i] = lp_s(p,rw);

while size(p,1) < l0
  p0 = rw(i);
  if imag(p0)
    p = [ p; p0; conj(p0) ];
  else
    p = [ p; p0];
  end
  [max_r,i] = lp_s(p,rw);
end
%% Choose further parameters
function [max_r,ind] = lp_s(p,set)
max_r = -1;
ind = 0;
for i = 1:length(set)
  x = set(i);
  rr = 1;
  for j = 1:length(p)
```

```
    rr = rr*abs(p(j)-x)/abs(p(j)+x);
  end
  if rr > max_r
    max_r = rr;
    ind = i;
  end
end
```

B.10 Algorithm 16

```
function p=heuristic_shift(A,E,l0,kl, ks)
n = size(A,1);
if kl >= n, error('kl must be smaller than n!'); end
if ks >= n, error('ks must be smaller than n!'); end
if (2*(l0) >= kl+ks), error('2*l0 must be smaller than
kl+ks!');
end
rw = [];
if kl > 0
 [Hl,Vl] = ritz_large(E,A,kl);
  rwp = eig(Hl(1:kl,1:kl));                        % =: R_+
  rw = [rw; rwp];
end
if ks > 0
    [Hs,Vs]=ritz_small(E,A,ks);
    ds = ones(ks,1)./eig(Hs(1:ks,1:ks));
      rw = [rw; ds];
end
  rw0 = rw; rw = [];
    for j = 1:length(rw0)
     if real(rw0(j))<0
       rw = [rw; rw0(j)];
     end
    end
  p=lp_mnmx(rw,l0);
return
%% large magnitude Ritz value
function [H,V] = ritz_large(E,A,k,rv)
na = nargin;
n = size(E,1);
if k >= n-1, error('k must be smaller than the order
of A!');
end
```

```
if na<7, rv = ones(n,1); end

V = zeros(n,k+1);
H = zeros(k+1,k);

V(:,1) = (1.0/norm(rv))*rv;

beta = 0;

for j = 1:k
     if j > 1
    H(j,j-1) = beta;
    V(:,j) = (1.0/beta)*rv;
     end
  x = V(:,j);
  w=E\(A*x);
  rv = w;

  for i = 1:j
    H(i,j) = V(:,i)'*w;
    rv = rv-H(i,j)*V(:,i);
  end

  beta = norm(rv);
  H(j+1,j) = beta;

end

V(:,k+1) = (1.0/beta)*rv;
%% small magnitude ritzvalue
function [H,V] = ritz_small(E,A,k,rv)
na = nargin;
n = size(A,1);
if k >= n-1, error('k must be smaller than the order
of A!');
end
if na<7, rv = randn(n,1); end
V = zeros(n,k+1);
H = zeros(k+1,k);
V(:,1) = (1.0/norm(rv))*rv;
beta = 0;
for j = 1:k
    if j > 1
     H(j,j-1) = beta;
     V(:,j) = (1.0/beta)*rv;
```

```
    end
    w = A\(E*V(:,j));
    rv = w;
    for i = 1:j
      H(i,j) = V(:,i)'*w;
      rv = rv-H(i,j)*V(:,i);
    end
    beta = norm(rv);
   H(j+1,j) = beta;
  end
V(:,k+1) = (1.0/beta)*rv;

%% solve the minimax problem
function p = lp_mnmx(rw,l0)
if length(rw)<l0
  error('length(rw) must be at least l0.');
end
max_rr = +Inf;
for i = 1:length(rw)
  max_r = lp_s(rw(i),rw);
  if max_r < max_rr
    p0 = rw(i);
    max_rr = max_r;
  end
end

if imag(p0)
  p = [ p0; conj(p0) ];
else
  p = p0;
end
[max_r,i] = lp_s(p,rw);

while size(p,1) < l0
  p0 = rw(i);
  if imag(p0)
    p = [ p; p0; conj(p0) ];
  else
    p = [ p; p0];
  end
  [max_r,i] = lp_s(p,rw);
end
%% Choose further parameters
function [max_r,ind] = lp_s(p,set)
max_r = -1;
```

```
ind = 0;
for i = 1:length(set)
  x = set(i);
  rr = 1;
  for j = 1:length(p)
    rr = rr*abs(p(j)-x)/abs(p(j)+x);
  end
  if rr > max_r
    max_r = rr;
    ind = i;
  end
end
```

B.11 Algorithm 19

```
function [Er,Ar,Br,Cr]=lr_srm(E,A,B,C,tol,l,restol,
maxitr)
Zc=lradi(E,A,B,l,maxitr,restol);
Zo=lradi(E',A',C',l,maxitr,restol);
Zon=E*Zo;
[Uc,S,Ub] = svd(full(Zc'*Zon),0);
s0=diag(S);
ks=length(s0);
K0=ks;
while (sum(s0(K0-1:ks))<tol/2)&&(K0>2)
  K0=K0-1;
end
r=K0;
sigma_r=diag(S(1:r,1:r));
Vc=Zc*Uc(:,1:r);
Vo=Zo*Ub(:,1:r);
TR=Vc*(diag(ones(r,1)./sqrt(sigma_r)));
TL=Vo*(diag(ones(r,1)./sqrt(sigma_r)));
Er=TL'*E*TR;
Ar=TL'*A*TR;
Br=TL'*B;
Cr=C*TR;
return
%% Solution of lyapunov equation
function Z = lradi(E,A,B,l,maxitr,restol)
i=1;
Z=[];ip=0;
res1=zeros(1,maxitr+2);
```

```
  m3=size(B,2);
W=B;
 bnorm=norm(W'*W,'fro');
%% Initial shift computation
   Bn=sprand(size(A,1),100,.8);
  [Q,~]=qr(full(Bn),0);
   p = adaptive_shift(E,A,Q,1);
   l=length(p);
 while i<=maxitr
  if ip<l
     ip=ip+1;
  else
   m2=size(Z,2);
   Zcn=Z(:,m2+1-l*m3:end);
   [Q,~]=qr(full(Zcn),0);
   p = adaptive_shift(E,A,Q,1);
   l=length(p);
   ip =1;
  end %
   pc=p(ip);
   V=(A+pc*E)\W;
   if isreal(pc)
   Z=[Z sqrt(-2*real(pc))*real(V)];
  W = W-2*pc*E*V;
   else
   beta=real(pc)/imag(pc);
   gam=2*sqrt(-real(pc));
   Z=[Z,...
       gam*(real(V)+beta*imag(V)),...
       gam*sqrt(beta^2+1)*imag(V)];
   V=real(V)+beta*imag(V);
   pc=conj(pc);
  W = W-4*real(pc)*E*V;
   i=i+1;
   res1(i)=norm(W'*W,'fro')/bnorm;
   fprintf(1,'step: %4d  normalized residual: %d\n',i,
   res1(i))
   if res1(i)<restol
   break;
   end
   end
   i=i+1;
end
  %% adaptive shifts computation
function p = adaptive_shift(E,A,V,l)
```

```
    An=V'*A*V;
    En=V'*E*V;
    rw=eig(full(An),full(En));
     rw0 = rw; rw = [];
    for j = 1:length(rw0)
     if real(rw0(j))<0
      rw = [rw;rw0(j)];
     end
    end
    p=lp_mnmx(rw,l);
%% solve the minimax problem
function p = lp_mnmx(rw,l0)
if length(rw)<l0
  error('length(rw) must be at least l0.');
end
max_rr = +Inf;
for i = 1:length(rw)
 max_r = lp_s(rw(i),rw);
 if max_r < max_rr
  p0 = rw(i);
  max_rr = max_r;
 end
end
if imag(p0)
  p = [ p0; conj(p0) ];
else
  p = p0;
end
[max_r,i] = lp_s(p,rw);
while size(p,1) < l0
 p0 = rw(i);
 if imag(p0)
 p = [ p; p0; conj(p0) ];
 else
 p = [ p; p0];
 end
  [max_r,i] = lp_s(p,rw);
end

function [max_r,ind] = lp_s(p,set)
max_r = -1;
ind = 0;
for i = 1:length(set)
  x = set(i);
  rr = 1;
```

```
  for j = 1:length(p)
    rr = rr*abs(p(j)-x)/abs(p(j)+x);
  end
  if rr > max_r
   max_r = rr;
   ind = i;
  end
end
```

B.12 Algorithm 21

```
function [Er,Ar,Br,Cr] = irka(E,A,B,C,r,maxiter,tol)
%% Initialization
n = size(A,1);
m = size(B,2);
p = size(C,1);
S = 100*rand(r,1);
b = randn(m,r);
c = randn(p,r);
for iter = 1:maxiter
    S_old = S;
   V = zeros(n,r);
   W = zeros(n,r);
    j = 1;
    while(j < r+1)
        x = (S(j)*E-A)\(B*b(:,j));
        y = (S(j)*E'-A')\(C'*c(:,j));
        if(abs(imag(S(j))) > 0)
            V(:,j) = real(x);
            W(:,j) = real(y);
            V(:,j+1) = imag(x);
            W(:,j+1) = imag(y);
            j = j + 2;
        else
            V(:,j) = real(x);
            W(:,j) = real(y);
            j = j + 1;
        end
    end
    [V,~] = qr(V,0);
    [W,~] = qr(W,0);
    Er = (W'*E*V);
    Ar = (W'*A*V);
```

```
    Br = (W'*B);
    Cr = C*V;
%% Update interpolation points/tangential directions
    [T,S] = eig(Ar,Er);
    S = - diag(S);
    b = (T\(Er\Br)).';
    c = Cr*T;

%% Check for convergence
   err = norm(sort(S)-sort(S_old))/norm(S_old);
%% comment out for non-verbose mode %%
   fprintf('IRKA step %d, conv. crit. = %e \n', iter, err)
   if(err < tol)
   break
  end
end
if (iter == maxiter && err > tol),
   fprintf('IRKA: No convergence in %d iterations.\n',
   maxiter)
end
```

Bibliography

[1] A. Yousuff, D. A. Wagie, and R. E. Skelton. Linear system approximation via covariance equivalent realizations. *J. Math. Anal. Appl.*, 106(1):91–115, 1985.

[2] J. Alan and Z. Daniel. *Table of Integrals, Series, and Products.* Academic press, San Diego, CA, 6th edition, 2000.

[3] G. Alì, N. Banagaaya, W.H.A. Schilders, and C. Tischendorf. Index-aware model order reduction for linear index-2 DAEs with constant coefficients. *SIAM Journal on Scientific Computing*, 35(3):A1487–A1510, 2013.

[4] L. Amodei and J.-M. Buchot. A stabilization algorithm of the Navier-Stokes equations based on algebraic Bernoulli equation. *Numer. Lin. Alg. Appl.*, 19(4):700–727, 2012.

[5] A. C. Antoulas, C. A. Beattie, and S. Gugercin. Interpolatory model reduction of large-scale dynamical systems. In Javad Mohammadpour and Karolos M. Grigoriadis, editors, *Efficient Modeling and Control of Large-Scale Systems*, pages 3–58. Springer US, 2010.

[6] A. C. Antoulas, D. C. Sorensen, and S. Gugercin. A survey of model reduction methods for large-scale systems. *Contemp. Math.*, 280:193–219, 2001.

[7] A.C. Antoulas. *Approximation of Large-Scale Dynamical Systems*, volume 6 of *Advances in Design and Control.* SIAM Publications, Philadelphia, PA, 2005.

[8] V. I. Arnold. *Mathematical Methods of Classical Mechanics.* Springer-Verlag, New York, 1989.

[9] U. M. Ascher and L. R. Petzold. *Computer Methods for Ordinary Differential Equations and Differential-Algebraic Equations.* SIAM, Philadelphia, PA, 1998.

[10] Z. Bai, J. Demmel, J. Dongarra, A. Ruhe, and H. van der Vorst, editors. *Templates for the Solution of Algebraic Eigenvalue Problems: A Practical Guide.* SIAM, Philadelphia, PA, 2000.

[11] Z. Bai and Y. Su. SOAR: A second-order Arnoldi method for the solution of the quadratic eigenvalue problem. *SIAM J. Matrix Anal. Appl.*, 26(3):640–659, 2005.

[12] Z. Bai and Y.-F. Su. Dimension reduction of large-scale second-order dynamical systems via a second-order Arnoldi method. *SIAM J. Sci. Comput.*, 26(5):1692–1709, 2005.

[13] M. J. Balas. Trends in large space structure control theory: Fondest hopes, wildest dreams. *IEEE Trans. Automat. Control*, 27:522–535, 1982.

[14] E. Bänsch, P. Benner, J. Saak, and H. Weichelt. Riccati-based boundary feedback stabilization of incompressible Navier-Stokes flow. Preprint SPP1253-154, DFG-SPP1253, 2013.

[15] E. Bänsch, P. Benner, J. Saak, and H. K. Weichelt. Riccati-based boundary feedback stabilization of incompressible Navier-Stokes flows. *SIAM J. Sci. Comput.*, 37(2):A832–A858, 2015.

[16] S. Barnett and C. Storey. Stability analysis of constant linear systems by Lyapunov second method. *Electronics Letters*, 2:165–166, 1966.

[17] S. Barrachina, P. Benner, and E. S. Quintana-Ortí. Efficient algorithms for generalized algebraic Bernoulli equations based on the matrix sign function. *Numer. Algorithms*, 46(4):351–368, 2007.

[18] R. H. Bartels and G. W. Stewart. Solution of the matrix equation $AX + XB = C$: Algorithm 432. *Comm. ACM*, 15:820–826, 1972.

[19] U. Baur. *Control-Oriented Model Reduction for Parabolic Systems*. Ph.D. Thesis, Technische Universität Berlin, Berlin, January 2008. ISBN 978-3639074178 Vdm Verlag Dr. Müller, available from `http://www.nbn-resolving.de/urn:nbn:de:kobv:83-opus-17608`.

[20] C. A. Beattie and S. Gugercin. Krylov-based minimization for optimal h 2 model reduction. In *Proceedings of the 46th IEEE Conference on Decision and Control*, pages 4385–4390. IEEE, 2007.

[21] C. A. Beattie and S. Gugercin. Interpolatory projection methods for structure-preserving model reduction. *Syst. Control Lett.*, 58(3):225–232, 2009.

[22] C. A. Beattie and S. Gugercin. Interpolatory projection methods for structure-preserving model reduction. *Syst. Cont. Lett.*, 58(3):225–232, 2009.

[23] T. Bechtold, G. Schrag, and L. Feng, editors. *System-Level Modeling of MEMS*. Advanced Micro & Nanosystems. Wiley-VCH, 2013.

[24] P. Benner, J. M. Claver, and E. S. Quintana-Ortí. Efficient solution of coupled Lyapunov equations via matrix sign function iteration. In A. Dourado et al., editor, *Proc. 3rd Portuguese Conf. on Automatic Control CONTROLO'98,* Coimbra, pages 205–210, 1998.

[25] P. Benner, M. Hinze, and E. J. W. ter Maten, editors. *Model Reduction for Circuit Simulation*, volume 74 of *Lecture Notes in Electrical Engineering*. Springer, Dodrecht, 2011.

[26] P. Benner, M. Köhler, and J. Saak. Fast approximate solution of the non-symmetric generalized eigenvalue problem on multicore architectures. In Michael Bader, Arndt Bodeand Hans-Joachim Bungartz, Michael Gerndt, Gerhard R. Joubert, and Frans Peters, editors, *Parallel Computing: Accelerating Computational Science and Engineering (CSE)*, volume 25 of *Advances in Parallel Computing*, pages 143–152. IOS Press, 2014.

[27] P. Benner, P. Kürschner, and J. Saak. Efficient handling of complex shift parameters in the low-rank Cholesky factor ADI method. *Numer. Algorithms*, 62(2):225–251, 2013.

[28] P. Benner, P. Kürschner, and J. Saak. An improved numerical method for balanced truncation for symmetric second order systems. *Math. Comput. Model. Dyn. Syst.*, 19(6):593–615, 2013.

[29] P. Benner, P. Kürschner, and J. Saak. Self-generating and efficient shift parameters in ADI methods for large Lyapunov and Sylvester equations. *Electron. Trans. Numer. Anal.*, 43:142–162, 2014.

[30] P. Benner, J.-R. Li, and T. Penzl. Numerical solution of large-scale Lyapunov equations, Riccati equations, and linear-quadratic optimal control problems. *Numer. Lin. Alg. Appl.*, 15(9):755–777, 2008.

[31] P. Benner, V. Mehrmann, and D. C. Sorensen. *Dimension Reduction of Large-Scale Systems*, volume 45 of *Lect. Notes Comput. Sci. Eng.* Springer-Verlag, Berlin/Heidelberg, Germany, 2005.

[32] P. Benner, H. Mena, and J. Saak. On the parameter selection problem in the Newton-ADI iteration for large-scale Riccati equations. *Electron. Trans. Numer. Anal.*, 29:136–149, 2008.

[33] P. Benner and E. S. Quintana-Ortí. Solving stable generalized Lyapunov equations with the matrix sign function. *Numer. Algorithms*, 20(1):75–100, 1999.

[34] P. Benner, E. S. Quintana-Ortí, and G. Quintana-Ortí. A portable subroutine library for solving linear control problems on distributed memory computers. In G. Cooperman, E. Jessen, and G. O. Michler, editors, *Workshop on Wide Area Networks and High Performance Computing,*

Essen (Germany), September 1998, Lecture Notes in Control and Information, pages 61–88. Springer-Verlag, Berlin/Heidelberg, Germany, 1999.

[35] P. Benner, E. S. Quintana-Ortí, and G. Quintana-Ortí. Balanced truncation model reduction of large-scale dense systems on parallel computers. *Math. Comput. Model. Dyn. Syst.*, 6(4):383–405, 2000.

[36] P. Benner, E. S. Quintana-Ortí, and G. Quintana-Ortí. Singular perturbation approximation of large, dense linear systems. In *Proc. 2000 IEEE Intl. Symp. CACSD, Anchorage, Alaska, USA, September 25–27, 2000*, pages 255–260. IEEE Press, Piscataway, NJ, 2000.

[37] P. Benner and J. Saak. Efficient balancing-based MOR for large-scale second-order systems. *Math. Comput. Model. Dyn. Syst.*, 17(2):123–143, 2011.

[38] P. Benner and J. Saak. Numerical solution of large and sparse continuous time algebraic matrix Riccati and Lyapunov equations: a state of the art survey. *GAMM Mitteilungen*, 36(1):32–52, August 2013.

[39] P. Benner, J. Saak, M. Stoll, and H. Weichelt. Efficient solution of large-scale saddle point systems arising in Riccati-based boundary feedback stabilization of incompressible Stokes flow. *SIAM J. Sci. Comput.*, 35(5):S150–S170, 2013. Special Section: 2012 Copper Mountain Conference.

[40] P. Benner, J. Saak, and M. M. Uddin. Second order to second order balancing for index-1 vibrational systems. In *7th International Conference on Electrical & Computer Engineering (ICECE) 2012*, pages 933–936. IEEE, 2012.

[41] P. Benner, J. Saak, and M. M. Uddin. Balancing based model reduction for structured index-2 unstable descriptor systems with application to flow control. *Numerical Algebra, Control and Optimization*, 6(1):1–20, 2016.

[42] P. Benner, J. Saak, and M. M. Uddin. Structure preserving model order reduction of large sparse second-order index-1 systems and application to a mechatronics model. *Math. Comput. Model. Dyn. Syst.*, 0(0):1–15, 2016.

[43] P. Benner and T. Stykel. Numerical solution of projected algebraic Riccati equations. *SIAM J. Numer. Anal*, 52(2):581–600, 2014.

[44] D. Billger. The butterfly gyro. 2005. Chapter 18 (pages 349–352) of [31].

[45] K. E. Brenan, S. L. Campbell, and L. R. Petzold. *Numerical Solution of Initial–Value Problems in Differential–Algebraic Equations.* Elsevier Science Publishing, North-Holland, 1989.

[46] A. Bryson and A. Carrier. Second-order algorithm for optimal model order reduction. *Journal of Guidance, Control, and Dynamics*, 13(5):887–892, 1990.

[47] A. Bunse-Gerstner, D. Kubalinska, G. Vossen, and D. Wilczek. ℓ_2-norm optimal model reduction for large scale discrete dynamical MIMO systems. *J. Comput. Appl. Math.*, 233(5):1202–1216, 2010.

[48] Y. Chahlaoui, K. A. Gallivan, A. Vandendorpe, and P. Van Dooren. Model reduction of second-order systems. Chapter 6 (pages 149–172) of [31].

[49] Y. Chahlaoui, D. Lemonnier, A. Vandendorpe, and P. Van Dooren. Second-order balanced truncation. *Linear Algebra Appl.*, 415(2–3):373–384, 2006.

[50] Y. Chahlaoui and P. Van Dooren. A collection of benchmark examples for model reduction of linear time invariant dynamical systems. SLICOT Working Note 2002–2, University of Manchester, February 2002. Available from `www.slicot.org`.

[51] C.-T. Chen. *Linear System Theory and Design.* Oxford University Press, 1984.

[52] K. A. Cliffe, T. J. Garratt, and A. Spence. Eigenvalues of block matrices arising from problems in fluid mechanics. *SIAM J. Matrix Anal. Appl.*, 15(4):1310–1318, 1994.

[53] R. R. Craig. *Structural Dynamics: an Introduction to Computer Methods.* John Wiley & Sons Inc, New York, 1981.

[54] R.R. Craig and A J. Kurdila. *Fundamentals of Structural Dynamics.* Wiley, 2006.

[55] B. N. Datta. Linear and numerical linear algebra in control theory: Some research problems. *Linear Algebra Appl.*, 197/198:755–790, 1994.

[56] B. N. Datta. *Numerical Methods for Linear Control Systems.* Elsevier Academic Press, 2004.

[57] B. N. Datta. *Numerical linear algebra and applications*, volume 116. SIAM, 2010.

[58] T. A. Davis. *Direct Methods for Sparse Linear Systems.* Number 2 in Fundamentals of Algorithms. SIAM, Philadelphia, PA, USA, 2006.

[59] James W. Demmel. *Applied numerical linear algebra.* Society for Industrial and Applied Mathematics (SIAM), Philadelphia, PA, 1997.

[60] A. Dou. Method of undetermined coefficients in linear differential systems and the matrix equation $YB - AY = F$. *SIAM J. Appl. Math.*, 14:691–696, 1966.

[61] W. Draijer, M. Steinbuch, and O. H. Bosgra. Adaptive control of the radial servo system of a compact disc player. *IFAC Automatica*, 28(3):455–462, 1992.

[62] Vladimir Druskin and Leonid Knizhnerman. Extended Krylov subspaces: Approximation of the matrix square root and related functions. *SIAM J. Matrix Anal. Appl.*, 19(3):755–771, 1998. DOI: 10.1137/S0895479895292400.

[63] I. S. Duff, A. M. Erisman, and J. K. Reid. *Direct methods for sparse matrices.* Clarendon Press, Oxford, UK, 1989.

[64] E. Eich-Soellner and C. Führer. *Numerical Methods in Multibody Dynamics.* European Consortium for Mathematics in Industry. B. G. Teubner GmbH, Stuttgart, 1998.

[65] R. Eid, B. Salimbahrami, and B. Lohmann. Equivalence of Laguerre-based model order reduction and moment matching. *IEEE Trans. Autom. Control*, 52(6):1104–1108, 2007.

[66] I. M. Elfadel and D. D. Ling. A block rational Arnoldi algorithm for multipoint passive model-order reduction of multiport RLC networks. In *Proceedings of the 1997 IEEE/ACM international conference on Computer-aided design*, pages 66–71. IEEE Computer Society, 1997.

[67] D. F. Enns. Model reduction with balanced realizations: An error bound and a frequency weighted generalization. In *Proc. 23rd IEEE Conf. Decision Contr.*, volume 23, pages 127–132, 1984.

[68] P. Feldmann and R. W. Freund. Efficient linear circuit analysis by Padé approximation via the Lanczos process. In *Proc. of EURO-DAC '94 with EURO-VHDL '94, Grenoble, France*, pages 170–175. IEEE Computer Society Press, 1994.

[69] P. Feldmann and R. W. Freund. Efficient linear circuit analysis by Padé approximation via the Lanczos process. *IEEE Trans. Comput.-Aided Design Integr. Circuits Syst.*, 14:639–649, 1995.

[70] K. V. Fernando and H. Nicholson. Minimality of siso linear systems. *Proc. IEEE*, 70(10):1241–1242, 1982.

[71] F. Freitas, J. Rommes, and N. Martins. Gramian-based reduction method applied to large sparse power system descriptor models. *IEEE Trans. Power Syst.*, 23(3):1258–1270, August 2008.

[72] R. W. Freund. Krylov-subspace methods for reduced-order modeling in circuit simulation. *J. Comput. Appl. Math.*, 123(1–2):395–421, 2000.

[73] Z. Gajić and M. Qureshi. *Lyapunov Matrix Equation in System Stability and Control.* Math in Science and Engineering. Academic Press, San Diego, CA, 1995.

[74] K. Gallivan, E. Grimme, and P. Van Dooren. A rational Lanczos algorithm for model reduction. *Numer. Algorithms*, 12:33–63, 1996.

[75] K. Gallivan, A. Vandendorpe, and P. Van Dooren. Model reduction of MIMO systems via tangential interpolation. *SIAM J. Matrix Anal. Appl.*, 26(2):328–349, 2004.

[76] M. Gerdin. *Identification and estimation for models described by differential-algebraic equations.* PhD thesis, Linköpings Universitet, Linköpings, Sweden, 2006.

[77] K. Glover. All optimal Hankel-norm approximations of linear multivariable systems and their L^∞ norms. *Internat. J. Control*, 39:1115–1193, 1984.

[78] G. H. Golub and C. F. Van Loan. *Matrix Computations.* Johns Hopkins University Press, Baltimore, third edition, 1996.

[79] G. H. Golub and C. F. Van Loan. *Matrix Computations.* Johns Hopkins University Press, Baltimore, fourth edition, 2013.

[80] M. Green and D. J. N Limebeer. *Linear Robust Control.* Prentice-Hall, Englewood Cliffs, NJ, 1995.

[81] E. J. Grimme. *Krylov projection methods for model reduction.* Ph.D. Thesis, Univ. of Illinois at Urbana-Champaign, USA, 1997.

[82] S. Gugercin. An iterative rational krylov algorithm (IRKA) for optimal $\mathcal{H}_2$ model reduction. In *Householder Symposium XVI, Seven Springs Mountain Resort, PA, USA*, 2005.

[83] S. Gugercin and A. C. Antoulas. A survey of model reduction by balanced truncation and some new results. *Internat. J. Control*, 77(8):748–766, 2004.

[84] S. Gugercin, A. C. Antoulas, and C. A. Beattie. Rational krylov methods for optimal $\mathcal{H}_2$ model reduction. In *In Proc. of 17th International Symposium on Mathematical Theory of Networks and Systems*, 2006.

[85] S. Gugercin, A. C. Antoulas, and C. A. Beattie. $\mathcal{H}_2$ model reduction for large-scale dynamical systems. *SIAM J. Matrix Anal. Appl.*, 30(2):609–638, 2008.

[86] S. Gugercin, A. C. Antoulas, and N. Bedrossian. Approximation of the international space station 1r and 12a models. In *Proc. of the 40th IEEE Conferences on Decision and Control*, pages 1515–1516, 2001.

[87] S. Gugercin, R. V. Polyuga, C. Beattie, and A. Van Der Schaft. Structure-preserving tangential interpolation for model reduction of port-Hamiltonian systems. *Automatica*, 48(9):1963–1974, 2012.

[88] S. Gugercin, D. C. Sorensen, and A. C. Antoulas. A modified low-rank Smith method for large-scale Lyapunov equations. *Numer. Algorithms*, 32(1):27–55, 2003.

[89] S. Gugercin, T. Stykel, and S. Wyatt. Model reduction of descriptor systems by interpolatory projection methods. *SIAM J. Sci. Comput.*, 35(5):B1010–B1033, 2013.

[90] Y. Halevi. Frequency weighted model reduction via optimal projection.

[91] S. J. Hammarling. Numerical solution of the stable, non-negative definite Lyapunov equation. *IMA J. Numer. Anal.*, 2:303–323, 1982.

[92] S. J. Hammarling. Numerical solution of the discrete-time, convergent, non-negative definite Lyapunov equation. *Syst. Cont. Lett.*, 17:137–139, 1991.

[93] M. Heinkenschloss, D. C. Sorensen, and K. Sun. Balanced truncation model reduction for a class of descriptor systems with application to the Oseen equations. *SIAM J. Sci. Comput.*, 30(2):1038–1063, 2008.

[94] M. Hochbruck and G. Starke. Preconditioned Krylov subspace methods for Lyapunov matrix equations. *SIAM J. Matrix Anal. Appl.*, 16(1):156–171, 1995. (See also: IPS Research Report 92–17, ETH Zürich, Switzerland (1992)).

[95] P. Hood and C. Taylor. Navier-Stokes equations using mixed interpolation. In J. T. Oden, R. H. Gallagher, C. Taylor, and O. C. Zienkiewicz, editors, *Finite Element Methods in Flow Problems*, pages 121–132. University of Alabama in Huntsville Press, 1974.

[96] D. Hyland and D. Bernstein. The optimal projection equations for model reduction and the relationships among the methods of wilson, skelton, and moore. *IEEE Transaction on*, 30(12): 1201–1211, 1985.

[97] I. M. Jaimoukha and E. M. Kasenally. Krylov subspace methods for solving large Lyapunov equations. *SIAM J. Numer. Anal.*, 31:227–251, 1994.

[98] I.M. Jaimoukha and E.M. Kasenally. Krylov subspace methods for solving large Lyapunov equations. *SIAM J. Numer. Anal.*, 31:227–251, 1994.

[99] K. Jbilou and A. J. Riquet. Projection methods for large Lyapunov matrix equations. *Linear Algebra Appl.*, 415(2–3):344–358, 2006.

[100] S. Jianbin, X. Dingbang, W. Xiong, C. Zhihua, and W. Xuezhong. Vibration sensitivity analysis of the butterfly-gyrostructure. *Microsystem technologies*, 20(7):1281–1290, 2014.

[101] B. Kranz. Zustandsraumbeschreibung von piezo-mechanischen Systemen auf Grundlage einer Finite-Elemente-Diskretisierung. In *ANSYS Conference & 27th CADFEM Users' Meeting*, Congress Center Leipzig, Germany, November 18-20 2009.

[102] D. Kressner. *Numerical methods for general and structured eigenvalue problems.* Springer-Verlag, New York, 2005.

[103] D. Kressner, V. Mehrmann, and T. Penzl. CTDSX - a collection of benchmark examples for state-space realizations of time-invariant continuous-time systems. SLICOT Working Note 1998-9, November 1998. Available from `www.slicot.org`.

[104] V. Kučera. *Analysis and Design of Discrete Linear Control Systems.* Academia, Prague, Czech Republic, 1991.

[105] A. Kumar and P. Daoutidis. Feedback control of nonlinear differential-algebraic-equation systems. *AIChE Journal*, 41(3):619–636, 1995.

[106] P. Kunkel and V. Mehrmann. *Differential-Algebraic Equations: Analysis and Numerical Solution.* Textbooks in Mathematics. EMS Publishing House, Zürich, Switzerland, 2006.

[107] S. Lall, P. Krysl, and J. E. Marsden. Structure-preserving model reduction for mechanical systems. *Phys. D*, 184(1-4):304–318, 2003. Complexity and nonlinearity in physical systems (Tucson, AZ, 2001).

[108] I. Lasiecka. *Mathematical Control Theory of Coupled PDEs*, volume 75 of *CBMS-NSF Regional Conference Series in Applied Mathematics.* SIAM Publications, Philadelphia, PA, 2002.

[109] A. J. Laub, M. T. Heath, C. C. Paige, and R. C. Ward. Computation of system balancing transformations and other applications of simultaneous diagonalization algorithms. *IEEE Trans. Autom. Control*, 32(2):115–122, 1987.

[110] J. R. Leigh. *Functional Analysis and Linear Control Theory.* Academic Press, New York, 1980.

[111] A. Lepschy, G. A. Mian, G. Pinato, and U. Viaro. Rational L_2 approximation: a non-gradient algorithm. In *Decision and Control, 1991, Proceedings of the 30th IEEE Conference on*, pages 2321–2323. IEEE, 1991.

[112] J.-R. Li. *Model Reduction of Large Linear Systems via Low Rank System Gramians.* Ph.D. Thesis, Massachusetts Institute of Technology, September 2000.

[113] J.-R. Li and J. White. Efficient model reduction of interconnect via approximate system Gramians. In *Proceedings of the IEEE/ACM international conference on Computer-aided design*, pages 380–384. IEEE Press, 1999.

[114] J.-R. Li and J. White. Low rank solution of Lyapunov equations. *SIAM J. Matrix Anal. Appl.*, 24(1):260–280, 2002.

[115] F. Lin. *Robust Control Design: An Optimal Control Approach.* AFI Press, 1997.

[116] Y. Liu and B. Do Anderson. Singular perturbation approximation of balanced systems. *International Journal of Control*, 50(4):1379–1405, 1989.

[117] A. Lu and E. L. Wachspress. Solution of Lyapunov equations by alternating direction implicit iteration. *Comput. Math. Appl.*, 21(9):43–58, 1991.

[118] David G. Luenberger. *Introduction to dynamic systems. Theory, models, and applications.* John Wiley & Sons, New York, first edition, 1979.

[119] E. Ma. A finite series solution of the matrix equation AX-XB=C. *SIAM J. Appl. Math.*, 14:490–495, 1966.

[120] M. Marcus and H. Minc. *A Survey of Matrix Theory and Matrix Inequalities.* Allyn and Bacon, Boston, 1964.

[121] R. März. Canonical projectors for linear differential algebraic equations. 31(4):121–135, 1996.

[122] V. Mehrmann and T. Stykel. Balanced truncation model reduction for large-scale systems in descriptor form. 2005. Chapter 20 (pages 357–361) of [31].

[123] L. Meier and D. Luenberger. Approximation of linear constant systems. *IEEE Trans. Autom. Control*, 12(5):585–588, 1967.

[124] D. G. Meyer and S. Srinivasan. Balancing and model reduction for second-order form linear systems. *IEEE Trans. Autom. Control*, 41(11):1632–1644, 1996.

[125] B. C. Moore. Principal component analysis in linear systems: controllability, observability, and model reduction. *IEEE Trans. Autom. Control*, AC–26(1):17–32, 1981.

[126] R. Neugebauer, B. Denkena, and K. Wegener. Mechatronic Systems for Machine Tools. *CIRP Annals - Manufacturing Technology*, 56:657–686, 2007.

[127] R. Neugebauer, B. Denkena, and K. Wegener. Adaptive spindle support for improving machining operations. *CIRP Annals - Manufacturing Technology*, 57:395–398, 2008.

[128] R. Neugebauer, W.-G. Drossel, A. Bucht, B. Kranz, and K. Pagel. Control design and experimental validation of an adaptive spindle support for enhanced cutting processes. *CIRP Annals - Manufacturing Technology*, 59:373–376, 2010.

[129] R. Neugebauer, W. G. Drossel, A. Bucht, B. Kranz, and K. Pagel. Control design and experimental validation of an adaptive spindle support for enhanced cutting processes. *CIRP Annals - Manufacturing Technology*, 59:373–376, 2010.

[130] R. Neugebauer, W.-G. Drossel, K. Pagel, and B. Kranz. Making of state space models of piezo-mechanical systems with exact impedance mapping and strain output signals. In *Mechatronics and Material Technologies*, volume 2, pages 73–80, Zurich, Switzerland, June 28-30 2010. Swiss Federal Institute of Technology ETH.

[131] D. Peaceman and H. Rachford. The numerical solution of elliptic and parabolic differential equations. *J. Soc. Indust. Appl. Math.*, 3:28–41, 1955.

[132] T. Penzl. *Numerische Lösung großer Lyapunov-Gleichungen*. Logos–Verlag, Berlin, Germany, 1998. Dissertation, Fakultät für Mathematik, TU Chemnitz, 1998.

[133] T. Penzl. A cyclic low rank Smith method for large sparse Lyapunov equations. *SIAM J. Sci. Comput.*, 21(4):1401–1418, 2000.

[134] T. Penzl. Algorithms for model reduction of large dynamical systems. *Linear Algebra Appl.*, 415(2–3):322–343, 2006. (Reprint of Technical Report SFB393/99-40, TU Chemnitz, 1999.).

[135] Z.-Q. Qu. *Model Order Reduction Techniques: with Applications in Finite Element Analysis*. Springer-Verlag, Berlin, 2004.

[136] T. Reis. Circuit synthesis of passive descriptor systems – a modified nodal approach. *Int. J. Circ. Theor. Appl.*, 38(1):44–68, 2010.

[137] T. Reis and T. Stykel. Balanced truncation model reduction of second-order systems. *Math. Comput. Model. Dyn. Syst.*, 14(5):391–406, 2008.

[138] R. Riaza. *Differential-Algebraic Systems. Analytical Aspects and Circuit Applications.* World Scientific Publishing Co. Pte. Ltd., Singapore, 2008.

[139] J. R. Rice. *Matrix Computation and Mathematical Software.* McGraw-Hill, New York, 1981.

[140] Axel Ruhe. Rational Krylov algorithms for nonsymmetric eigenvalue problems, II: Matrix pairs. *Linear Algebra Appl.*, 197/198:283–295, 1994.

[141] Y. Saad. Numerical solution of large Lyapunov equation. In M. A. Kaashoek, J. H. van Schuppen, and A. C. M. Ran, editors, *Signal Processing, Scattering, Operator Theory and Numerical Methods*, pages 503–511. Birkhäuser, 1990.

[142] Y. Saad. *Numerical Methods for Large Eigenvalue Problems.* Manchester University Press, Manchester, UK, 1992.

[143] Y. Saad. *Iterative Methods for Sparse Linear Systems.* SIAM, Philadelphia, PA, USA, 2003.

[144] J. Saak. Effiziente numerische Lösung eines Optimalsteuerungsproblems für die Abkühlung von Stahlprofilen. Diplomarbeit, Fachbereich 3/Mathematik und Informatik, Universität Bremen, D-28334 Bremen, September 2003.

[145] J. Saak. *Efficient Numerical Solution of Large Scale Algebraic Matrix Equations in PDE Control and Model Order Reduction.* Dissertation, TU Chemnitz, July 2009. available from `http://nbn-resolving.de/urn:nbn:de:bsz:ch1-200901642`.

[146] M. G. Safonov and R. Y. Chiang. A Schur method for balanced-truncation model reduction. *IEEE Trans. Autom. Control*, 34(7):729–733, 1989.

[147] B. Salimbahrami. *Structure Preserving Order Reduction of Large Scale Second Order Models.* Dissertation, Technische Universität München, München, 2005.

[148] B. Salimbahrami and B. Lohmann. Order reduction of large scale second-order systems using Krylov subspace methods. *Linear Algebra Appl.*, 415(2–3):385–405, 2006.

[149] W. H. A. Schilders, H. A. van der Vorst, and J. Rommes. *Model Order Reduction: Theory, Research Aspects and Applications.* Springer-Verlag, Berlin, Heidelberg, 2008.

[150] J. Sherman and W. J. Morrison. *Ann. Math. Statist.*, 21(1):124–127, 1950.

[151] B. Simeon, C. Führer, and P. Rentrop. The Drazin inverse in multibody system dynamics. *Numer. Math.*, 64:521–539, 1993.

[152] V. Simoncini. A new iterative method for solving large-scale Lyapunov matrix equations. *SIAM J. Sci. Comput.*, 29(3):1268–1288, 2007.

[153] R. E. Skelton. *Dynamic systems control: linear systems analysis and synthesis.* John Wiley & Sons, Inc., New York, NY, 1988.

[154] R. A. Smith. Matrix equation $XA + BX = C$. *SIAM J. Appl. Math.*, 16(1):198–201, 1968.

[155] E. D. Sontag. *Mathematical Control Theory.* Springer-Verlag, New York, NY, 2nd edition, 1998.

[156] D. C. Sorensen and A. C. Antoulas. On model reduction of structured systems. Chapter 4 (pages 117–130) of [31].

[157] J. T. Spanos, M. H. Milman, and D. L. Mingori. A new algorithm for L_2 optimal model reduction. *Automat.*, 28(5):897–909, 1992.

[158] T. Stykel. *Analysis and Numerical Solution of Generalized Lyapunov Equations.* Dissertation, TU Berlin, 2002.

[159] T. Stykel. *Analysis and Numerical Solution of Generalized Lyapunov Equations.* Ph.D. Thesis, Technische Universität Berlin, Berlin, 2002.

[160] T. Stykel. Balanced truncation model reduction for semidiscretized Stokes equation. *Linear Algebra Appl.*, 415(2–3):262–289, 2006.

[161] F. Tisseur. Backward error and condition of polynomial eigenvalue problems. *Linear Algebra Appl.*, 309(1):339–361, 2000.

[162] F. Tisseur and K. Meerbergen. The quadratic eigenvalue problem. *SIAM Rev.*, 43(2):235–286, 2001.

[163] M. S. Tombs and I. Postlethwaite. Truncated balanced realization of a stable non-minimal state-space system. *Internat. J. Control*, 46(4):1319–1330, 1987.

[164] N. Truhar and K. Veselić. An efficient method for estimating the optimal dampers' viscosity for linear vibrating systems using Lyapunov equation. *SIAM J. Matrix Anal. Appl.*, 31(1):18–39, 2009.

[165] M. M. Uddin. Model reduction for piezo-mechanical systems using Balanced Truncation. Master's thesis, Stockholm University, Stockholm, Sweden, 2011.

[166] M. M. Uddin. *Computational methods for model reduction of large-scale sparse structured descriptor systems.* Ph.D. Thesis, Otto-von-Guericke-Universität, Magdeburg, Germany, 2015.

[167] M. M. Uddin. Gramian-based model-order reduction of constrained structural dynamic systems. *IET Control Theory & Applications*, 2018.

[168] M. M. Uddin, J. Saak, B. Kranz, and P. Benner. Computation of a compact state space model for an adaptive spindle head configuration with piezo actuators using balanced truncation. *Production Engineering*, 6:577–586, 2012.

[169] M. M. Uddin, J. Saak, B. Kranz, and P. Benner. Efficient reduced order state space model computation for a class of second order index one systems. *Proc. Appl. Math. Mech.*, 12(1):699–700, 2012.

[170] H. A. Van der Vorst. *Iterative Krylov Methods for Large Linear Systems.* Cambridge University Press, Cambridge, 2003.

[171] P. Van Dooren, K. Gallivan, and P.-A. Absil. $\mathcal{H}_2$-optimal model reduction of MIMO systems. *Appl. Math. Lett.*, 21:1267–1273, 2008.

[172] A. Varga. A note on Hammarling's algorithm for the discrete Lyapunov equation. *Syst. Cont. Lett.*, 15(3):273–275, 1990.

[173] De C. Villemagne and R. E. Skelton. Model reduction using a projection formulation. *Internat. J. Control*, 46:2141–2169, 1987.

[174] M. Voigt. *On Linear-Quadratic Optimal Control and Robustness of Differential-Algebraic-Systems.* Dissertation, Otto-von-Guericke-Universität, Magdeburg, Germany, July 2015.

[175] E. L. Wachspress. Iterative solution of the Lyapunov matrix equation. *Appl. Math. Letters*, 107:87–90, 1988.

[176] E. L. Wachspress. The ADI minimax problem for complex spectra, *Appl. Math. Lett.*, 1(3):311–314, 1988.

[177] E. L. Wachspress. *The ADI Model Problem.* Springer, New York, 2013.

[178] D. Wang and M. N. Harris. Stability analysis of the equilibrium of a constrained mechanical system. *Internat. J. Control*, 60(5):733–746, 1994.

[179] D.S. Watkins. *The matrix eigenvalue problem: GR and Krylov subspace methods.* SIAM, Philadelphia, PA, USA, 2007.

[180] D. S. Weile, E. Michielssen, E. Grimme, and K. Gallivan. A method for generating rational interpolant reduced order models of two-parameter linear systems. *Appl. Math. Lett.*, 12(5):93–102, 1999.

[181] K. Willcox and G. Lassaux. Model reduction of an actively controlled supersonic diffuser. 2005. Chapter 20 (pages 357–361) of [31].

[182] J. L. Willems and J. C. Willems. The application of Lyapunov methods to the computation of transient stability regions for multimachine power systems. *IEEE Trans. Power Systems*, (5):795–801, 1970.

[183] D. A. Wilson. Optimum solution of model-reduction problem. *Proceedings of the Institution of Electrical Engineers*, 117(6):1161–1165, 1970.

[184] T. Wolf, H. K. F. Panzer, and B. Lohmann. ADI iteration for Lyapunov equations: a tangential approach and adaptive shift selection. e-print 1312.1142, arXiv, December 2013.

[185] Max A. Woodbury. The stability of out-input matrices. *Chicago, IL*, 9, 1949.

[186] Max A. Woodbury. Inverting modified matrices. *Memorandum Rept.*, 42(106):336, 1950.

[187] P. M. R. Wortelboer, M. Steinbuch, and O. H. Bosgra. Closed-loop balanced reduction with application to a compact disc mechanism. *Selected Topics in Identification, Modeling and Control*, 9:47–58, 1996.

[188] B. Yan, S. X.-D. Tan, and B. McGaughy. Second-order balanced truncation for passive-order reduction of RLCK circuits. *IEEE Trans. Circuits Syst. II, Exp. Briefs*, 55(9):942–946, September 2008.

[189] W. Y. Yan and J. Lam. An approximate approach to $\mathcal{H}_2$ optimal model reduction. *IEEE Trans. Autom. Control*, 44(7):1341–1358, 1999.

[190] A. Yousuff and R. E. Skelton. Covariance equivalent realizations with applications to model reduction of large-scale systems. *Control and Dynamic Systems*, 22:273–348, 1985.

[191] F. Zhang. *The Schur Complement and its Applications*, volume 4. Springer Science + Business Media, New York, 2006.

[192] F. Zhang. *The Schur Complement and Its Applications.* Springer, New York, 2006.

[193] K. Zhou, J. C. Doyle, and K. Glover. *Robust and Optimal Control.* Prentice-Hall, Upper Saddle River, NJ, 1996.

[194] K. Zhou, G. Salomon, and E. Wu. Balanced realization and model reduction for unstable systems. *Internat. J. Robust and Nonlinear Cont.*, 9(3):183–198, 1999.

[195] Kemin Zhou. *Essentials of Robust Control.* Prentice-Hall, Upper Saddle River, NJ, 1997.

[196] D. Žigić, L. T. Watson, and C. Beattie. Contragredient transformations applied to the optimal projection equations. *Linear Algebra Appl.*, 188:665–676, 1993.

Index